TEXT BOOK
OF
ROTATIONAL MECHANICS

DPH PHYSICS SERIES

TEXT BOOK OF ROTATIONAL MECHANICS

By

D.K. Jha

DISCOVERY PUBLISHING HOUSE
NEW DELHI-110002

Published by:
Namit Wasan
DISCOVERY PUBLISHING HOUSE PVT. LTD.
4383/4B, Ansari Road, Darya Ganj
New Delhi-110 002 (India)
Phone : +91-11-23279245; 23253475; 43596065
E-mail : discoverybooksindia@gmail.com
discoverypublishinghouse@gmail.com
namitwasan9@gmail.com
web : www.discoverypublishinggroup.com

***First Published:* 2005**

***Reprinted:* 2021**

ISBN: 978-81-7141-997-5

Text Book of Rotational Mechanics

Printed at:
Infinity Imaging Systems
Delhi

Preface

This book "Text book of Rotational Mechanics" cover the syllabi of B.Sc. (Pass & Honours) and engineering students of various universities in India. The mathematical description of the book is based on the vector analysis provided an efficient short hand for writing physics and at the same time makes it possible to visualise the physical meaning of concepts and laws distinctly and exactly. Hence, the vactor treatment becomes necessary. The emphasis is on the basic physics with some instructive, stimulative and useful applications.

Keeping the view the recent trend of various examinations particularly competitive once, illustrative objective questions are added at the end of each chapter in addition to problems and reviews questions.

I wish to express my severe thanks to all who have helped me in one way or other in the preparation of this book criticism and suggestions for further improvement shall be gratefully acknowledge.

D.K. Jha

Contents

1

DYNAMICS OF RIGID BODIES

RIGID BODY—TRANSLATIONAL AND ROTATIONAL MOTION

A rigid body is defined as a solid body, in which the particles are arranged in such a way that the inter-moleculer distance are small and is not disturbed by any external forces applied on it. This type of body preserves its shape or configuration intact and does not bend stretch or vibrate when in motion.

There are two ways in which a body can more. One type of movement, which we have already investigated is called linear or tranaslational, and is a mere change in position. The other type is called angular or rotational movement in which the body rotates about a line called an axis. Many motions, of course are a combination of the two—a spiral point spins as it sources through the air and the wheels of a car rotate as they travel along the road.

In translation motion our basic measure was the distance traveled, in rotation we must start with the angle through which the body turns it might seen good enough to measure thise angle in degree or in terms of complete revoluations, but in practice the relationship and equations are much similar and easier to comprehend if we use another unit called radian.

The rotational motion, on the other hand, can be quite a complicated affair but we shall restrict ourselves here to the *simplest* case of rotation, *viz., the rotation of a rigid body about a line or axis fixed in the reference frame in which we are making our observation,* this line or axis being referred to as the axis of rotation of the body. A given point in the body thus obviously moves in a plane perpendicular to this axis of rotation and the rotation is referred to as *plane rotation or rotation in two dimensions.*

Let us consider a rigid body free to rotate about the axis OZ through O, with the axis fixed in an *inertial reference frame* (Fig. 1.1).

All the particles of the body describe circles about the axis of rotation (OZ) in a plane perpendicular to it and have their centres lying on it, with their radii equal to their respective perpendicular distances from it (except, of course, those that lie on the axis itself) and the radius of each obviously sweeps through the same angle $\delta\theta$, say, in the same time δt.

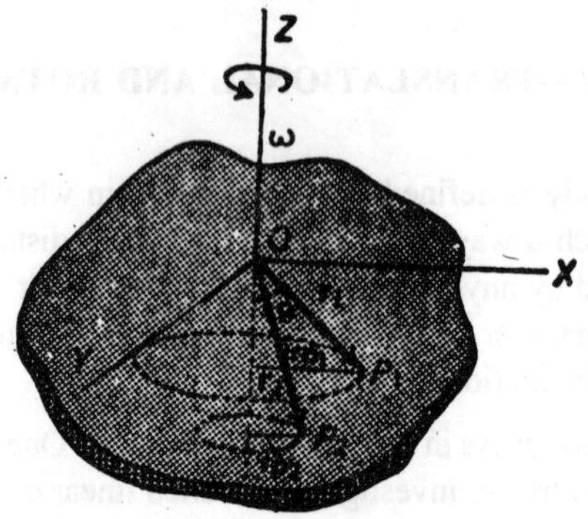

Fig. 1.1

Thus, for example, particles P_1 and P_2, with position vectors r_1 and r_2 with respect to origin O and distant r_{p1} and r_{p2} from the axis, describe circles of radii $P_1A = r_{p1}$ and $PB = r_{p2}$ with their centres at A and B respectively, where P_1A and P_2B are perpendiculars dropped from P_1 and p_2 on to the axis, and both sweep through the same angle $\delta\theta$ in time δt.

The angular velocity of both P_1 and P_2, and in fact of all the particles, and hence of the body, as a whole, is thus the same, *viz.* $\omega = \delta\theta/\delta t$ which, in the limit $\delta t \rightarrow 0$ gives $\omega = d\theta/dt$. Since infinitesimally small angular displacements are vectors, we have $\omega = d\theta/dt$ (a vector quantity $d\theta$ divided by a scalar quantity dt) also a vector, the direction of which (in accordance with the right hand rule) is along the axis of rotation (or the Z-axis), upward, if the rotation of the body be anticlockwise about the axis. This is indicated in the figure by the thick arrow, the length of which is proportional to the magnitude of ω.

It may be noted that nothing actually moves in the direction of the angular velocity vector; *it merely represents conventionally the angular velocity of the rotational motion* that is taking place in the plane perpendicular to it (*i.e.*, in the x-y plane, in the case shown). The *angular acceleration* of the body is given by $\alpha = d\omega/dt$ which, again, is a vector, since, as before, a vector ($d\omega$) is divided by a scalar (dt).

The *linear speed* of a particle P is, as we know, given by $v = \omega r_p$ (*i.e., its angular velocity × radius of the circle in which it rotates*), where v and ω are respectively the *magnitudes* of the linear speed and the angular velocity.

Now, in Fig. 1.2, if **r** be the position vector of a particle *P* of a rigid body with respect to the origin O, it obviously rotates about the axis in a circle of radius $r_p = r \sin POA = r \sin \theta$, where θ is the angle that the position vector **r** makes with the axis. So that, $v = \omega r \sin\theta$. And therefore, *linear velocity vector* $v = \omega \times r$, its direction being *tangential* to the *circular path at* P and *perpendicular to the position vector* r, as shown.

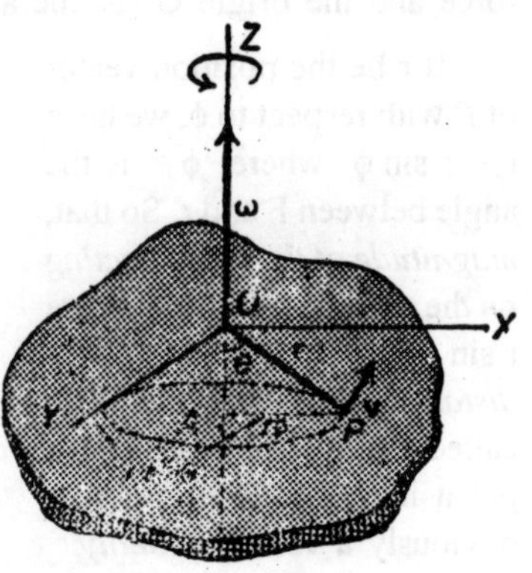

Fig. 1.2

As mentioned already, we are concerned here with only this plane rotational motion of a rigid body and the study of its relationship with the properties of the body or the causes of its rotation, and we call it the study of the *rotational dynamics* of the rigid body.

As will be readily seen, however, and as will be further observed as we proceed along, the analogous translational and rotational parameters, like linear and angular displacements (r and θ), linear and angular velocities (v and ω), linear and angular acceleration (a and $\alpha = d\omega/dt$) etc., differ dimensionally in most cases only by a length factor r; for example, $v = r\omega$, $a = r\, d\omega/dt$ and so on.

TORQUE

The one important difference between translational and rotational motion that is clearly apparent is that for linear acceleration, in the case of the former, we need to know only the magnitude of the force applied; for angular acceleration in the case of the latter, we must know also the actual point of application of the force and the way it is directed, or, in other words, the *moment of the force* which, in the terminology used in the context of rotational motion, we refer to as the *torque* (from the Latin word *torquere* meaning *to twist*).

Thus, in Fig. 1.3, if F be the *force* acting on a particle *P* of a rigid body, the *moment of the force* about the origin ϕ in an inertial reference frame, or the *torque* τ acting on the particle, with respect to O, is F × r_p, where r_p is the perpendicular distance between the line of action of the force and the origin O (or the axis passing through O).

If r be the position vector of *P* with respect to ϕ, we have $r_p = r \sin\phi$ where ϕ = is the angle between F and r. So that, *magnitude of the torque acting on the particle* = Fr sin θ where r sin $\phi = r_p$, *the perpendicular distance between* F and O, is called the *moment arm.* Or, to put it in vector form, for it is obviously a *vector quantity,*

we have $\tau = r \times F$.

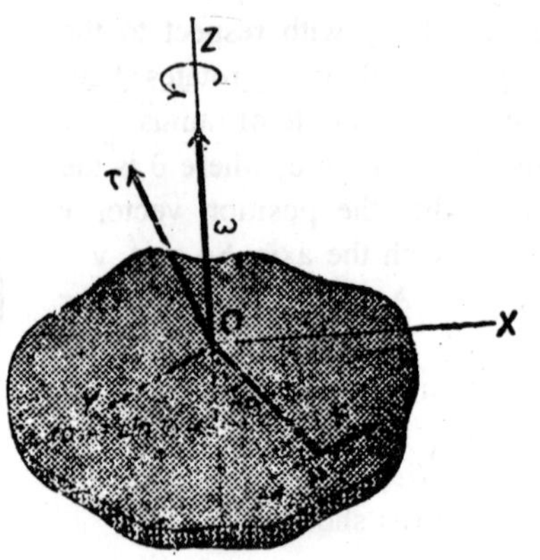

Fig. 1.3

The *direction* or *sense* of the torque is as shown, given by the *right-handed screw rule,* according to which if we curl the fingers of our right hand in the direction in which r must be swung to move into the position of F, through the smaller angle between them, the extended thumb gives the direction or the sense of the vector τ.

Obviously, if r = 0, *i.e.,* if the point *P* lies at the origin O, the torque is zero. In other words, the torque about the origin itself is zero.

It will also be seen that the magnitude of the torque may be put either as $\tau = F(r \sin\phi)$ or Fr ⊥ or as $\tau = r\,(F \sin\phi)$ or rF⊥, where r sin ϕ or r⊥ represents the component of r perpendicular to the line of action of F and *F* sin θ or F⊥ the component of F perpendicular to r.

It is thus clear that *only the component of* F *perpendicular to* r is *responsible for the torque.* For, if θ = 0 or π, this perpendicular component of F, and hence also the torque, vanishes; for, then F sin θ = 0 and the line of action of F thus passes through the origin O.

It may as well be noted that reversing either F or r reverses only the direction of the torque, its magnitude remaining unaffected (because a reversal of the direction of one vector in a vector product means a reversal of direction of the product itself). If, however; both r and F be reversed simultaneously, both magnitude and direction of τ remain unaffected (because a reversal of the direction of both the vectors in a vector product leaves the direction of the product unaffected).

And, obviously, *for a rigid body,* consisting of a large number of particles, the torque

$$\tau = \Sigma \, r \times F$$

It will be seen at once that *torque has the same dimensions as work* (both being force times distance), viz., ML^2T^{-2}. The two are, however, entirely different physical quantities; *whereas work is a scalar quantity, torque is a vector*. To distinguish between the two, therefore, we express work in units like ergs (dyne × cm) in the C.G.S. system, joules (N × . m) in the MKS or SI system and foot-pound (ft × lb) in the F.P.S. system, and torque in *dyne-cm, N-metre* and lb-ft in the three systems respectively.

ANGULAR MOMENTUM–ANGULAR IMPULSE

Although, we have already dealt with angular momentum in connection with angular momentum in may recapitulate some salient points here.

Angular momentum, in rotational motion, is the analogue of linear momentum, in translational motion, and is defined as the *moment of linear momentum.*

Now, linear momentum for a particle of mass m, moving with velocity v is given by p = mv and for system of particles or a rigid body, by $P = M \times v_{c.m.}$, where M is the mass of the whole system or body and $V_{c.m.}$, the velocity of its *centre of mass.*

Thus, considering a particle of a rigid body, free to turn about the fixed axis OZ through O (Fig. 1.4), if its linear momentum p be directed as shown, such that the momentum vector makes an angle θ with its position vector r (with respect to the origin O) in an initial frame of reference, its regular momentum is given by

$$J_p = r \times p = r \times mv = m \times (r\omega)$$

It is obviously a vector and its magnitude is given by $J_p = pr \sin \theta$, with its sense or direction given by the usual right handed screw rule for the vector product of two vectors as in the case of torque in 1.3 above. So that, in the case shown, it is perpendicular to the plane of r and p, directed as indicated.

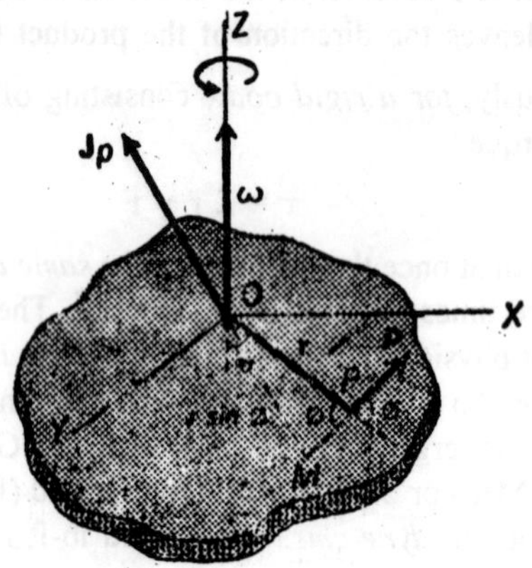

Fig. 1.4

This direction of the angular momentum vector J_p, it will be noted, is not, in general, the same as that of the angular velocity vector ω, for the simple reason that J_p is not the product of a scalar with ω.

Again, as in the case of the torque, so also here, we may write the magnitude of J_p as $J_p = p\ (r \sin \theta)$ or pr ⊥, or, as $J_p = r\ (p \sin \theta)$ or rp ⊥, where $r \sin \theta = r\perp$, called the *moment arm*, is the component of r at right angles to the line of action of p and $p \sin \theta = p\perp$, the component of p at right angles to r.

It follows, therefore, that *only the component of p perpendicular to r is responsible for the angular momentum.* For, if angle θ between r and p be *zero* or π the line of action of p passes through the origin O, and p sin θ being zero, the angular momentum J_p is also zero.

Obviously enough, *the angular momentum of the whole rigid body (J) about the point O will be the vector sum of the angular momenta of all the particles constituting it about the same point (O), i.e.,*

$$J = \Sigma mr \times v = \Sigma mr \times (\omega \times r). \qquad ...(i)$$

its direction, as pointed out earlier, being not, in general, the same as that of ω.

As shown under 6.9, *the rate of change of angular momentum, i.e.,*

$$\frac{dJ}{dt} = \Sigma r \times \frac{d}{dt}(mv) = \Sigma r \times F = \tau, \quad ...(ii)$$

where $\Sigma r \times F$ is the sum of the torques, or the total torque, *due to all the external forces acting on the system or the rigid body*. The reason why we explicity use the word *external* as explained already under 6.9, is that internal forces all form *collinear pairs of opposite forces* (of *action* and *reaction* in accordance with Newton's third law of motion), having equal and opposite moments about the given point, so that their sum is zero and they produce no effect.

This relation (ii) is the *fundamental equation o motion* of a rigid body about a fixed axis through a given point O (or about O) and *applies irrespective of whether the particles constituting the system are in motion relative to each other or in fixed spatial relationship with each other, as in a rigid body.*

Further, as will be easily seen, this relation is the rotational analogue of the familiar relation F = dp/dt in translatory motion (*i.e.*, Newton's second law of motion). For, just as the rate of change of linear momentum (dp/dt) of a particle (or a body) is equal to the force (F) acting upon it, so also, the rate of change of angular momentum (dJ/dt) of a particle (or a body) is equal to the torque (τ) acting upon it, *i.e., angular momentum has here been substituted for linear momentum and torque for force.*

Now, since the direction of the angular momentum vector J is not, in general, the same as that of the angular velocity vector ω, the torque vector τ will not, in general, lie along the vector ω, *i.e.*, along the axis of rotation. Only a component of it will thus lie along the axis, with its other component perpendicular to it. The former component alone, however, produces rotation of the body about the axis, the latter merely tending to turn the axis away from its fixed position.

Let us, therefore, proceed to obtain the value of this *axial component of J and hence also that of the axial component of torque* τ.

Simplifying the triple vector product in expression I for J above, with the help of the vector identity A × (B × C) = B (A.C) – C (A.B), where A may be replaced by r, B by ω and C by r, we have

$$J = \Sigma m\ [r \times (\omega \times r)] = \Sigma m\ [w(r.\rho r) - r(r.r\ \omega)]$$

$$\text{Or, } J = \Sigma m\ [r^2\omega - (r.\omega)r]$$

Since the angle between r and ω is φ (Fig. 1.4), the magnitude of the component of r along the direction of ω, *i.e.*, along the axis, is r cos φ, as shown separately in Fig. 1.5 for the sake of clarity. And, therefore, ***magnitude of the component of J along the axis, i.e., its axial component***, is given by

$$J_A = \Sigma m[r^2\omega - (r\omega \cos \phi)\, r \cos \phi)]$$

$$= \Sigma m r^2 \omega\, (1 - \cos^2 \phi) = \Sigma m r^2 \omega \sin^2 \phi.$$

Or, since $r \sin \phi = r_p = PA$ and ω is the same for all particles of the body, we have $J_A = \Sigma m r_p^2$.

This summation $\Sigma m r_p^{\,2}$ is called *rotational inertial* or the *moment of inertia* of the body about the axis of rotation and is represented by the symbol I. So that, *axial component of angular momentum*, $J_A = I\omega$). Since torque $\tau = dJ/dt$, we have *axial component of the torque,*

i.e., $\tau_A = \dfrac{dJ_A}{dt} = \dfrac{d}{dt} I\omega$.

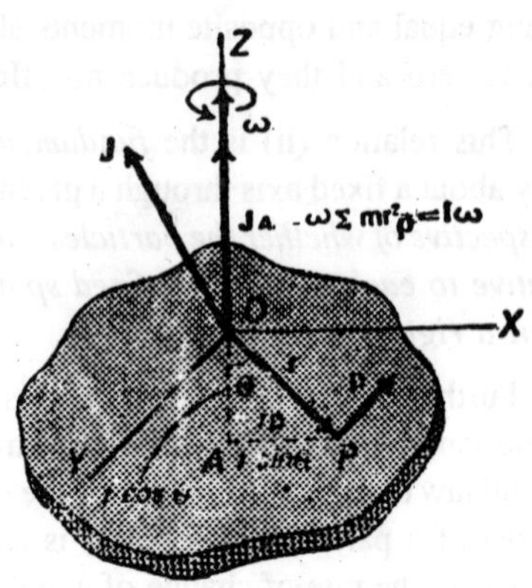

Fig. 1.5

Or, since I is constant for the body for a given axis of rotation, we have $\tau_A = \dfrac{I d\omega}{dt}$.

If, however, the axis of rotation passes through the centre of mass of the body, and coincides with its axis of symmetry, the components of the angular momenta of the various particles, and hence also those of the torque perpendicular to the axis, form equal and collinear pairs and thus cancel out. The entire angular momentum vector J, and hence also the entire torque vector τ, then lies along the angular velocity vector. There is thus no component of the torque tending to turn the axis away from its fixed position and the whole of it is, there we, effective in producing rotation of the body.

In such a case, in which the axis of rotation passes through the centre of mass and coincides with the axis of symmetry of the body, we drop the subscript (A) from J and τ and simply write

$$J = I\omega \quad \text{and} \quad \tau = \frac{Id\omega}{dt}.$$

With the axis of rotation passing through the centre of mass of the body, these relations for J and τ still hold good even if the axis of rotation be not fixed in an inertial frame but be a moving one, *provided that it always moves parallel to itself, i.e.,* its position at one instant is parallel to that at another.

To summarise, then, the relation $\tau = I\ dw/dt$ is applicable (i) when the axis of rotation passes through the centre of mass of the body and is itself its axis of symmetry (even though it may be moving parallel to itself) or (ii) when the axis of rotation is a fixed one.

Angular Impulse

We have the relation $I\ d\omega/dt = \tau$, whence, we have $I\ d\omega = \tau\ dt$, where τ is the torque acting on the body.

Integrating with respect to t, we have *angular momentum* = $I\omega = \int_0^t \tau\, dt$, the expression being true *howsoever* τ *may vary with time.*

If, however, τ be constant, we have $I\omega = \int_0^t \tau\, dt = \tau t$.

If t be small and τ quite large, the expression $\int_0^t \tau\, dt$ is called the *angular impulse* given to the body and, as we have just seen, with τ constant, it is equal to τt.

Also if I remains constant during the time t, we may express it as

$$\int_0^t \tau\, dt = \int I \frac{d\omega}{dt} dt = \int_{\omega_1}^{\omega_2} I\, d\omega = I(\omega_2 - \omega_1).$$

GENERAL THEOREMS ON MOMENT OF INERTIA

There are two important theorems on moment of inertia which, in some cases, enable the moment of inertia of a body to be determined about an axis, if its moment of inertia about some other axis be known. Let us now proceed to discuss these.

(i) The Theorem (or Principle) of Perpendicular Axes

(a) *For a plane laminar body :* The theorem states that *the moment of inertia of a plane lamina about an axis perpendicular to the plane of the lamina is equal to the sum of the moments of inertia of the lamina about two axes perpendicular to each other, in its own plane, and intersecting each other at the point where the perpendicular axis passes through it.*

Thus, if I_x and I_y be the moments of inertia of a plane lamina (Fig. 1.6), about the perpendicular axis OX and OY respectively, which lie *in the plane of the lamina* and intersect each other at O, its moment of inertia (I) about an axis passing through O and *perpendicular to its plane* is given by $I = I_x + I_y$.

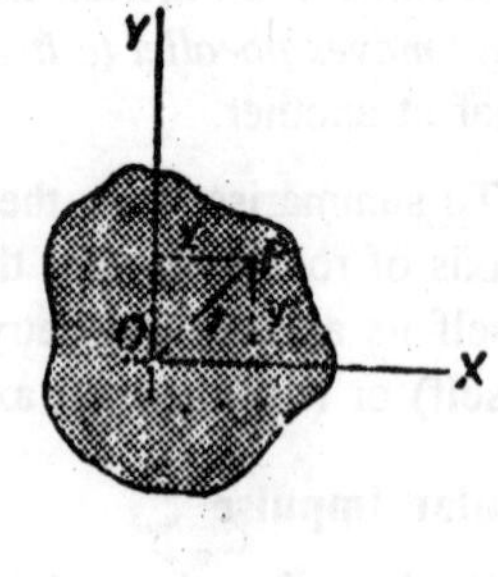

For, considering a particle of mass m at P, distant r from O and x and y from the axes OY and OX respectively, we have

Fig. 1.6

$$I = \Sigma mr^2,\ I_x = \Sigma my^2 \text{ and } I_y = \Sigma mx^2.$$

So that, $I_x + I_y = \Sigma my^2 + \Sigma mx^2 = \Sigma m(y^2 + x^2) = \Sigma mr^2,$

$$[\because y^2 + x^2 = r^2]$$

i.e., $\quad I_x + I_y = I.$

(b) *For a three-dimensional body :* In this case, the theorem demands that *the sum of the moments of inertia of a three-dimensional body about its three mutually perpendicular axes is equal to twice the summation Σmr^2 about the origin.* This may be seen from the following.

Suppose we have a three-dimensional or a cubical body, shown dotted in Fig. 1.7, with OX, OY and OZ as its three mutually perpendicular axes, representing its *length, breadth and height respectively.*

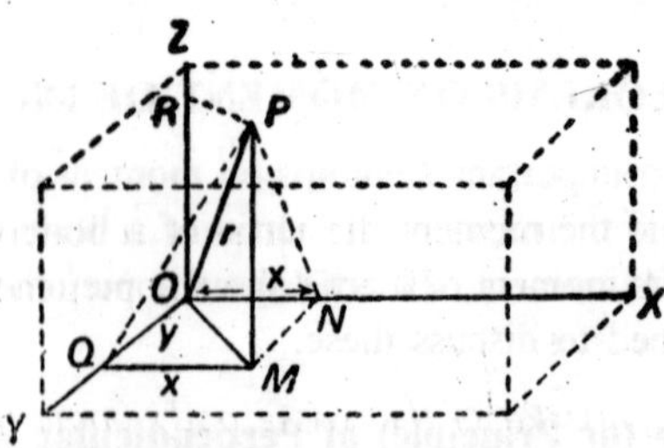

Fig. 1.7

Consider a small element of mass m of the body at a point P somewhere inside it. Drop a perpendicular PM from P on the x – y plane to meet it in M. Join OM and OP and from M draw MQ parallel to the x-axis and MN parallel to the y-axis. Also, from P draw PR parallel to OM, Then, clearly, the coordinates of the point P are x = ON = QM, y = OQ = NM and z = MP = OR.

Since plane x – y is perpendicular to the z-axis, any straight line drawn on this plane is also perpendicular to it. So that, both OM and PR are perpendicular to the z-axis (PR being parallel to OM).

Clearly, therefore, ∠OMP is a *right angle* [∵ OM ∥ PR and PM ∥ OR]. We, therefore, have

$OM^2 + MP^2 = OP^2$. Or, $OM^2 + z^2 = r^2$. ...(i) [∵ OP = r.]

But $OM^2 = QM^2 + OQ^2 = x^2 + y$.

[∵ OQM is a right angle, being the angle between the axes of x and y].

∴ substituting the value of OM^2 in relation (i) above, we have

$$x^2 + y^2 + z^2 = r^2. \quad ...(ii)$$

Join PN and PQ. Then, PN and PQ are the respective normals to the axes of x and y. For ∠PMN being a *right angle* (the angle between the axes of y and z), we have $PN^2 = MN^2 + PM^2 = y^2 + z^2$ and, therefore, $x^2 + PN^2 = x^2 + y^2 + z^2 = r^2$. Or, $ON^2 + PN^2 = r^2$ (∵ = x = ON). Angle PNO is thus a right angle and, therefore, PN perpendicular to the x-axis.

Similarly, in the right-angled ΔPMQ, we have

$$PQ^2 = MQ^2 + PM^2 = x^2 + z^2.$$

Again, because $x^2 + y^2 + z^2 = r^2$, we have $PQ^2 + y^2 = r^2$.

Or, $PQ^2 + OQ^2 = OP^2$, [∵ y = OQ and r = OP]

Angle PQO too is thus a right angle and, therefore, PQ perpendicular to the y-axis.

Now, moment of inertia of the element at P about the z-axis = m × PR^2 = m × OM^2, because PR = OM is the perpendicular distance of the mass from the axis.

∴ *moment of inertia of the whole body about the z-axis, i.e.,*

$$I_x = \Sigma m.OM^2 = \Sigma m(x^2 + y^2).$$

Similarly, *moment of inertia of the body about the y-axis. i.e.,*

$$I_y = \Sigma m.PQ^2 = \Sigma m(x^2 + z^2).$$

because PQ is the perpendicular distance of mass m from this axis. And, *moment of inertia of the body about the x-axis, i.e.,*

$$I_x = \Sigma m\, PN^2 = \Sigma m(y^2 + z^2),$$

because PN is the perpendicular distance of mass m from the x-axis.

∴ adding up the moments of inertia of the body about the three axes, we have

$I_x + I_y + I_z = \Sigma m(y^2 + z^2) + \Sigma m(x^2 + z^2) + \Sigma m(x^2 + y^2) = 2\Sigma m(x^2 + y^2 + z^2) = 2\Sigma mr^2.$

(ii) The Theorem (or Principle) of Parallel Axes

This theorem, due to *Steiner, is true for both a plane laminar body and a thin three-dimensional body, and states that the moment of inertia of a body about any axis is equal to its moment of inertia about a parallel axis through its centre of mass, plus the product of the mass of the body and the square of the distance between the two axes.*

(a) *For a plane laminar body :* Let AB be the axis in the plane of the paper about which the moment of inertia (I) of the *plane lamina,* shown in Fig. 1.8, is to be determined and CD, an axis parallel to AB, through the *centre of mass* O of the lamina, at a distance r from AB.

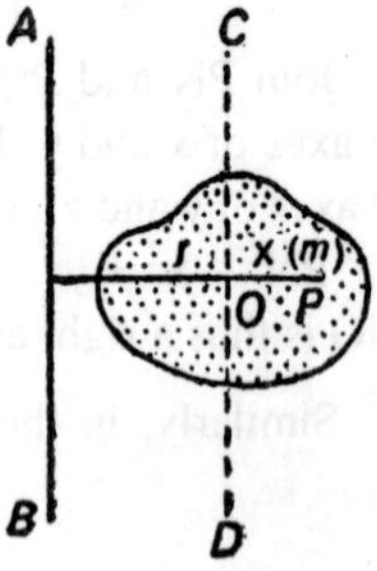

Fig. 1.8

Considering a small element of mass m of the lamina at the point P, distant x from CD, we have *moment of inertia of the element about the axis AB* = m (x + r)^2.

And, therefore, *moment of inertia of the whole lamina about the axis AB, i.e.,* $I = \Sigma m(x + r)^2$.

Or, $\quad I = \Sigma mx^2 + \Sigma mr^2 + 2\Sigma mxr.$

Obviously, $\Sigma mx^2 + I_{c.m.,}$ the moment of inertia of the lamina about the axis CD, through its centre of mass. So that,

$$I = I_{c.m.,} + \Sigma mr^2 + 2\Sigma mxr.$$

Now, $\Sigma mr^2 = r^2\Sigma m$ (r being constant) $= Mr^2$, where M is the mass of the whole lamina, and Σmx = the sum of the moments of all the particles of the lamina about the axis CD, passing through its centre of mass and, therefore, equal to zero; for, as we know, a body always balances about its c.m., showing that the algebraic sum of the moments of all its particles about the c.m. is zero.

We, therefore, have $I = I_{c.m.} + Mr^2$, *i.e., the moment of inertia of the lamina about the axis AB = its moment of inertia about a parallel axis CD through its centre of mass + mass of the lamina × (distance between the two axes)^2.*

(b) *For a three-dimensional body :* Let AB be the axis about which the moment of inertia of a cubical or a three-dimensional body, shown dotted in Fig. 1.9, is to be determined.

Draw a parallel axis CD through the centre of mass O of the body at a distance r from it.

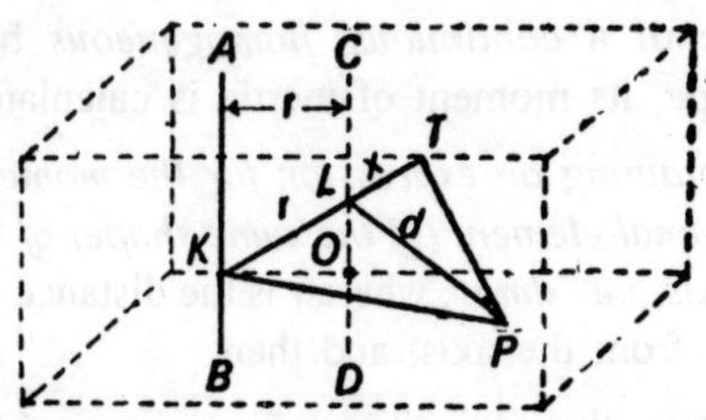

Fig. 1.9

Imagine an element of the body, of mass m, at a point P outside the plane of the axes AB and CD and let PK and PL be perpendiculars drawn from P to AB and CD respectively and PT, the perpendicular dropped from P on to KL produced.

Put PL = d, LK = r, LT = x and $\angle PLK = \theta$.

Then, if I be the moment or inertia of the body about the axis AB and $I_{c.m.}$, its moment of inertia about the axis CD (through its centre of mass O), we have

$$I = \Sigma m.PK^2 \quad \text{and} \quad I_{c.m.} = \Sigma mPL^2 = \Sigma md^2.$$

Now, from the geometry of the figure, we have $PK^2 = PL^2 + LK^2 - 2PL.LK \cos PLK = d^2 + r^2 - 2d.r \cos\theta$.

And, in the right-angled triangle PTL, we have cos PLT = LT/PL, where $\angle PLT = 180° - \angle PLK = 180° - \theta$. So that, $\cos(180 - \theta) = x/d$. Or, $-\cos\theta = x/d$, whence $d \cos\theta = -x$.

Substituting this value of d cos θ in the expression for PK^2 above, we have $PK^2 = d^2 + r^2 + 2\,rx$. And, therefore, $I = \Sigma m\, PK^2 = \Sigma m(d^2 + r^2 + 2rx) = \Sigma md^2 + \Sigma mr^2 + 2r\Sigma mx$.

Or, $I = I_{c.m.} + Mr^2 + 2r\, Smx$

[$\because$ Σm = M, mass of the body and r is constant.]

Clearly Σmx = 0, being the total moment about an axis passing through the centre of mass of the body. We, therefore, have

$$I = I_{c.m.} + Mr^2,$$

the same result as for a plane laminar body.

CALCULATION OF MOMENT OF INERTIA

In the case of a *continuous, homogeneous* body of a definite geometrical shape, its moment of inertia is calculated by :

(i) *First obtaining an expression for the moment of inertia of an infinitesimal element (of the same shape) of mass dm about the given axis, i.e., dm.r*^{2}, where r is the distance of the infinitesimal element from the axis, and then

(ii) *Integrating this expression over appropriate limits so as to cover the entire body*, making full use of the theorems of perpendicular and parallel axes, where necessary, as we shall shortly.

The moments of inertia of certain bodies of a regular geometrical shape *about any one of their three perpendicular axes of symmetry. passing through their centres of mass* may, however, be directly obtained by the application of what is known as *Routh's rule*. According to this rule:

(i) *The moment of Inertia of a rectangular lamina or a parallelopiped* about any one of these axes (passing through its centre of mass) is given by

I = mass of the lamina × 1/3 (sum of the squares of the other two semi-axes).

Thus, considering a rectangular lamina of mass M, length *l* and breadth b, (Fig. 1.10), its moment of inertia about an axis passing through its centre and perpendicular to its plane is given by

$$I = M\left[\frac{(l/2)^2 + (b/2)^2}{3}\right] = M\left(\frac{l^2 + b^2}{12}\right),$$

the two *semi-axes* being $\frac{l}{2}$ and $\frac{b}{2}$.

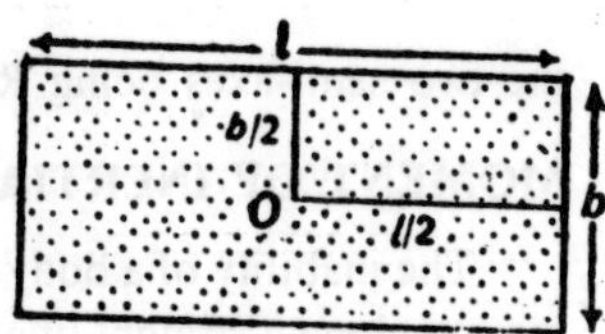

Fig. 1.10

(ii) *The moment of inertia of a circular or elliptical lamina* about any one of the three axes of symmetry (and passing through its centre mass) is given by

= *mass of the lamina* × *1/4(sum of the squares of the other two semi-axes)*

Thus, in the case of a circular lamina of mass M and radius R Fig. 1.11(a) its moment of inertia about an axis passing through its centre and perpendicular to its plane is given by

$$I = M\left(\frac{R^2 + R^2}{4}\right) = M\frac{R^2}{2},$$

the two *semi axes* being R and R.

And in the case of an *elliptical lamina* of *mass M and major* and *minor axes* 2a and 2b respectively, [Fig. 1.10 (b)], its moment of inertia about an axis passing through its centre and perpendicular to its plane is given by

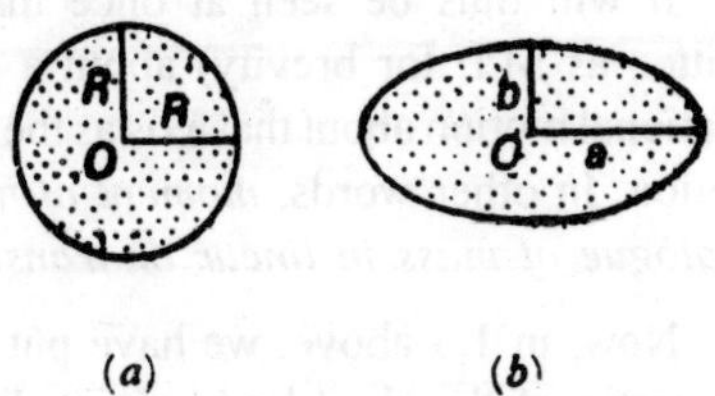

Fig. 1.11

$$I = M\left(\frac{a^2 + b^2}{4}\right),$$

the two semi-axes being a and b.

(iii) *The moment of inertia of a sphere or spheroid* about any axis of symmetry (passing through the centre of mass) is given by

I = mass of the sphere or spheroid × 1/5 (*sum of the squares of the other two semi-axes*)

Thus, the moment of inertia of a solid sphere, of mass M and radius R_4 about any axis of symmetry, *i.e.,* about any diameter, is given by

$$I = M\left(\frac{R^2 + R^2}{5}\right) = \frac{2}{5}MR^2,$$

the two semi-axes being R and R.

MOMENT OF INERTIA–RADIUS OF GYRATION

The inability of a body to change by itself its state of rest or uniform motion along a straight line (Newton's first law of motion) is an inherent property of matter and is called *inertia.* The greater the mass of the body, the greater the resistance offered by it to any change in its state of rest or linear motion. So that, *mass is taken to be a measure of inertia for linear or translatory motion.*

In exactly the same manner, a body free to rotate about an axis opposes any change in its state of rest or uniform rotation. In other words, it possesses inertia for rotational motion, *i.e.,* it opposes the torque or the moment of the couple applied to it to change its state of rotation, for only a couple can oppose another couple. Hence the name *moment of inertia* given to it.

It will thus be seen at once that the moment of inertia (usually written as M.I. for brevity) about a given axis plays the same part an rotational motion about that axis as the mass of a body does in translational motion. In other words, *moment of inertia, in rotational motion, is the analogue of mass in linear or translational motion.*

Now, in 1.3 above, we have put $\Sigma m r_p^2 = I$, where I is the moment of inertia of the rigid body about the axis of rotation.

This means, obviously, that if r_1, r_2, r_3 etc. be the perpendicular distances from the axis, of particles of respective masses m_1, m_2, m_3, etc., we have $I = \Sigma mr^2 = m_1r_1^2 - m_2r_2^2 + m_3r_3^2 + ...$

We may, therefore, define the *moment of inertia of a rigid body about a given axis of rotation as the sum of the products of the masses of the various particles of the body and the squares of their respective distances from the axis.* We usually put it as

$$I = \Sigma mr^2 = MK^2,$$

where M is the *total mass of the body* (*i.e.,* equal to $m_1 + m_2 + m_3$ etc.) and K, the *effective distance of its particles from the axis,* called its *radius of gyration about the axis of rotation.*

In the case of a body which does not consist of separate, discrete particles but has a continuous and homogeneous distribution of matter in it, the summation becomes an integration, the integral being taken over the entire body. So that, $I = \int r^2 dm = MK^2$, where dm is the mass of an infinitesimally small element of the body at distance r from the axis.

Also, since for a body free to rotate about a fixed axis of rotation, we have $\tau = I\, d\omega/dt = 1$, $I = \tau$, *i.e., the moment of inertia of a body about a fixed axis of rotation may also be defined as the torque that must be applied to the body to produce unit angular Acceleration in it about that axis.*

The kinetic energy of rotation of a body rotating about a fixed axis $= \frac{1}{2} I\omega^2$. So that, if ω be *unity*, the kinetic energy of rotation of the body $= \frac{1}{2} I$ and, therefore, I = 2 (*kinetic energy of rotation*).

This gives us another method of defining moment of inertia, *viz., that the moment of inertia of a body about a fixed axis of rotation is twice the kinetic energy of relation of the body, rotating about that axis with unit angular velocity.*

Radius of Gyration

It follows from the above discussion that K, the *radius of gyration* of a body about its axis of rotation, which we have called the *effective distance* of its particles from the axis, is actually *that distance from the axis at which if its entire mans (M) be supposed to be concentrated, its moment of inertia about the given axis would be the same as with its actual distribution of mass.*

Clearly, therefore, a change In the position or inclination of the axis of rotation of a body will bring about a change in the relative distances of its particles and hence in their *effective distance* or the radius of gyration of the body about the axis. And so it will a change in the distribution of mass about the axis, *e.g.*, the transference of a portion of matter (or mass) of the body from one part to another, despite the fact that the total mass of the body remains unaltered.

Thus, whereas the mass of a body remains the same irrespective of the location or inclination of its axis of rotation, its radius of gyration about the axis depends upon (i) *the position and inclination of the axis and (ii) the distribution of mass of the body about the axis.*

This may be illustrated by a simple example: As we shall presently see, the M.I. of a circular hoop or ring about an axis passing through its centre and perpendicular to its plane (MR^2) is twice the M.I. of a circular disc of *exactly* the *same mass and radius and about an identical axis, i.e.,* an axis passing through its centre and perpendicular to its plane, ($MR^2/2$). The masses in the two cases being the same, obviously, K^2 for the hoop or ring is twice as large as K^2 for the disc, due, no doubt, to the average, and hence the effective, distance of the particles from the axis of rotation being greater in the case of the hoop or ring than in that of the disc.

DIMENSIONS AND UNITS OF MOMENT OF INERTIA

Since the moment of inertia of a body about a given axis of rotation is MK^2, where M is the *mass* of the body and K, its *radius of gyration* about the axis, *its dimensions* are 1 *in mass*, 1 *in length* and *zero in time.* Its dimensional formula is thus $ML^2\ T^0$ or ML^2.

Obviously, therefore, it is expressed in *gm-cm²* in the C.G.S. system, kg-m² in the MKS (or SI) *system* and lb-ft² in the F.P.S. *system.*

Further, it may be noted that since the moment of inertia of a body about a given axis remains unaffected by reversing its direction of rotation about that axis, it is a *scalar quantity*, once its axis of rotation is fixed.

It follows, therefore, that the total M I of a number of bodies about a given axis is equal to the sum of the moments of inertia of the individual bodies about that axis, *i.e.,* $I = I_1 + I_2 + I_3 + ...$, where I is the total M.I. of the bodies and I_1, I_2, I_3 etc. their individual moments of inertia about the given axis.

For the same reason, if a body has a M.I. about a given axis equal to I, and a portion of it, having a moment of inertia I' about the same axis be removed from it, the M.I. of the remaining part of the body about the given axis will be I' – I.

ANALOGOUS PARAMETERS IN TRANSLATIONAL AND ROTATIONAL MOTION

We have already come across some analogous parameters in translational and rotational motion, like *linear and angular velocity, linear and angular acceleration.* There are, of course, many others, the important ones among which are given here in tabular form for purposes of ready reference, along with the symbols commonly used for them and their various relationships.

Translational motion	Rotational motion
1. Mass (or translational inertia)–m or M	Moment of inertia – $I = MK^2$ (or rotational inertia)
2. Distance covered (or linear displacement)–s	Angle described–θ (radians) (or angular displacement)
3. Linear velocity–$v = ds/dt$	Angular velocity–$\omega = d\theta/dt$ (radian/sec)
4. Linear acceleration–$a = dv/dt = d^2s/dt^2$	Angular acceleration–$\alpha = d\omega/dt = d^2\theta/dt^2$
5. Linear momentum–$p = mv$	Angular momentum–$J = I\omega$
6. Force (or rate of change of linear momentum)–$F = dp/dt = ma$	Torque or moment of couple (or rate of change of angular momentum)– τ or $C = dJ/dt = I\, d\omega/dt$
7. Work done by force, $W = \int F ds = Fs$	Work done by torque or couple– $W = \int \tau d\theta$ or $\int C d\theta$ or $\tau\theta$ or $C\theta$.
8. Power–$P = Fv$	Power–$P = \tau\omega$ or $C\omega$
9. Translational kinetic energy $= \frac{1}{2}mv^2$	Rotational kinetic energy $= \frac{1}{2}I\omega^2$
10. Equations of translational motion:	Equations of rotational motion:
(i) $v = u + at$	(i) $\omega_2 = \omega_1 + \alpha t = \omega_1 + (d\omega/dt)\, t$
(ii) $s = ut + \frac{1}{2}at^2$	(ii) $\theta = \omega_1 t + \frac{1}{2}\alpha t^2 = \omega_1 t + \frac{1}{2}\left(\frac{d\omega}{dt}\right)t^2$
(iii) $v^2 - u^2 = 2\, as$	(iii) $\omega_2^2 - \omega_1^2 = 2\alpha\theta = 2\left(\frac{d\omega}{dt}\right)\theta$

Further, corresponding to Newton's three laws of translational motion, we have the following *three laws of rotational motion*:

1. Unless an external torque is applied to it, the state of rest or uniform rotational motion of a body about its fixed axis of notation remains unaltered.
2. The rate of change of angular momentum (or the rate of change of rotation) of a body about a fixed axis of rotation is directly

proportional to the torque applied and lakes place in the direction of the torque.

3. When a torque is applied by one body on another, an equal and opposite torque is applied by the latter on the former about the same axis of rotation.

PARTICULAR CASES OF MOMENT OF INERTIA

Let us now see how to obtain expressions for moments of inertia in some typical and important cases, with regular geometrical shapes.

1. Moment of Inertia of a Uniform Rod

(i) *about an axis through its centre and perpendicular to its length.* Let AB be a *thin uniform* rod, of *length l* and mass M, free to rotate about an axis YOY' passing through its centre O and perpendicular to its length, as shown in Fig: 1.12. Since the rod is uniform, its *mass per unit length* is clearly M/*l*.

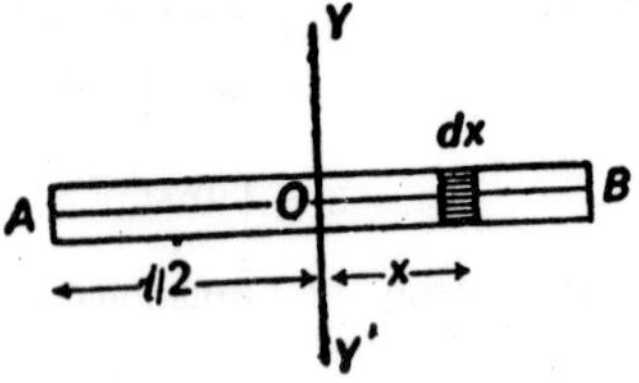

Fig. 1.12

Considering a small element of the rod, of length dx at a distance x from the axis through O, we have *mass of the element* = (M/*l*).dx and, therefore, its M.I. about the axis (YOY') through O = (M/*l*).dx.x^2.

The moment of inertia (I) of the whole rod about the axis YOY' is thus given by the integral of the above expression between the limits x = – *l*/2 and x = + *l*/2 or by twice its integral between the limits x = 0 and x = *l*/2, *i.e.,*

$$I = 2\int_0^{l/2} \frac{M}{l} x^2 dx = \frac{2M}{l}\left[\frac{x^3}{3}\right]_0^{l/2} = \frac{2M}{l}\cdot\frac{l^3}{24} = \frac{Ml^2}{12}$$

(ii) *about an axis through one end of the rod and perpendicular to its length.*

Proceeding as in case (i) above, we obtain the moment of inertia of the rod about the axis, now passing through one end A of the rod, by integrating the expression for the M.I. of the element dx of the rod, between the limits x = 0 at A and x = l at B, *i.e.,*

$$I = \int_0^l \frac{M}{l} x^2 dx = \frac{M}{l}\left[\frac{x^3}{3}\right]_0^l = \frac{Ml^2}{3}$$

Alternatively, we could obtain the same result by an application of the principle of parallel axes, according to which

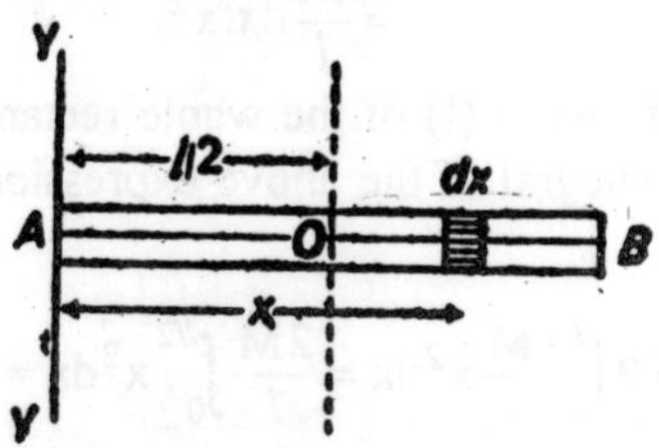

Fig. 1.13

M.I. of the rod about the axis Y A Y' = its M.I. about a parallel axis through O (mass of the rod × square of the distance between the two axes).

So that, $I = \frac{Ml^2}{12} + M\left(\frac{l}{2}\right)^2 + \frac{Ml^2}{12} + \frac{Ml^2}{4} = \frac{Ml^2}{3}$.

2. Moment of Inertia of a Rectangular Lamina (or Bar)

(i) *about an axis through its centre and parallel to one side.* Let ABCD be a rectangular lamina, of *length l, breadth b* and *mass* M and let YOY' be the axis through its centre O and parallel to the side AD or BC about which its moment of inertia is to be determined.

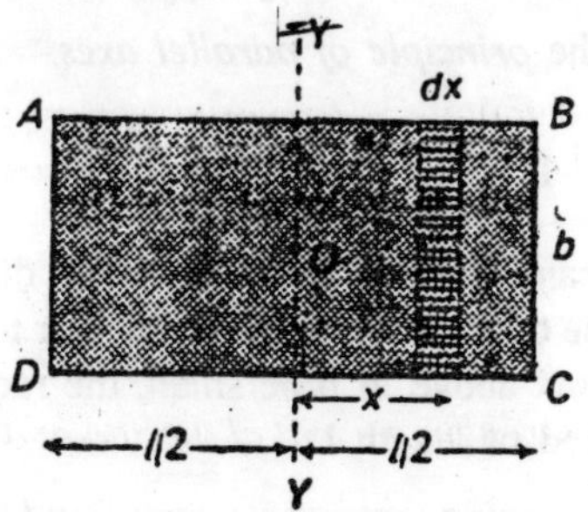

Fig. 1.14

Consider an element, or a small rectangular strip of the lamina, parallel to, and at a distance x from, the axis. The area of this strip or element = dx × b. And, since the *mass per unit area of the lamina* = $M/(l \times b)$, we have *mass of the strip or element* = $\frac{M}{l \times b} \times dx \times b = \frac{M}{l} dx$.

And, therefore, *M.I. of the element about the axis YOY'*

$$= \frac{M}{l} dx.x^2.$$

The moment of inertia (I) of the whole rectangular lamina is thus given by twice the integral of the above expression between the limits x = 0 and x = $l/2$.

i.e.,
$$I = 2\int_0^{l/2} \frac{M}{l} x^2 dx = \frac{2M}{l}\int_0^{l/2} x^2 dx = \frac{2M}{l}\left[\frac{x^3}{3}\right]_0^{l/2}$$

$$= \frac{2M}{l} \cdot \frac{l^3}{24} = \frac{Ml^2}{12}.$$

As will be readily seen, if b be small, the rectangular lamina becomes a *rod* of length l whose M.I. about the axis YOY′ passing through its centre and perpendicular to its length would be $Ml^2/12$, as obtained under case 1, (i) above.

(ii) *About one side:* In this case, since the axis coincides with AD or BC, we integrate the expression for the M.I. of the element of length dx and distance x from the axis, *i.e.,* $(M/l)dx.x^2$, between the limits x = 0 at AD and x = l at BC. So that M.I. of the lamina about side AD or BC is given by

$$I = \int_0^l \frac{M}{l} x^2 dx = \frac{M}{l}\left[\frac{x^3}{3}\right]_0^l = \frac{Ml^2}{3}.$$

Alternatively, by the *principle of parallel axes,*

$$I = \frac{Ml^2}{12} + M\left(\frac{l}{2}\right)^2 = \frac{Ml^2}{3}$$

This is again the same case as that of the M.I. of a rod about an axis passing through one of its ends and perpendicular to its length [case I (ii)]; for, as pointed out above, if b be small, the rectangular lamina too reduces to a thin rod of length I.

(iii) *About an axis passing through its centre and perpendicular to its plane :* This may be easily obtained by an application of the principle of perpendicular axes to case (i) above, according to which, *M.I. of the lamina about an axis through O and perpendicular to its planes = M.I. of the lamina about an axis through O parallel to b + M.I. of the lamina about an axis through O parallel to l, i e.,*

$$I = \frac{Ml^2}{12} + \frac{Mb^2}{12} = \frac{M(l^2 + b^2)}{12}.$$

This relation is true for both *thin (i.e., laminar)* as well as *thick* plates or bars, because no stipulation has been made in deducing it as to the thickness of the lamina. And, in fact, a *thick rectangular plate or bar may be regarded as a combination of thin or laminar plates or bars, piled up one over the other.*

(iv) *About an axis passing through the mid-point of one side and perpendicular to its plane :* In this case the axis passes through the mid-point of side AD or BC, say, and perpendicular to the plane of the lamina, so that it is parallel to the axis through O (the c.m. of the lamina) in case (iii).

In accordance with the *principle of parallel axes*, therefore, the M.I. of the lamina about this axis is given by

$$I = M\frac{(l^2 + b^2)}{12} + M\left(\frac{l}{2}\right)^2 = M\left(\frac{l^2 + b^2}{12} + \frac{l^2}{4}\right) = M\left(\frac{l^2}{3} + \frac{b^2}{12}\right).$$

And, if the axis passes through the mid-point of AB or DC, we, similarly, have

$$I = M\left(\frac{l^2 + b^2}{12}\right) + M\left(\frac{b}{2}\right)^2 = M\left(\frac{l^2 + b^2}{12} + \frac{b^2}{4}\right) = M\left(\frac{l^2}{12} + \frac{b^2}{3}\right)$$

(v) *About an axis passing though one of its corners and perpendicular to its plane :* Let the axis pass through the corner D of the lamina. Since it is perpendicular to the plane of the lamina, it is parallel to the axis through its centre of mass O in case (iii). Again, therefore, by the *principle of parallel axes*, we have moment of inertia of the rectangular lamina about this axis through D given by

$$I = M\frac{(l^2 + b^2)}{12} + Mr^2,$$ where r is the distance between the two axes.

Clearly, $$r^2 = \left(\frac{l}{2}\right)^2 + \left(\frac{b}{2}\right)^2 = \frac{(l^2 + b^2)}{4}.$$

So that, $$I = M\left(\frac{l^2 + b^2}{12}\right) + M\left(\frac{l^2 + b^2}{4}\right) = M\left(\frac{l^2 + b^2 + 3l^2 + 3b^2}{12}\right)$$

$$= M\left(\frac{l^2 + b^2}{3}\right).$$

3. Moment of Inertia of a Hoop or a Thin Circular Ring

(i) *About an axis through its centre and perpendicular to its plane.* Let the radius of the hoop or the thin circular ring be R and its mass M, (Fig. 1.15).

Consider a particle of mass m of the hoop or the ring. Clearly, its M.I. about an axis passing through the centre O of the hoop or the ring and perpendicular to its plane = mR^2.

∴ *M.I. of the entire hoop or ring about this axis passing through its centre and perpendicular to its plane,* $I = \Sigma mr^2 = Mr^2$

[$\Sigma m = M$, the mass of the hoop or ring].

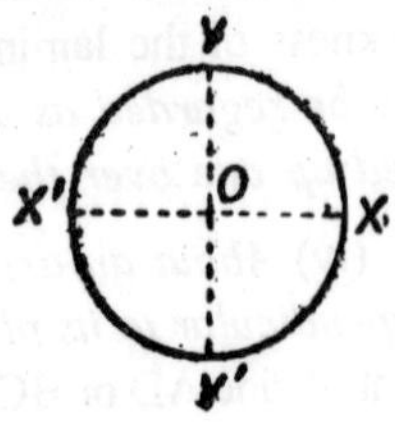

Fig. 1.15

(ii) *About its diameter* : Obviously, due to symmetry, the M.I. of the hoop or the ring will be the same about one diameter as about another. Thus, if I be its M.I. about the diameter XOX′ (Fig. 1.16), it will also be I about the diameter YOY′ perpendicular to XOX′.

By the principle of perpendicular axes, therefore, the M.I. of the hoop or the ring about the axis through its centre O and *perpendicular to its plane* is equal to the sum of its moments of inertia about the perpendicular axes XOX′ and YOY′ in its own plane and intersecting at *O*, *i.e.*,

$$I + I = M.R^2. \text{ Or, } 2I = MR^2, \text{ whence, } I = \frac{MR^2}{2}.$$

4. Moment of Inertia of a Circular Lamina or Disc

(i) *About an axis through its centre and perpendicular to its plane.* Let M be the mass of the disc and R, its radius, so that its *mass per unit area* is equal to $M/\pi R^2$.

Considering a ring of the disc, of width dx, and distant x from the axis passing through O and perpendicular to the plane of the disc, we have

area of the ring = circumference × width
= $2\pi x dx$, and hence *its mass*

$$= \left(\frac{M}{\pi R^2}\right) \times 2\pi x dx = \frac{2Mxdx}{R^2}.$$

Fig. 1.16

And, therefore, its M.I. about the perpendicular axis through O

$$= \frac{2Mxdx}{R^2} x^2 = \frac{2Mx^3dx}{R^2}.$$

Since the whole disc may be supposed to be made up at such like concentric rings of radii ranging from 0 to R, the M.I. of, *the whole disc* about the axis through O and perpendicular to its plane, *i.e.*, I, is obtained by integrating the above expression for the M.I of the ring, between the limits x = 0 and x = R. Thus,

$$I = \int_0^R \frac{2M}{R^2} x^3 dx = \frac{2M}{R^2}\left[\frac{x^4}{4}\right]_0^R = \frac{2M}{R^2} \cdot \frac{R^4}{4} = \frac{MR^2}{2}$$

(ii) *About a diameter* : Here, again, due to symmetry, the M.I. of the disc about one diameter is the same as about another. So that, if I be the M.I. of the disc about each of the perpendicular diameters XOX′ and YOY′, (Fig. 1.17), we have, by the *principle of perpendicular axes*,

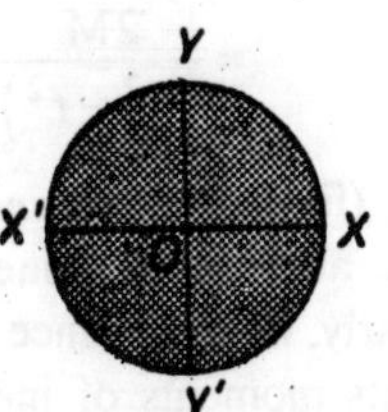

Fig. 1.17

I + I = M.I. of the disc about an axis through O and perpendicular to its plane,

i.e., $2I = \frac{MR^2}{2}$, whence, $I = \frac{MR^2}{4}$.

5. Moment of Inertia of an Annular Ring or Disc

(i) *About an axis through its centre and perpendicular to its plane:* An annular disc is just an ordinary disc, with a smaller coaxial disc removed from it, leaving a concentric circular hole in it, as shown in Fig. 1.18.

Fig. 1.18

If R and r be the outer and inner radii of the disc and M, its mass, we have *mass per unit area of the disc = mass/area =* $M/\pi(R^2 - r^2)$.

Now, the disc may be imagined to be made up of a number of circular rings, with their radii ranging from r to R. So that, considering one such ring of *radius* x and *width* dx, we have *face area of the ring* $= 2\pi x\ dx$ and $\therefore$ its mass

$$= 2\pi x\ dx\left[\frac{M}{\pi\left(R^2 - r^2\right)}\right] = \left[\frac{2Mx}{\left(R^2 - r^2\right)}\right]dx.$$

And hence its *M.I. about the axis through O and perpendicular to its plane* $= \frac{2Mx}{\left(R^2 - r^2\right)}dx.x^2 = \frac{2Mx^3}{\left(R^2 - r^2\right)}dx$

The M.I. of the whole annular disc *i.e.,* I, is, therefore, given by the integral of the above expression between the limits x = r and x = R.

$$\text{Or, } I = \int_r^R \frac{2Mx^3}{\left(R^2 - r^2\right)}dx = \frac{2M}{\left(R^2 - r^2\right)}\int_r^R x^3 dx = \frac{2M}{\left(R^2 - r^2\right)}\left[\frac{x^4}{4}\right]_r^R$$

$$= \frac{2M}{\left(R^2 - r^2\right)}\left[\frac{\left(R^4 - r^4\right)}{4}\right] = \frac{M\left(R^2 + r^2\right)}{2}$$

(ii) *About a diameter* : Due to symmetry, the M.I. of the annular disc about one diameter is the same as about another, say, I. Then, clearly, in accordance with the *principle of perpendicular axes*, the sum of its moments of inertia about two perpendicular diameters must be equal to its M.I. about the axis passing through its centre (where the two diameters intersect) and perpendicular to its plane, *i.e.,*

$$I + I = M(R^2 + r^2)/2.$$

Or, $$2I = M(R^2 + r^2)/2,$$

whence, $$I = M\ (R^2 + r^2)/4.$$

6. Moment of Inertia of a Solid Cylinder

(i) *About its own axis of cylindrical symmetry* : A solid cylinder is just a thick circular disc or a number of thin circular discs (all of the same radius) piled up one over the other, so that its axis of cylindrical symmetry is the same as the axis passing through the centre of the thick disc (or the pile of thin discs) and perpendicular to its plane.

$\therefore$ *M.I. of the solid cylinder about its axis of cylindrical symmetry, i.e., I = M.I. of the thick disc (or the pile of thin discs) of the same mass*

and radius about the axis through its centre and perpendicular to its plane.

Or, $$\frac{MR^2}{2}.$$

(ii) *About the axis through, its centre and perpendicular to its axis of cylindrical symmetry.* If R be the *radius*, I, the *length* and M, the mass of the solid cylinder, supposed to be uniform and of a homogeneous composition, we have its *mass per unit length* = M/*l*.

Now, imagining the cylinder to be made up of a number of discs each of radius R, placed adjacent to each other, and considering one such disc of thickness dx and at a distance x from the centre O of the cylinder, (Fig. 1.19), we have

$$\textit{mass of the disc} = \left(\frac{M}{l}\right)dx, \text{ and } \textit{radius} = R.$$

And ∴ $$\textit{M.I. of the disc about its diameter AB} = \frac{M}{l}dx.\frac{R^2}{4},$$

and its M.I. about the parallel axis YOY', passing through the centre O of the cylinder and perpendicular to its axis of cylindrical symmetry (or its length), in accordance with the principle of parallel axes.

$$= \frac{M}{l}dx\frac{R^2}{4} + \frac{M}{l}dx.x^2.$$

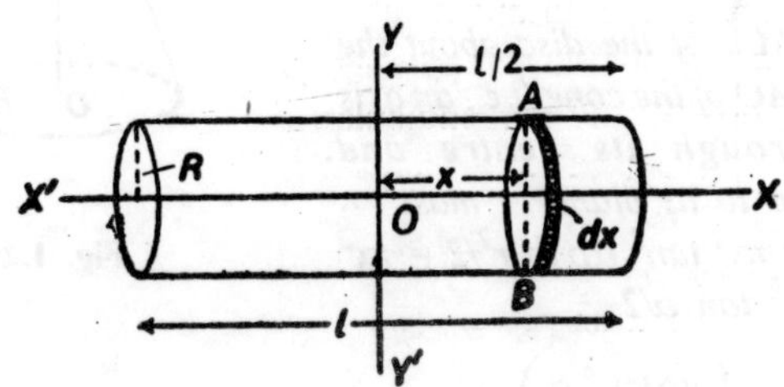

Fig. 1.19

Hence, *M.I. of the whole cylinder about this axis, i.e., I = twice the integral of the above expression between the limits*

x = 0 and x = l/2,

$$\textit{i.e.,}\quad I = 2\int_0^{l/2}\left(\frac{M}{l}\cdot\frac{R^2}{4}dx + \frac{M}{l}x^2dx\right) = \frac{2M}{l}\int_0^{l/2}\left(\frac{R^2}{4}dx + x^2dx\right)$$

$$= \frac{2M}{l}\left[\frac{R^2x}{4} + \frac{x^3}{3}\right]_0^{l/2}$$

Or, $I = \frac{2M}{l}\left[\frac{R^2}{4}\cdot\frac{l}{2} + \frac{l}{8\times 3}\right] = \frac{2M}{l}\left(\frac{R^2 l}{8} + \frac{l^3}{24}\right) = M\left(\frac{R^2}{4} + \frac{l^2}{12}\right).$

7. Moment of Inertia of a Solid Cone

(i) *About its vertical axis :* Let M be the *mass* of the solid cone, h its *vertical height* and R, the *radius of its base* (Fig. 1.20).

Clearly, *volume of the cone* = $\frac{1}{3}\pi R^2 h$ and if ρ be the density of its material its *mass*

$$M = \frac{1}{3}\pi R^2 h\rho, \text{ whence, } \rho = \frac{3M}{\pi R^2 h}.$$

Now, the cone may be imagined to consist of a number of discs of progressively decreasing radii, from R to 0, piled up one over the other.

Considering one such disc of thickness dx and at a distance x from the vertex (A) of the cone, we have *radius of the disc*, r = x tan α, where α is the semi-vertical angle of the cone, and, therefore, its *volume* = $\pi r^2\,dx = \pi x^2 \tan^2 \alpha dx$ and its *mass* = $\pi x^2 \tan^2 \alpha\rho\, dx$.

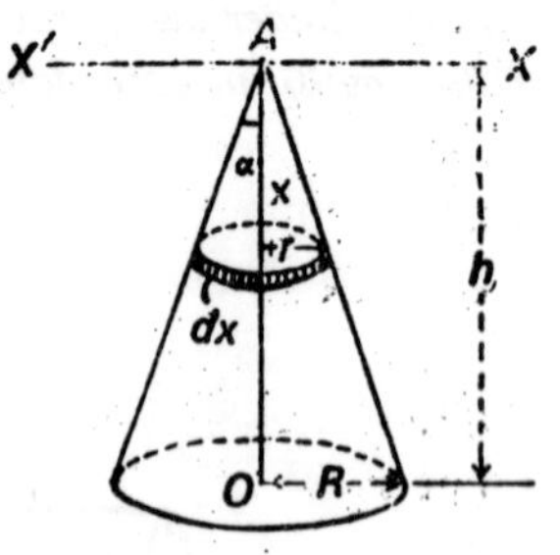

Fig. 1.20

Hence, *M.I. of the disc about the vertical axis AO of the cone (i.e., an axis passing through its centre and perpendicular to its plane) = mass × (radius)²/2* = $\pi x^2 \tan^2 \alpha\rho\, dx.r^2/2 = \pi x^2 \tan^2 \alpha\rho\, dx.x^2 \tan \alpha/2$

$$= \left(\frac{\pi\rho\tan^4\alpha}{2}\right)x^4 dx.$$

And ∴ *M.I. of the entire cone about its vertical axis AO is given by*

$$I = \int_0^h \frac{\pi\rho\tan^4\alpha}{2} x^4 dx = \frac{\pi\rho\tan^4\alpha}{2}\int_0^h x^4 dx = \frac{\pi\rho\tan^4\alpha}{2}\left[\frac{x^5}{5}\right]_0^h$$

$$= \frac{\pi\rho\tan^4\alpha}{2}\cdot\frac{h^5}{5} = \frac{\pi\rho R^4}{2h^4}\cdot\frac{h^5}{5},$$

substituting R/h for tan α.

Or, substituting the value of ρ obtained above, we have

$$I = \frac{\pi.3M}{\pi R^2 h}\cdot\frac{R^4}{2h^4}\cdot\frac{h^5}{5} = \frac{3MR^2}{10}$$

(ii) *About an axis passing through the vertex and parallel to its base:* Again, considering the disc at a distance x from the vertex of the cone, we have its *M.I. about its diameter = mass × (radius)²/4*

$$= \pi x^2 \tan^2 \alpha\rho\ dx.\frac{r^2}{4}$$

$$= \pi x^2 \tan^2 \alpha\rho\ dx.x^2\ \frac{\tan^2\alpha}{4} = \left(\frac{\pi\rho\tan^4\alpha}{4}\right)x^4 dx$$

∴ its M.I. about the parallel axis XX′, distant x from it

$$= \left(\frac{\pi\rho\tan^4\alpha}{4}\right)x^4 dx + \pi x^2\tan^2\alpha\rho x^2 dx$$

$$= \left(\frac{\pi\rho\tan^4\alpha}{4}\right)x^4 dx + \pi\rho\tan^2\alpha x^4 dx.$$

Hence M.I. *of the entire cone* about the axis XX′, *parallel to its base* is given by

$$I = \int_0^h \frac{\pi\rho\tan^4\alpha}{4} x^4 dx + \pi\rho\tan^2\alpha\, x^4 dx$$

$$= \frac{\pi\rho\tan^4\alpha}{4}\int_0^h x^4 dx + \pi\rho\tan^2\alpha\int_0^h x^4 dx$$

$$= \frac{\pi\rho}{4}\cdot\frac{R^4}{h^4}\left[\frac{x^5}{5}\right]_o^h + \pi\rho.\frac{R^2}{h^2}\left[\frac{x^5}{5}\right]_o^h = \frac{\pi\rho R^4}{4h^4}\cdot\frac{h^5}{5} + \frac{\pi\rho R^2}{h^2}\cdot\frac{h^5}{5}$$

$$\left[\because \tan\alpha + \frac{R}{h}\right]$$

Or, substituting the value of ρ, we have *M.I. of the cone about the axis XX′, i.e.,.*

$$I = \frac{\pi.3M}{4\pi R^2 h}\cdot\frac{R^4}{h^4}\cdot\frac{h^5}{5} + \frac{\pi.3M}{\pi R^2 h}\cdot\frac{R^2}{h^2}\cdot\frac{h^5}{5}.$$

or,
$$I = \frac{3MR^2}{20} + \frac{3Mh^2}{5}.$$

8. Moment of Inertia of a Hollow Cylinder

(i) *About its axis of cylindrical symmetry :* A hollow cylinder may be considered to be a thick annular disc or a combination of thin annular discs, each of the same external and internal radii, placed adjacent to each other, the axis of the cylinder (*i.e.,* its axis of cylindrical symmetry) being the same as the axis passing through the centre of the thick annular disc (or the combination of thin annular discs) and perpendicular to its plane.

The M.I. of the hollow cylinder about its own axis is, therefore, the same as that of a thick annular disc (or a combination of thin annular discs) of the same mass M and external and internal radii R and r respectively about the axis passing through its centre and perpendicular to its plane, *i.e.,*

$$I = \frac{M\left(R^2 + r^2\right)}{2}$$

Alternatively, we may obtain the same result directly as follows: Let R and r be the external and internal radii respectively of a hollow cylinder of length l and mass M, (Fig. 1.21). Then, *face-area* of the cylinder = $\pi(R^2 - r^2)$ and its volume = $\pi(R^2 - r^2)\, l$ and, therefore, its *mass per unit volume*

$$= \frac{M}{\pi\left(R^2 - r^2\right)l}.$$

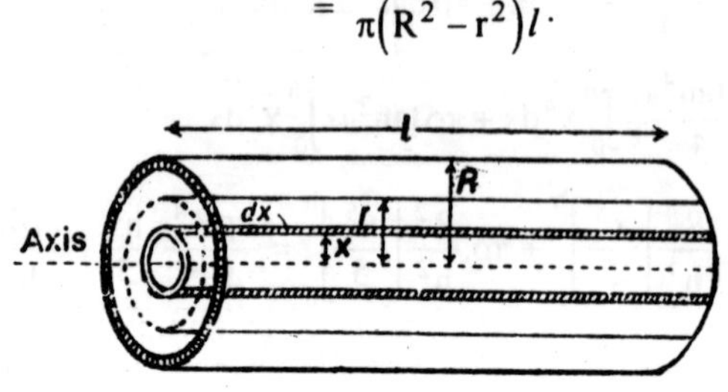

Fig. 1.21

Imagining the cylinder to be made up of a large number of thin, coaxial cylinders, with their radii varying from r to R, and considering one such cylinder of radius x and thickness dx, we have its *face area* = $2\pi x dx$ and its *volume* = $2\pi x dx . l$ and hence its mass = $2\pi x dx . l \times M/\pi(R^2 - r^2)l = 2Mxdx/R^2 - r^2)$.

And, therefore, its *M.I. about the axis of the cylinder*

$$= \frac{2Mxdx}{(R^2 - r^2)} \cdot x^2 = \frac{2M}{(R^2 - r^2)} x^3 dx$$

Hence ***M.I. of the entire cylinder about its axis, i e., I,*** **is given by**

$$I = \int_r^R \frac{2M}{(R^2 - r^2)} x^3 dx = \frac{2M}{(R^2 - r^2)} \int_r^R x^3 dx = \frac{2M}{(R^2 - r^2)} \left[\frac{x^4}{4} \right]_r^R$$

$$= \frac{2M}{(R^2 - r^2)} \left(\frac{R^4 - r^4}{4} \right) = M\left(\frac{R^2 + r^2}{2} \right)$$

(ii) *About an axis passing through its centre and perpendicular to its own axis (Fig. 1.22).* Again, if R and r be the *external and internal radii* respectively of the hollow cylinder, *l*, its length and M, its mass, we have *mass per unit volume of the cylinder*

$$= \frac{M}{\pi(R^2 - r^2)l}.$$

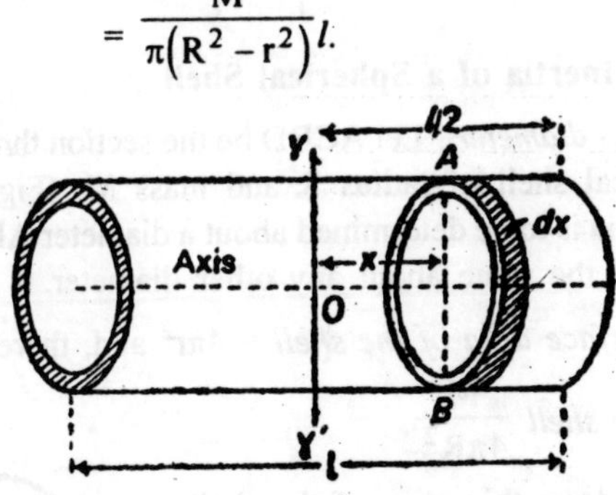

Fig. 1.22

Imagining the hollow cylinder to be made up of a large number of annular discs, of external and internal radii R and r respectively, placed adjacent to each other, and considering one such disc at a distance x from the axis YOY′ passing through the centre O of the cylinder and perpendicular to its own axis, we have *surface area of the disc* = $\pi(R^2 - r^2)$, its *volume* = $(R^2 - r^2)dx$ and, therefore, its *mass* = $\pi(R^2 - r^2)dx \times M/\pi(R^2 - r^2)l = Mdx/l$.

So that, *M.I. of the disc about its diameter (AB)* = $\frac{Mdx}{l} \cdot \frac{(R^3 + r^2)}{4}$.

And, therefore, *its M.I. about the parallel axis YOY′* distant x from it, in accordance with the *principle of parallel axes*,

$$= \frac{M}{l}dx.\left(\frac{R^2+r^2}{4}\right)+\frac{M}{l}dx.x^2$$

Hence, *M.I. of entire hollow cylinder about the axis YOY'* is equal to twice the integral of the above expression between the limits x = 0 and $x = \frac{l}{2}$, *i.e.,*

$$I = 2\int_0^{l/2}\left[\frac{M(R^2+r^2)}{4l}dx + \frac{M}{l}x^2dx\right] = \frac{2M}{l}\int_0^{l/2}\left[\frac{R^2+r^2}{4} + dx + x^2dx\right]$$

$$= \frac{2M}{l}\left[\frac{(R^2+r^2)x}{4} + \frac{x^3}{3}\right]_0^{l/2}$$

Or, $I = \frac{2M}{l}\left[\frac{(R^2+r^2)l}{4\times2} + \frac{l^3}{8\times3}\right] = M\left[\frac{(R^2+r^2)}{4} + \frac{l^2}{12}\right]$

9. Moment of Inertia of a Spherical Shell

(ii) *About its diameter* : Let ACBD be the section through the centre O, of a spherical shell of radius R and mass M, (Fig. 1.23), whose moment of inertia is to be determined about a diameter AB, say, its value being obviously the same about any other diameter.

Clearly, *surface area of the shell* = $4\pi r^2$ and, therefore, *mass per unit area of the shell* $\frac{M}{4\pi R^2}$.

Now, consider a thin slice of the shell, lying between two planes EF and GH, perpendicular to the diameter AB at distances x and x + dx respectively from its centre O. This slice is, obviously, a *ring* of radius PE and width EG (and not PQ which is equal to dx, the distance between the two planes).

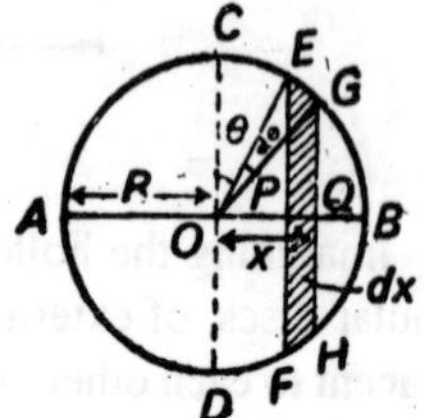

Fig. 1.23

Clearly, *area of the ring = circumference × width* = $2\pi PE \times EG$ and hence its mass = $2\pi\, PE \times EG \times \frac{M}{4\pi R^2}$.

Join OE and OG and let angle COE be equal to θ and angle EOG = dθ.

Then, $PE = OE \cos OEP = R \cos\theta$,

because $OE = R$ and $< OEP =$ *alternate*$< COE = \theta$.

And, $OP = OE \sin OEP$. Or, $x = R \sin\theta$ and

$\therefore \quad \frac{dx}{d\theta} = R\cos\theta$. [$\because$ $OP = x$ and $OE = R$.]

Or, $dx = R\cos\theta \, d\theta = PE.d\theta$ (because $R\cos\theta = PE$)

and $EG = OE \, d\theta = Rd\theta$.

$\therefore$ *mass of the ring* $= 2\pi \, PE \times R \, d\theta \times \frac{M}{4\pi R^2} = \frac{Mdx}{2R}$

[$\because$ $PE.d\theta = dx$.]

Hence, *M.I. of the ring about diameter AB of the* shell *(i.e., an axis passing through the centre of the ring and perpendicular to its plane*

$= mass \times (radius)^2 = \left(\frac{Mdx}{2R}\right)(R^2 - x^2)$,

[because $PE^2 = OE - OP^2 = R^2 - x^2$.]

And, therefore, M.I. of the whole spherical shell about the diameter (AB) is equal to twice the integral of this expression between the limits $x = 0$ and $x = R$, *i.e.*,

$$I = 2\int_o^R \frac{M}{2R}(R^2 - x^2)dx = \frac{M}{R}\int_o^R (R^2 - x^2)dx = \frac{M}{R}\left[R^2x - \frac{x^3}{3}\right]_o^R$$

$$= \frac{M}{R}\left[R^3 - \frac{R^3}{3}\right] = \frac{M}{R}\cdot\frac{2}{3}R^3$$

Or, $$I = \frac{2}{3}MR^2.$$

10. Moment of Inertia of a Solid Sphere

(i) *About a diameter* : Figure 1.24 represents a section, through the centre, of a solid sphere of radius R and mass M, whose moment of inertia is to be determined about a diameter AB, say, its value being the same about any other diameter.

Since the volume of sphere $= \frac{4\pi R^3}{3}$, its *mass per unit volume* (or *density*) $= \frac{M}{4/3\pi R^3} = \frac{3M}{4\pi R^3}$.

Considering a thin circular slice of the sphere at a distance x from its centre O and of thickness dx, we have surface area of the slice (which is obviously a disc of radius $\sqrt{R^2 - x^2}$) = $\pi(R^2 - x^2)$ and *its volume* = *area* × *thickness* = $\pi(R^2 - x^2)dx$ and hence its *mass* = *volume* × *density*

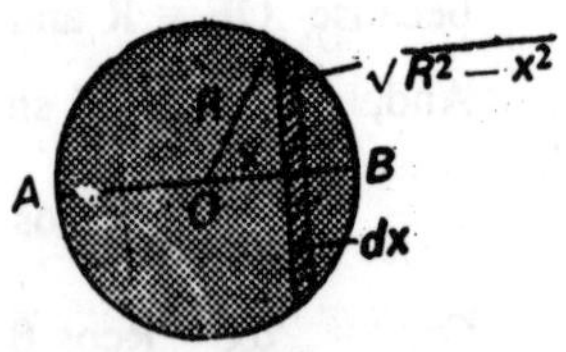

Fig. 1.24

$$= \pi(R^2 - x^2)dx \times \frac{3M}{4\pi R^2} = \frac{3M\left(R^2 - x^2\right)dx}{4R^3}.$$

And ∴ *M.I. of this slice or disc about AB (i.e., an axis passing through its centre and perpendicular to its plane)* = *its mass* × *(radius)²/2*

$$= \frac{3M\left(R^2 - x^2\right)}{4R^3}dx.\frac{R^2 - x^2}{2} = \frac{3M\left(R^2 - x^2\right)^2}{8R^3}dx.$$

Hence, M.I. of the whole sphere about its diameter (AB) is equal to twice the integral of the above expression between the limits x = 0 and x = R, *i.e.*,

$$I = 2\int_o^R \frac{3M\left(R^2 - x^2\right)^2}{8R^3}dx = \frac{2 \times 3M}{8R^3}\int_o^R \left(R^2 - x^2\right)^2 dx$$

$$= \frac{3M}{4R^3}\int_o^R \left(R^4 - 2R^2x^2 + x^4\right)dx$$

Or, $$I = \frac{3M}{4R^3}\left[R^4x - 2R^2\frac{x^3}{3} + \frac{x^5}{5}\right]_o^R = \frac{3M}{4R^3}\left(R^5 - \frac{2}{3}R^5 + \frac{1}{5}R^5\right)$$

$$= \frac{3M}{4R^3} \times \frac{8R^5}{15} = \frac{2}{5}MR^2.$$

(ii) *About a tangent* : A tangent drawn to the sphere at any point will clearly be parallel to one or the other diameter of it (*i.e.*, an axis passing through its centre or centre or mass) and at a distance equal to the radius of the sphere, R, from it. We therefore have, by the principle of parallel axes, *moment of inertia of the sphere about a tangent, i.e.*,

$$I = \frac{2}{5}MR^2 + MR^2 = \frac{7}{5}MR^2.$$

11. Moment of Inertia of a Hollow Sphere or a Thick Shell

(i) *About its diameter :* A hollow sphere (or a thick shell) is just a solid sphere from the inside of which a small concentric solid sphere has been removed.

Since the M.I. about a given axis is a scalar quality, we have *moment of inertia of hollow sphere about a diameter = moment of inertia of the solid sphere minus moment of inertia of the smaller solid sphere removed from it, both about the same diameter.*

Let R and r be the external and internal radii of the hollow sphere, *i.e.,* the radius of the bigger solid sphere and the smaller solid sphere (removed from it) respectively. Then, if ρ be the density of the material of the given hollow sphere, we have

$$\textit{mass of the bigger sphere} = \frac{4}{3}\pi R^3 \rho \text{ and}$$

$$\textit{mass of the smaller sphere} = \frac{4}{3}\pi r^3 \rho.$$

so that mass of the hollow sphere, $M = \frac{4}{3}\pi(R^3 - r^3)\rho$, and. therefore,

$$\rho = \frac{3M}{4\pi(R^3 - r^3)}.$$

And, clearly, moments of inertia of the bigger and the smaller spheres about a given diameter are respectively $\frac{2}{5}\left(\frac{4}{3}\pi R^3 \rho\right)R^2$ and $\frac{2}{5}\left(\frac{4}{3}\pi r^3 \rho\right)r^2$

And, therefore, M.I. *of the hollow sphere about the same diameter* is given by

$$I = \frac{2}{5}\left(\frac{4}{3}\pi R^3 \rho\right)R^2 - \frac{2}{5}\left(\frac{4}{3}\pi r^3 \rho\right)r^2 = \frac{2}{5}\cdot\frac{4}{3}\pi\rho\left(R^5 - r^5\right)$$

Or, substituting the value of ρ obtained above, we have

$$I = \frac{2}{5}\cdot\frac{4}{3}\pi\frac{3M}{4\pi(R^3 - r^3)}\left(R^5 - r^5\right) = \frac{2}{5}M\left(\frac{R^5 - r^5}{R^3 - r^3}\right).$$

(ii) *About a tangent :* A tangent to the sphere at any point being parallel to one diameter or the other (*i.e.,* the axis passing through the centre of mass) of the sphere and at a distance equal to its external radius R from it, we have, by the principle of parallel axes,

M.I. of hollow sphere about a tangent, i.e.,

$$I = \frac{2}{5}M\left(\frac{R^5 - r^5}{R^3 - r^3}\right) + MR^2$$

THE GYROSTAT

A gyrostat, in actual practice, usually takes the form of a *heavy circular disc,* free to rotate at a high speed about an axle passing through its centre of mass and perpendicular to its plane, *i.e.,* about its axis of symmetry, and is so mounted that the disc, along with the axle can turn freely about any one of the three mutually perpendicular axes.

Thus, in Fig. 1.25, the disc or the gyrostat G is free to rotate about the axle, coincident with the axis of x and is carried by a circular ring R_1 which is itself tree to turn about the axis of y inside another ring R_2, which is free to rotate about the axis of z, within a rigid framework F. The three axes being obviously perpendicular to each other, intersect in the same point O and therefore, for any given position of F, the disc or the gyrostat G can turn freely about any one of them.

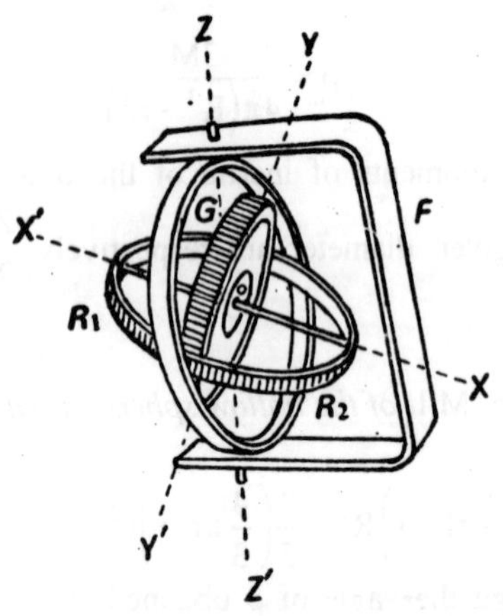

Fig. 1.25

Now, from the discussion under 1.12 above, it is clear that *with the disc rotating fast, (i) any rotation of the framework leaves the position and direction of the axis of rotation of the disc unaffected with respect to it and (ii) any torque applied to the axis of rotation of the disc results in its precession.*

Since the rate of precession is inversely proportional to the angular momentum $I\omega$ of the disc, the greater its M.I. and the higher its angular velocity about the axis of rotation, the smaller the precession of the latter. Obviously, therefore, a massive disc rotating at a high speed makes for a greater stability of its axis of rotation. Hence, the use of gyrostats for steadying of motions and ensuring stability of direction.

KINETIC ENERGY OF ROTATION

Here, we must distinguish between : (i) *pure rotation, in which the centre of mass of the rotating body has zero linear velocity, e.g.,* the rotation of a body about an axis through its centre of mass, and (ii) *rotation in which the centre of mass of the rotating body has also a linear velocity, e.g.,* the rolling of a body along a plane or an inclined surface. Let us consider the two cases separately.

(i) *Kinetic energy of a body rotating about an axis through its centre of mass (i.e., in the case of pure rotation).* Consider a body of mass M, rotating with angular velocity ω about an axis AB, passing through its *centre of mass,* O (Fig. 1.26), so that *the centre of mass has zero linear velocity.* It is thus a case of *pure rotation.*

Obviously, the body possesses kinetic energy in virtue of its motion of rotation which is, therefore, aptly called its *kinetic energy of rotation.* Let us obtain an expression for it.

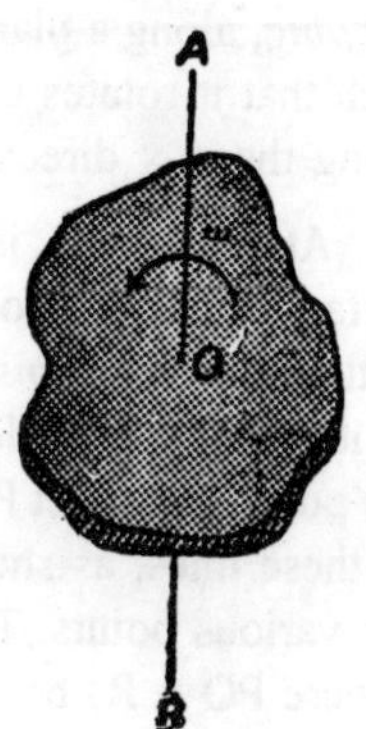

Fig. 1.26

The body is clearly made up of a large number of particles of masses m_1, m_2, m_3...etc. at respective distances r_1, r_2, r_3...etc. from the axis AB through O. Since their angular velocity is the same (ω), their linear velocities are respectively $r_1\omega = v_1$, $r_2\omega = v_2$, $r_3\omega = v_3$...etc. and hence their respective kinetic energies equal to

$$\frac{1}{2}m_1v_1^2 = \frac{1}{2}m_1r_1^2\omega, \frac{1}{2}m_2v_2^2 = \frac{1}{3}m_2r_2^2\omega^2,$$

$$\frac{1}{2}m_3v_3^2 = \frac{1}{2}m_3r_3^2\omega^2 \text{...etc.}$$

$\therefore$ *total K.E. of all the particles, i.e., the K.E. of the body itself*

$$= \frac{1}{2}m_1r_1^2\omega^2 + \frac{1}{2}m_2r_2^2\omega^2 + \frac{1}{2}m_3r_3^2\omega^2 + ...$$

$$= \frac{1}{2}\omega^2\left(m_1r_1^2 + m_2r_2^2 + m_3r_3^2 + ...\right)$$

$$= \frac{1}{2}\omega^2 \Sigma mr^2 = \frac{1}{2}\omega^2 MK^2,$$

where $\Sigma mr^2 = MK^2$, with M, as the mass of the body and K, its radius of gyration about the axis of rotation AB.

Since $MK^2 = I$, the *moment of inertia* of the body about the axis AB, we have

Kinetic energy of rotation of the body about the axis (AB) through its centre of mass $= \frac{1}{2}I\omega^2$. It is naturally expressed in *ergs* in the *C.G.S. system*, in *joules* in the M.K.S. (*or the SI*) *system* and in *foot-poundals* or *foot-pounds* in the *F.P.S. system*.

(ii) *Kinetic energy of a rotating body whose centre of mass has also a linear velocity.*

(a) *Case of a body rolling along a plane surface* : Let us consider a body, like a circular disc, a cylinder, a sphere etc. (*i.e.*, a body with a circular symmetry), of *mass M, radius* R and with its *centre of mass* at O, (Fig. 1.27), rolling, *without slipping*, along a plane or a level surface, such that it rotates clockwise and moves along the + x direction, as indicated.

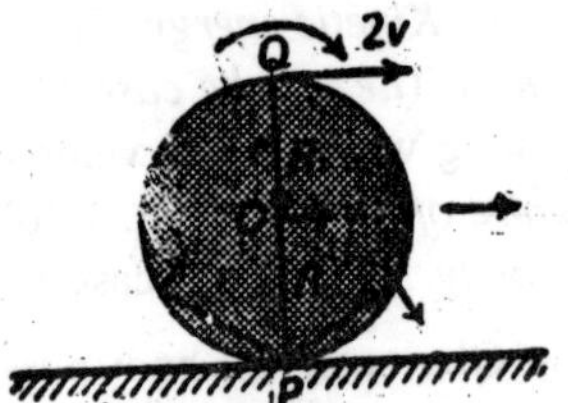

Fig. 1.27

At any given instant, the point P, where the body touches the surface, is at rest, so that an axis through P, perpendicular to the plane of the paper, is *its instantaneous axis of rotation* and the linear velocities of its various particles are perpendicular to the lines joining them with the point of contact P, their magnitudes being proportional to the lengths of these lines, as shown by the directions and lengths of the arrows at the various points. Thus, if the linear velocity of the centre of mass O (where PO = R) be v, that of the particle at Q (where PQ = 2R) is 2v.

This means clearly that the particles have all the *same* angular velocity *with respect to the point P* or that *the body is rotating about the fixed axis through P with an angular velocity ω, say, given by v/R,* where v is the linear velocity of the centre of mass.

The motion of the body is thus equivalent to one of pure rotation about the axis through P, with an angular velocity ω. The whole of the kinetic energy of the body is, therefore, the same as its kinetic energy of rotation about this axis and hence equal to $I_p\omega^2$, where $I_p\omega^2$ is the M.I. of the body about the axis through P.

Clearly, if $I_{c.m.}$ be the moment of inertia of the body about a parallel axis through its centre of mass, we have, by the principle of parallel axes, $I_P = I_{c.m.} + MR^2$.

So that, *K.E. of the rolling body* $= \frac{1}{2}(I_{c.m.} + MR^2)\omega^2$

$$= \frac{1}{2}I_{c.m.}\omega^2 + \frac{1}{2}MR^2\omega^2 = \frac{1}{2}I_{c.m.}\omega^2 + \frac{1}{2}Mv^2, \quad \text{...(i)}$$

where v is the linear speed of its centre of mass with respect to P.

Now, $I_{c.m.} = MK^2$, where K is the *radius of gyration* of the body about the axis through its centre of mass, and $\omega = v/R$. So that,

K.E. of the rolling body

$$= \frac{1}{2}MK^2\frac{v^2}{R^2} + \frac{1}{2}Mv^2 = \frac{1}{2}Mv^2\left(\frac{K^2}{R^2} + 1\right) \quad \text{...(ii)}$$

It will be readily seen that in expression I above for K.E. of a rotating body, the first term gives its *K.E. of pure rotation* about the centre of mass, *i.e.*, its K.E. when it is simply rotating with angular velocity ω about the axis through its centre of mass, without executing any translational motion (*i.e.*, with the linear velocity of its centre of mass zero). And, the second term gives its *K.E. of pure translation*, *i.e.*, its K.E. when it is simply moving with linear velocity v (or the linear speed of the centre of mass) without performing any rotational motion (*i.e.*,. with its angular velocity zero) Thus,

K.E. of a rolling body rotating with angular velocity ω and moving with linear velocity v (= Rω)	=	*its K.E. of pure rotation (with the same angular velocity ω) about its centre of mass*
		+
		its K.E. of pure translation, with its centre of mass moving with linear velocity v.

This, in fact, applies to all bodies, rolling or otherwise, which are simultaneously executing a translational motion and a rotational motion about an axis perpendicular to their planes of motion.

It follows at once from the above that :

rotation of the body about the fixed axis through P, with angular velocity ω = *its pure rotation about the c.m. (O) with the same angular velocity ω*
\+
the pure translational motion of its c.m. with linear velocity v(= Rω).

This may be very interestingly seen from Fig. 1.28, where Fig. (a) represents the rotation of the body about the fixed axis through the point of contact P, when, as we have seen, the linear velocity of the particle at P is zero, that of O, the centre of mass, +v along the +x direction and that of the particle at Q, +2v in the +x direction.

Figure 1.28 (b) represents the pure rotation of the body about the axis through its centre of mass O, when, obviously, the centre of mass (O) is at rest, *i.e.*, its linear velocity is zero, that of the particle at P, – v = – ωR along the –x direction and that of the particle at Q, +v = ωR along the +x direction.

Figure 1.28 (c) represents the pure translational motion of the body, with the linear velocity of the centre of mass O equal to +v, so that the linear velocities of all other particles at P, Q etc. are also the same, *viz.*, +v.

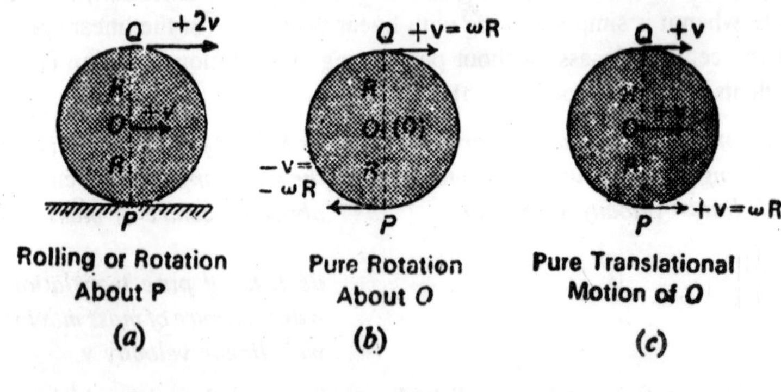

Rolling or Rotation About P
(*a*)

Pure Rotation About *O*
(*b*)

Pure Translational Motion of *O*
(*c*)

Fig. 1.28

Now, if really the rotation of the body about the fixed axis through P is equivalent to its pure rotation about O *plus* the translational motion of its centre of mass (O), the respective velocities of the particles at P, O and Q in Fig. 1.28 (b) and (c) must add up to give the same values as in Fig. 1.28 (a), as indeed they do, thereby fully confirming the conclusion arrived at above.

(b) *Case of a body rolling down an inclined plane* : Its acceleration along the plane. Let a body of circular symmetry (*e.g.*, a *disc, sphere, cylinder etc.*,), of *mass* M, roll freely down a plane, inclined to the horizontal at an angle θ, (Fig. 1.29) and rough enough to prevent slipping and hence any work done by friction.

If v be the linear velocity acquired by the body on covering a distance S *along the plane*, obviously, its vertical distance of descent = S sin θ.

∴ *potential energy lost by the body* = Mg. S sin θ.

This must obviously be equal to the kinetic energy gained by the body, *i.e.*, equal to its K.E. of rotation *plus* its K.E. of translation.

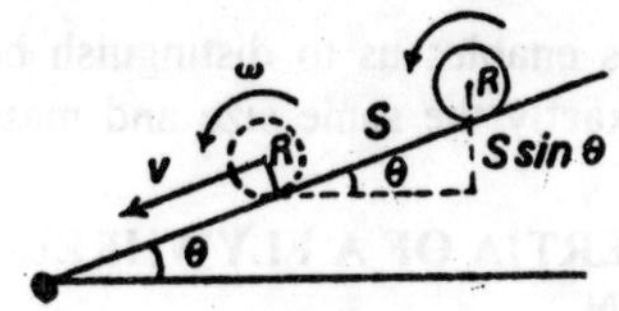

Fig. 1.29

Now, K.E. *of rotation of the body* = $\frac{1}{2}I\omega^2$, where ω is its *angular velocity* about a perpendicular axis through its centre of mass, and its *K.E. of translation* = $\frac{1}{2}Mv^2$, because its centre of mass has a linear velocity v.

∴ *total K.E. gained by the body* $= \frac{1}{2}I\omega^2 + \frac{1}{2}Mv^2 = \frac{1}{2}MK^2 \frac{R}{R^2} + \frac{1}{2}Mv^2$, because $I = MK^2$, where K is the radius of gyration of the body about the axis through its centre of mass, and $\omega = \frac{v}{R}$.

Or, *total K.E. gained by the body* $= \frac{1}{2}Mv^2\left(\frac{K^2}{R^2+1}\right)$

∴ equating this gain of K.E. against the loss of P.E., we have

$\frac{1}{2}Mv^2\left(\frac{K^2}{R^2+1}\right)$ = Mg S sin θ. Or, $\frac{v^2\left(K^2+R^2\right)}{R^2} = 2g \sin \theta . S$, whence, $$v^2 = 2\frac{R^2}{K^2+R^2}g\sin\theta.S.$$

Comparing this with the kinematic relation $v^2 = 2aS$ for a body starting from rest, we have *acceleration of the body along the plane,*

i.e., $$a = \frac{R^2}{K^2+R^2}g\sin\theta,$$

i.e., the acceleration is proportional to $\frac{R^2}{(K^2+R^2)}$ for a given angle of inclination (θ) of the plane.

Clearly, therefore, (i) *the greater the value of K, as compared with R, the smaller the acceleration of the body rolling down along the plane* and hence the greater the time taken by it in reaching the bottom of the plane, and (ii) *the acceleration, and hence the time of decsent, is quite independent of the mass of the body.*

Incidentally this enables us to distinguish between a solid and a hollow sphere of exactly the same size and mass.

MOMENT OF INERTIA OF A FLYWHEEL—EXPERIMENTAL DETERMINATION

A flywheel is just a *large, heavy wheel, with a long cylindrical axle passing through* its centre (Fig. 1.30), and with its *e.g.,* lying on its axis of rotation, so that, when properly mounted on ball bearings (to minimise friction), *it may continue to be at rest in any desired position.*

In order that its M.I. about its axis of rotation may be large, *practically the whole of its mass is concentrated on its rim.* In fact, the larger its M.I., the better its performance in making the running of a machine, coupled to the engine, smoother and steadier. It therefore, finds extensive use in the case of stationary engines, where, as we know power is delivered to the shaft (to which the machine is connected not *continuously* but only once in a cycle of four strokes. Without the flywheel, therefore, (also connected to the shaft), the machine would be driven in jerks. The flywheel, so to speak, absorbs the power delivered to the shaft and makes it available to the machine during the intervening three strokes. For, on account of its large M.I., the flywheel gets hardly accelerated by the power transmitted to it intermittently (after every three strokes) and it, therefore, continues to rotate with the same speed throughout, thus making for a smooth running of the machine.

For our purpose here, however, we may regard a flywheel to be just a thick disc (or a short cylinder) from which a smaller concentric disc (or cylinder) has been removed. In other words, we may take it to be an annular ring, whose moment of inertia is to be determined about an axis passing through its centre and perpendicular to its plane, or a hollow cylinder whose M.I. is to be determined about its axis of symmetry. So that, if R and r be the *external and internal* radii respectively of the annular ring (or the hollow cylinder) and M, its mass, its M.I. about *this* axis = $M(R^2 + r^2)/2$.

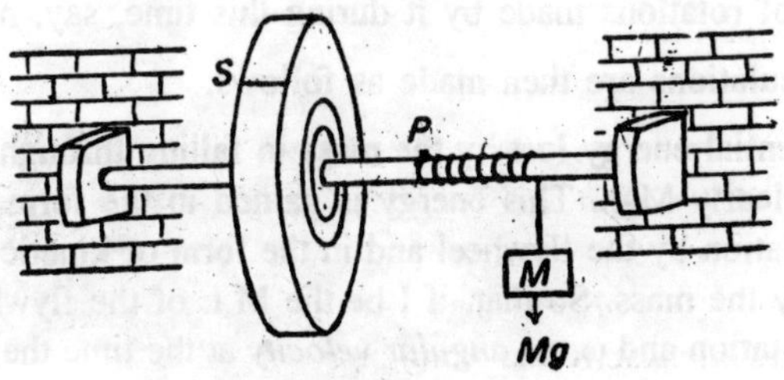

Fig. 1.30

The axle, again, is a disc (or a solid cylinder) of radius r and its M.I. about the axis passing through its centre and perpendicular to its plane (or about the axis of symmetry of the solid cylinder) = $mr^2/2$, where m is the mass of axle.

So that, *M.I. of the flywheel*, as a whole, (*i.e.*, of the wheel and the axle) *about the axis of rotation, i.e.*,

$$I = M.I.\ of\ the\ wheel + M.I.\ of\ the\ axle = M\frac{(R^2+r^2)}{2} + m\frac{r^2}{2}.$$

Experimental determination : The axle of ihe flywheel, whose moment of inertia is to be determined, is arranged to lie in a horizontal position, as shown, in Fig. 1.30, at a convenient height from the floor.

A small loop at the end of a piece of fine cord is slipped on to a tiny peg P on the axle and almost the whole length of it wound *evenly* round the latter, with a suitable mass M suspended from its other free end, as shown.

This mass is initially kept properly supported at a known height h from the floor, the length of the cord being so adjusted that when the mass is allowed to fall, resulting in the rotation of the wheel, *the cord just gets completely unwound as the mass touches the floor, and the loop slips off the peg.* The number of rotations made by the wheel during this interval of time is counted (with the help of a *reference mark* S made on the wheel) and is obviously equal to the number of turns of the cord round the axle, say, N.

The wheel having acquired kinetic energy due to the falling mass continues to rotate until brought to rest by the frictional forces at the bearings. The time t taken by it to come to rest after the mass gets detached from the axle is noted with the help of a stop watch, as also the number of rotations made by it during this time, say, n.

The calculations are then made as follows:

The potential energy lost by the mass in falling through a vertical height h is clearly Mgh. This energy is gained in the form of kinetic energy of rotation by the flywheel and in the form of kinetic energy of translation by the mass. So that, if I be the M.I. of the flywheel about the axis of rotation and ω, its *angular velocity* at the time the cord slips off the peg, the *K.E. of rotation acquired by the flywheel* $= \frac{1}{2}I\omega^2$. And, if v be the linear velocity of the mass on falling through vertical height h, the *K.E. of translation acquired by the mass* $= \frac{1}{2}Mv^2$. We, therefore, have

$$Mgh = \frac{1}{2}I\omega^2 + \frac{1}{2}Mv^2 + \textit{work done against friction.}$$

Now, if the frictional force be taken to be uniform and the work done against it per rotation of the wheel to be w, clearly, work done against friction during N rotations made by the wheel before the mass gets detached from the axle = Nw. So that, we have

$$Mgh = \frac{1}{2}I\omega^2 + \frac{1}{2}Mv^2 + Nw.$$

Again, since, after the mass gets detached from the axle, the wheel makes *n rotations* before coming to rest, the work done against friction during n rotations = nw. This must obviously be equal to the K.E. of rotation of the wheel and, therefore,

$$nw = \frac{1}{2}I\omega^2, \text{ whence, } w = \frac{1}{2}I\omega^2/n.$$

Substituting this value of w in the energy equation above, we have

$$Mgh = \frac{1}{2}I\omega^2 + \frac{1}{2}Mv^2 + \frac{1}{2}\frac{NI\omega^2}{n} = \frac{1}{2}Mv^2 + \frac{1}{2}I\omega^2\left(1 + \frac{N}{n}\right).$$

Or, $2Mgh = Mv^2 = I\omega^2\left(1+\frac{N}{n}\right)$.

Or, $2Mgh - Mv^2 = I\omega^2\left(1+\frac{N}{n}\right)$,

whence, $I = \frac{2Mgh - Mv^2}{\omega^2\left(1+\frac{N}{n}\right)} = \frac{2Mgh - Mr^2\omega^2}{\omega^2\left(1+\frac{N}{n}\right)} = \frac{M\left(\frac{2gh}{\omega^2} - r^2\right)}{\left(1+\frac{N}{n}\right)}$,

because $v = r\omega$, where r is the radius of the axle.

To determine the value of ω, we note that the wheel comes to rest in time t after the mass gets detached from the axle. Hence, if the retardation due to friction be considered to be uniform, the *average angular velocity of the wheel during* time t may be taken to be $(\omega + 0)/2 = \omega/2$. And, since the wheel makes n rotations before coming to rest, it describes an angle $2\pi n$ in time t, and its average angular velocity is, therefore, also equal to $2\pi n/t$. So that,

$$\frac{\omega}{2} = \frac{2\pi n}{t}, \text{ Or, } \omega = \frac{4\pi n}{t}.$$

Substituting this value of ω in the expression for I above, we have

$$I = \frac{M\left[\left(\frac{2ght^2}{16\pi^2 n^2}\right) - r^2\right]}{1+\frac{N}{n}} = \frac{M\left[\left(\frac{ght^2}{8\pi^2 n^2}\right) - r^2\right]}{\left(\frac{n+N}{n}\right)}$$

Or, $I = M\left(\frac{n}{n+N}\right)\left(\frac{ght^2}{8\pi^2 n^2} - r^2\right) = \frac{M}{n+N}\left(\frac{ght^2}{8\pi^2 n} - nr^2\right)$

whence, the value of I, the *moment of inertia of the flywheel about its axis* of rotation, may be easily obtained.

GYROSTATIC APPLICATIONS

The following are a few example of the use of a gyroscope for steadying of motions and ensuring directional stability, the underlying principle in each case being that the gyroscope is made to process in such a manner that the reaction couple due to the rotating disc or wheel offsets exactly the effect of the disturbing couple.

(i) *The gyrostatic or the Gyro-Compass* : Used in aeroplanes and ships and, particularly in submarines, it is a disc or a flywheel of a large moment of inertia (*i.e.*, a. gyrostat), G, with its axle PQ mounted in a horizontal ring R_1 free to rotate about its axis AB inside a vertical ring R_2 which can turn freely about the axis CD within a framework F, carried on horizontal gimbals, of which only one pair G_1G_1 is shown here, (Fig. 1.31). This ensures that the disc or the flywheel has three degrees of freedom, irrespective of the position of the framework F.

Rotated at a high speed by means of an electric motor, with its axis in the magnetic meridian (*i.e.*, in the vertical plane passing through the geographic north and south of the earth), the conservation of momentum of the disc or the wheel and its directional stability ensure that its axis POQ always lies in the direction of the meridian, *i.e.*, along the geographic north and south.

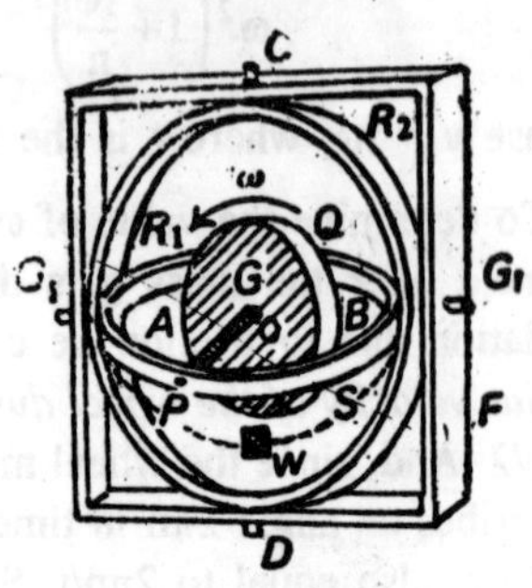

Fig. 1.31

In view of the complete freedom of movement of the disc, irrespective of the position of the supporting frame (F), any movement of the latter produces no deflecting couple or torque on it and this *particular (North-South) direction of the axis thus continues to be maintained in space all the time, despite any changes in the direction of the ship or the submarine or any tossings or pitchings of it.*

This compass, appropriately called the *gyro-compass*, is therefore more dependable than the ordinary compass, the more so because *it remains quite unaffected by any type of magnetic disturbances.*

(ii) *Rifles of barrels ef guns and rifles* : This is another familiar example of the directional stability of a rapidly spinning body. For, if the shot or the bullet be given a rapid spin about an axis along its direction of motion, it is less responsive to small deflective forces through air and its uniformity of flight is thus greatly improved.

This is achieved by what is called '*rifling*' the barrel of the gun or the rifle, *i.e.*, by cutting spiral grooves inside it, so that before emerging out, the shot or the bullet first moves along these, thereby acquiring the necessary spin to ensure its almost linear motion on emergence.

(iii) *Riding of bicycles and rolling of hoops or discs* : Both these are cases of what is called *statical instability*, for neither of the two, when at rest, can possibly remain in equilibrium in the position in which it does when in motion.

Here, again, the gyroscopic action does the trick by appropriately deflecting their axes of rotation, thereby changing their planes of rotation to counteract the disturbing effect due to gravity.

Thus, if a person riding a bicycle, without holding the handle, wishes to turn to one side, all he has to do is to tilt a little to that side, thus producing a couple about the horizontal direction of motion of the front wheel (here functioning as a rotating gyrostat) which then turns the axle of the wheel about the vertical, and hence its plane of rotation, in the desired direction.

The same is true far a hoop or a disc, projected with its plane vertical to roll over a horizontal surface.

(iv) *Stabilising of ships* : Although a ship is intrinsically stable, it is subject to disturbing oscillations in a rough or heavy sea, both about its transverse and longitudinal axes. The former type of oscillation is called pitching and the latter type, rolling. The amplitude in the case of rolling being much greater than that in the case of pitching, it is the more troublesome of the two. Efforts are, therefore, directed towards minimising rolling rather than pitching.

Now, rolling is caused by the torque generated as a result of the difference in the buoyancy on the two sides of the centre line of the ship when it happens to lie on a wave slope. This torque has its maximum value when the ship lies on either side of a wave at the point of maximum slope and, the minimum or zero value when it lies on either the crest or the trough of a wave.

To overcome this disturbing torque, a large gyroscope, here called a gyrostabiliser, is used. It is so mounted as to rotate or spin about a vertical shaft or axle which can be tilted only forwards and backwards but not sideways. The shaft is, in fact, enclosed in a casing which is supported in bearings fixed to the frame of the ship, thus allowing, for its precessional motion through about 60° on either side of the vertical, and is driven by an electric motor. The arrangement is such that during a roll of the ship one way or the other (*i.e.,*. clockwise or anticlockwise), the shaft of the gyro-stabiliser is automatically titled forwards or backwards, such that its precession gives rise to a torque in the opposite direction to that producing the roll.

(v) *Precession of the equinoxes* : The equatorial plane of the earth and the plane of its orbit round the sun (or the elliptic) are, as we know, inclined to each other at an angle of 23.5°, so that they intersect each other along a line, called the line of *equinoxes*. The earth, therefore, crosses this line twice in one full round about the sun, near about the 21st of March and the 22nd of September every year, its position in the orbit at these two points being referred to as the vernal and the *Autumnal equinox* respectively.

Now, the earth, as is well known, is not an exact sphere but slightly bulges out at the equator (its equatorial radius being about 13 miles greater than its polar radius) and thus has the form of a *flattened ellipsoid of revolution*. Further, the sun and the moon do not usually lie in its equatorial plane but rather in the plane of its orbit or the elliptic. In consequence, the gravitational attraction due to the sun and the moon on the equatorial bulge of the earth gives rise to a torque, bringing about the precession of its axis. And since the earth acts like a gigantic top, its precession axis describes a circular cone relative to the fixed stars, *e.g.*, the pole star. This precession of the earth's axis brings about a corresponding change in the direction of the line of equinoxes and the phenomenon is called *precession of the equinoxes*.

The torque on the earth due to the attractive force of the sun and the moon is, however, so small that it takes 25800 years for the earth's axis to describe the complete cone, at which rate of rotation the star *Vega* will be the pole star in about 1200 years hence.

(vi) *Other applications* : Among other applications of gyrostatic action may be mentioned the modern aircraft appliances like the *automatic pilot*, the *steering of torpedoes*, the *artificial horizon*, the *bomb sights* and the *turn and bank indicators* etc.

The function of all these instruments is to record the effects of a change of orientation between a relatively fixed plane (provided by a fast rotating gyrostat), serving as the *datum plane*, and some other movable plane in the machine. This they do with far greater precision and reliability than is possible by human judgement alone, however trained and practised.

PRECESSION

We know that when a body rotates about a *fixed* axis, under the action of a torque, the axial component of the torque alone is effective in producing rotation (1.3), its component perpendicular to the axis being neutralised by an equal and opposite torque due to the constraining forces applied to the axis to maintain its original position, *i.e.*, to keep it fixed.

In the absence of any such constraining forces, obviously, the axis of rotation, and hence also the plane of rotation, of the body will turn from its initial position under the action of the *component of the torque perpendicular to the axis*, which we may denote by the symbol τ_1.

Since this torque τ_1 (*i.e.*, the axis of this torque) is perpendicular to the axis of rotation of the body, the magnitude of the angular velocity, and hence also that of the angular momentum, of the body remains unaltered, with only its direction changed in exactly the same manner that in a linear motion, a constant force (viz., the centripetal force), acting on a body in a direction perpendicular to that of its velocity, simply changes its direction and not its magnitude. So that, if the torque τ_1 be a constant one, the rotation-axis and the plane of rotation of the rotating body continue to change their direction at a constant rate, with the axis of the torque always perpendicular to the axis of rotation of the body.

This change in the direction of the rotation-axis, or in the direction of the plane of rotation, of the rotating body under the action of a constant torque perpendicular to the axis of rotation is called precession.

The torque τ_1 which brings this about is, therefore, called the *precessional torque* and the rate of rotation of the axis of rotation or the plane of rotation of the body is called the *rate of precession*, usually denoted by the symbol ϕ.

Let us obtain the value of this *precessional torque* (τ_1) and the *rate of precession* (ϕ).

We shall consider here only the case of a symmetrical body rotating about one of its axes of symmetry, for in such a case alone, does the position of the axis of rotation remain the same, or very nearly the same, with respect to the rotating body itself.

Thus, let us consider the case of a uniform circular disc D rotating with a constant angular velocity ω, in the anticlockwise direction, about its geometric axis OY (*i.e.*; about the axis passing through its centre (or centre of mass) and perpendicular to its plane), in its own plane and hence perpendicular to OY, (Fig. 1.32).

Clearly, if I be the M.I. of the disc about its axis of rotation (OY), its angular momentum J will be equal to Iω.

Let it be represented by the vector OA, perpendicular to its plane of rotation or along the axis of rotation OY.

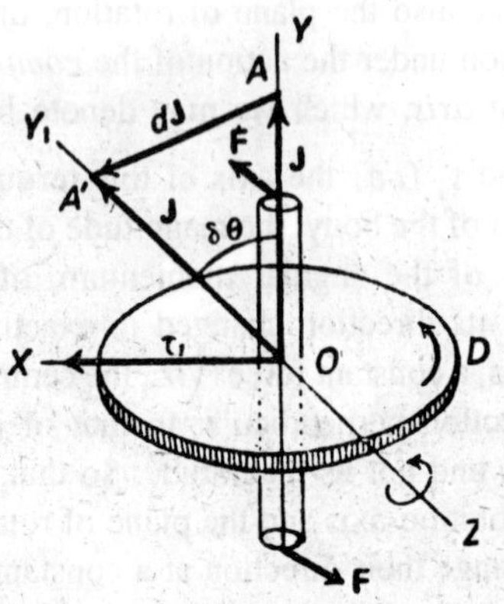

Fig. 1.32

Under the action of the component of the torque (F, F) perpendicular to it, (*i.e.*, with the axis of the torque perpendicular to it and hence in the plane, of the disc, along OX), let the axis of rotation OY, and hence also the plane of rotation of the disc, turn or precess through a small angle δθ, in a small interval of time δt, about the axis OZ, through O and perpendicular to both OX and OY, *i.e.*, perpendicular to the plane of the paper.

The angular momentum of the disc will now be represented by the vector OA', perpendicular to its new plane of rotation, or along its new axis of rotation OY_1. It will be seen that the torque vector being perpendicular to the angular momentum vector, the angular momentum of the disc remains unaltered in magnitude, with only its direction changing through angle AOA = δθ and, therefore, OA′ = QA = J.

Since even a change in its direction constitutes a change in momentum, we have, in accordance with the principle of addition of vectors, the change in angular momentum dJ represented by the vector AA′ which forms the third side of the triangle of vectors OAA′. This means that the torque τ_1 = dJ/δt, also directed along AA′ or OX, is again perpendicular to the axis of rotation in its new position (OY_1) and the axis, therefore, goes on precessing at a constant rate (ϕ) about the axis OZ through O, *in the direction of the torque.*

To obtain the value of ϕ, we note that $\tau_1 = \dfrac{dJ}{\delta t} = \dfrac{I\,d\omega}{\delta t}$

So that, *change in angular momentum* $dJ = \tau_1 \delta t$.

Since from ihe figure, $\delta\theta = \dfrac{AA'}{OA} = \dfrac{dJ}{J}$, we have $\delta\theta = \dfrac{\tau_1 \delta t}{J}$, whence,

the *rate of precession* $\phi = \dfrac{\delta\theta}{\delta t} = \dfrac{\tau_1}{J} = \dfrac{\tau_1}{I\omega}$. ...(i)

[$\because J = I\omega$.]

And, therefore, *precessional torque* $\tau_1 = \phi \times (I\omega) = \phi \times J$, its magnitude being equal to $\phi \times J \times$ *sine of the angle between them* (which is clearly 90° here). ...(ii)

Thus, *the precessional torque vector τ, is the cross product of the precessional rate vector ϕ and the angular momentum vector* J. So that, if we know the directions of vectors ϕ and J, the direction of the torque vector may be easily obtained from the *right hand rule for a vector product.*

It will thus be seen that *if the axis of rotation of a body lies along OY and the axis of the applied torque along OX, the body precesses about the third mutually perpendicular axis OZ.*

As can be seen from relation (i) above, *the rate of precession (ϕ) of the rotation-axis is proportional directly to the torque (τ_1) applied and inversely to the angular momentum ($J = I\omega$), i.e.., to the moment of inertia and the angular velocity of the rotating body about the axis of rotation, the direction of precession being the direction of the change of angular momentum or the torque.* So that, the angular momentum vector, so to speak, changes the torque vector.

For a given torque, obviously, the larger the moment of inertia of a body and the higher its angular velocity about the axis of rotation, the smaller the rate of precession. This is the principle underlying a *gyrostat*, (1.13).

ROTATIONAL ENERGY STATES OR A DIATOMIC MOLECULE

As we have just seen in 1.15 above, the M.I. of a diatomic molecule about an axis through its centre of mass and perpendicular to the distance r_0 (the bond length) between the two atoms is μr_0^2, where μ is its reduced mass. If, therefore, ω be the angular velocity of the molecule about its axis of rotation, its *angular momentum* j = Iω, and the *angular momentum vector* j lies in the plane perpendicular to the plane of rotation or along the axis of rotation.

Thus, $j^2 = I^2\omega^2$. Or, $\omega^2 = \frac{j^2}{I^2}$.

Now, as we know, the kinetic energy of rotation of a rigid body about an axis, about which its M.I. is I, is given by $E = \frac{1}{2}I\omega^2$, where ω is its angular velocity about the axis.

So that, substituting the value of ω^2 as obtained above, we have

$$E = \frac{1}{2}I\frac{j^2}{I^2} = \frac{j^2}{2I} = \frac{j^2}{2\mu r_0^2}$$

This is the relation for the kinetic energy of rotation of a diatomic molecule on the basis of classical mechanics, where j and, therefore, E may have any possible values and not only certain specified ones.

As we have seen under 1.14, however, the angular momentum in the case of small particles is *quantised* and, as deduced on the basis of wave mechanics, may have only discrete values $\sqrt{j(j+1)}$ in terms of $\hbar$ or h/2π units, where j is the *total angular momentum quantum* number, equal to 0, 1, 2, 3 etc., *i.e.*, the *magnitude of the angular momentum of the molecule* $= \sqrt{j(j+1)}\frac{h}{2\pi}$.

Substituting this in place of j, therefore, in the expression for rotational kinetic energy of the diatomic molecule, we have

$$E = \frac{j(j+1)h^2}{8\pi^2 I},$$

an expression which is also borne out by actual experiment.

This means, then, *that the rotational kinetic energy of a diatomic molecule is also quantised and may have only discrete values corresponding to j = 0, 1, 2, 3, etc. and no other intermediate values.*

Putting $\frac{h^2}{8\pi^2 I} = A$, we have $E = Aj(j+1)$.

So that, for j = 0, E = 0; for j = 1, E = 2A;

for j = 2, E = 6A and so on.

These are called the *rotational energy states, (or rotational energy levels) of the diatomic molecule*, indicating that it can have no rotational energy between 0 and 2A, between 2A and 6A and so on.

For convenience, these are usually shown in what is called an *energy level diagram*, as indicated in Fig. 1.33, which consists of a set of parallel lines, one above the other, representing the various energy levels, the lowest or the base line representing the *ground state* of the molecule, with E = 0 and hence also E = 0. The upper lines correspond to j = 1, j = 2 etc. and, therefore, E = 2A, 6A etc., the distances of the lines (or energy levels) from the base line (or the ground state) being proportional to the values of energy represented by them.

At or near the absolute zero, the molecule is in the ground state, with E = 0 corresponding to j = 0. As the temperature rises, the energy state or the energy level of the molecule-also rises as shown, so that E attains the values 2A, 6A, 12A, 20A, 30A etc. corresponding to j = 1, 2, 3, 4 and 5 etc. respectively, *i.e., there is a transition of the molecule from a lower to a higher energy state* as it absorbs more and more energy.

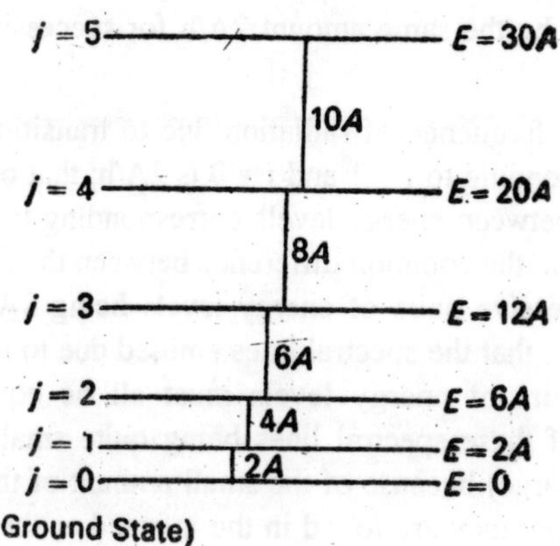

Fig. 1.33

On the other hand, if the transition occurs from a higher to a lower energy state, the molecule gives out its surplus energy which appears in the form of an electromagnetic radiation. These transitions are indicated by vertical lines in the figure along with the surplus energy emitted as electromagnetic radiation in each successive transition?.

In either case, however, only those transitions are possible for which the value of j changes by $\Delta j = \pm 1$, *i.e.*, from j to (j + 1) or from j to (j – 1).

The change in energy corresponding to a change $\Delta j = \pm 1$ in the value of j can be easily obtained from the difference in energy at energy levels corresponding to j and (j – 1).

Thus, the change in energy,

$$\Delta E = E_{(j)} - E_{(j-1)}$$
$$= j(j+1)A - (j-1)jA = 2Aj.$$

Or, $$\Delta E = 2\frac{h^2}{8\pi^2 I}j = \frac{h^2}{4\pi^2 I}j \quad ...(i)$$

So that, $h\nu = \frac{h^2}{4\pi^2 I}j$ and,

therefore, $$\nu = \frac{h}{4\pi^2 I}j = \frac{2Aj}{h}. \quad ...(ii)$$

This shows that the frequencies (ν) of the transition radiations emitted differ by the same amount 2A/h for successive pairs of energy levels.

Thus, the frequency of radiation due to transition between energy levels corresponding to j = 1 and j = 0 is 2A/h; that of the radiation due to transition between energy levels corresponding to j = 2 and j = 0 is 4A/h and so on, the common difference between the frequencies emitted between successive pairs of energy levels being 2A/h all along. This means, clearly, that the spectral lines emitted due to transitions between successive pairs of energy levels must all be equally spaced. The frequencies of these spectral lines being quite small (and hence their wavelengths large) because of the small amount of the energy involved in the transition, they are found in the infrared or the micro-radio wave region of the electromagnetic spectrum.

It will be readily seen that this common difference (2A/h) in the frequencies of the successive spectral lines enables us to obtain the value of the M.I. of a diatomic molecule about its own axis of rotation. For,

$$\frac{2A}{h} = \frac{h}{4\pi^2 I},$$

whence, $$I = \frac{h^2}{8\pi^2 A}.$$

And, since $I = \mu r_0^2$, where μ is the reduced mass of the molecule and r_0; the internuclear distance or bond length, we can also obtain the value of r_0.

Now, *defining wave number v as the number of wavelengths per cm,* we have

$$v = \frac{1}{\lambda} = \frac{v}{c} \text{ per cm,}$$

where c is the velocity of light or electromagnetic radiation, in general.

If, therefore, E be the energy of radiation, we have

$$E = hv \text{ or, } v = \frac{E}{h},$$

Or, $v = \frac{1}{\lambda} = E$, where $\frac{E}{hc} = \epsilon$ is referred to as *energy in wave numbers.*

We can thus put our equation I above in terms of ϵ, when it becomes

$$v = \frac{E_{(j)}}{hc} - \frac{E_{(j-1)}}{hc} = E_{(j)} - E_{(j-1)} = \frac{2A}{hc} j = \frac{hj}{4\pi^2 Ic} = \frac{hj}{4\pi^2 \mu r_0^2 c}.$$

This shows at once that the common spacing between the spectral lines emitted is $h/4\pi^2 Ic = h/4\pi^2 \mu {r_0}^2 c$. So that, knowing the spacing between the spectral lines we can again obtain the M.I. of the diatomic molecule and its intranuclear distance or bond length.

Further, knowing the energy in wave numbers (ϵ) for two successive energy levels, say, corresponding say and (j – 1), we can straightaway obtain the wave number v and hence, the wavelength (λ) of the spectral line emitted during transition between the two energy levels. For $v = E_{(j)} - E_{(j-1)}$.

MOMENT OF INERTIA OF A DIATOMIC MOLECULE

In a diatomic molecule, in its stable equilibrium position, the two atoms are a certain distance r_0 apart, where r_0 is called its *internuclear distance or bond length.* For our present purpose, however, we may imagine it to consist of two tiny spheres at either end of a thin weightless rod, as shown in Fig. 1.34.

Let C be the centre of mass of the molecule and r_1 and r_2, the respective distances of the two atoms from it.

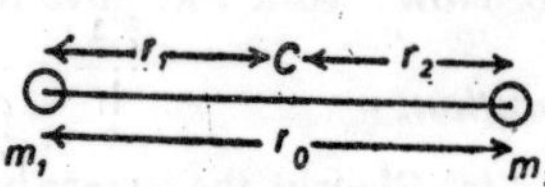

Fig. 1.34

Then, clearly, $r_1 + r_2 = r_0$ ...(i)

and $m_1 r_1 = m_2 r_2$, ...(ii)

where, m_1 and m_2 are the masses of the two atoms respectively.

From relations (i) and (ii), therefore, we have

$$r_1 = \frac{m_2}{m_1 + m_2} r_0$$

and $$r_2 = \frac{m_1}{m_1 + m_2} r_0.$$

Now, the moment of inertia of the molecule (*i.e.*, of the two atoms) about an axis passing through the centre of mass C and perpendicular to r is clearly $I = m_1 r_1^2 + m_2 r_2^2$.

Or, substituting the values of r_1 and r_2, we have

$$I = m_1\left(\frac{m_2}{m_1 + m_2}\right)^2 r_0^2 + m_2\left(\frac{m_1}{m_1 + m_2}\right)^2 r_0^2.$$

Or, $$I = \left[\frac{m_1 m_2^2 + m_2 m_1^2}{(m_1 + m_2)^2}\right] r_0^2 = \left[\frac{m_1 m_2 (m_1 + m_2)}{(m_1 + m_2)^2}\right] r_0^2 = \frac{m_1 m_2}{m_1 + m_2} r_0^2.$$

But, as we know, $m_1 m_2/(m_1 + m_2) = \mu$, the reduced mass of the molecule. So that, $I = \mu r_0^2$.

Or, *M.I. of diatomic molecule = (reduced mass of the molecule) × (internuclear distance or bond length)*^{2}.

SOLVED EXAMPLES

Example 1:

(a) A solid sphere of mail 100 gm and radius 2.5 cm roils without sliding with a uniform velocity of 10 cm/sec along a straight line on a smooth horizontal table. Calculate its total energy in ergs.

(b) A hoop of radius 100 cm and mass 19 kg is rolling a along a horizontal surface, so that its centre of mass has a velocity of 20 cm/sec. How work will have to be done to stop it?

Solution:

(a) Clearly, the sphere possesses only kinetic energy, so that its total energy is equal to its K.E.

Now, *K.E. of a rolling body*, as we know, is given by

$$\frac{1}{2}Mv^2\left(1+\frac{K^2}{R^2}\right).$$

For a *solid sphere,* $K^2 = \frac{2}{5}R^2$. And, therefore,

K.E. or the total energy of the rolling sphere,

$$E = \frac{1}{2}Mv^2\left(1+\frac{2}{5}\frac{R^2}{R^2}\right)$$

$$= \frac{1}{2}Mv^2\left(1+\frac{2}{5}\right) = \frac{7}{10}Mv^2.$$

Or, substitutıng the values of M and v, we haze

$$E = \frac{7}{10} \times 100 \times 10^2 = 7 \times 10^3 \text{ ergs.}$$

(b) Obviously, *work required to be done to stop the rolling hoop*

$$= \textit{its K.E.} = \frac{1}{2}Mv^2\left(1+\frac{K^2}{R^2}\right) = \frac{1}{2}Mv^2\left(1+\frac{R^2}{R^2}\right)$$

$$= Mv^2 = 10^4(20)^2 = 4 \times 10^6 \text{ ergs.}$$

Example 2:

A small solid sphere rolls without slipping along a loop-the-loop track, shown in Fig. 1.35 from a height OR from the bottom of the track, where R is the radius of the circular part of the track. Calculate the horizontal and the vertical forces acting on the sphere when it rises up to the point P in a level with the centre O of the circular part.

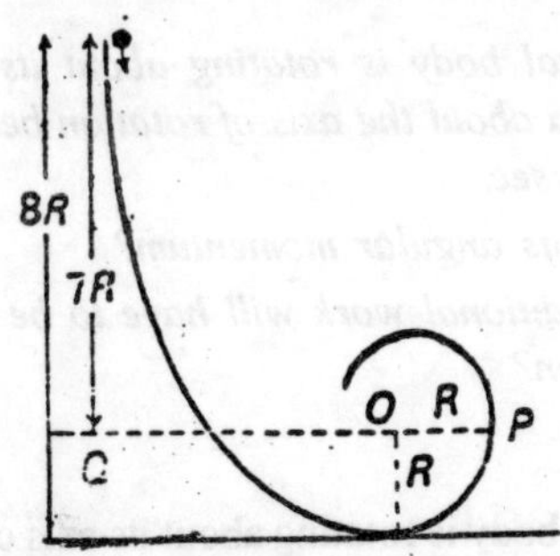

Fig. 1.35

Solution:

Clearly, the kinetic energy acquired by the sphere when it arrives at P in ihe circular part of ihe loop will be the same as that at Q in a level with P and the centre O of the circular part, *i.e.*, at a distance (8 R – R) = 7R from the top of the track. If, therefore, v be the velocity of the sphere at Q or P, *its K.E. of rotation* $= \frac{1}{2}Mv^2\left(\frac{1+K^2}{r^2}\right)$, where M is its *mass, r, its radius* and K, its *radius of gyration about its diameter* (about which it is rotating as it is rolling down).

Since $K^2 = \frac{2}{5}r^2$ in the case of a sphere, we have *K.E. acquired by the sphere on falling through a height 7R*

$$= \frac{1}{2}Mv^2\left(1+\frac{2}{5}\frac{r^2}{r^2}\right) = \frac{1}{2}Mv^2\left(\frac{7}{5}\right) = \frac{7}{10}Mv^2$$

And *P.E. lost by the sphere on falling through this height 7R = 7 MgR.* Since gain in K.E. of the sphere = loss in its P.E., we have

$$\frac{1}{10}Mv^2 = 7MgR,$$

whence, $$v^2 = 10gR.$$

Obviously, this Velocity v of the sphere at the point P is tangential to the circular part of *the track at P. The horizontal force acting on it, therefore, towards O is the centripetal force*

$$\frac{Mv^2}{R} = \frac{M(10gR)}{R} = 10\ Mg.$$

And, the *vertical force acting on it is clearly its weight Mg.*

Example 3:

A symmetrical body is rotating about its axis of symmetry, its., moment of inertia about the axis of rotation being 1 kg-m² and its rate of rotation 2 rev/sec.

(a) What is its angular momentum?

(b) What additional work will have to be done to double its rate of rate of rotation?

Solution:

(a) Since the body is rotating about its axis of symmetry, *the angular momentum vector coincides with its axis of rotation* and we have its *angular momentum.* J = Iω.

And since its K.E. of rotation, $E = \frac{1}{2} I\omega^2$,
we have $I\omega^2 = 2E$, and, therefore,

$$I^2\omega^2 = 2EI, \text{ whence, } J = J\omega = \sqrt{2EI}\,.$$

Here, $I = 1$ kg-m^2 and $\omega = 2$ rev/sec $= 2 \times 2\pi$ or 4π radian/sec,

So that, $E = \frac{1}{2}$ (1) $(4\pi)^2 = 8\pi^2$ joules and, therefore

$$J = \sqrt{2EI} = \sqrt{2 \times 8\pi^2 \times 1} = \sqrt{16\pi^2}$$
$$= 4\pi = 12.57 \text{ kg-m}^2\text{/sec.}$$

(b) When doubled, the rate of rotation of the body will be 4 rev/sec. $= 8\pi$ radian/sec.

$\therefore$ *its K.E. of rotation* will be $\frac{1}{2} I(8\pi)^2 = \frac{1}{2}(1)(64\pi^2) = 32\pi^2$ joules.

Hence, *additional work required to be done = its final K.E. of rotation–its initial K.E. of rotation* $= 32\pi^2 - 8\pi^2 = 24\pi^2 = 236.8$ joules.

Example 4:

A particle of mass m is released from rest at the point A, a distance x = b from the origin O so as to fall vertically parallel to the y-axis, as shown in Fig. 1.36.

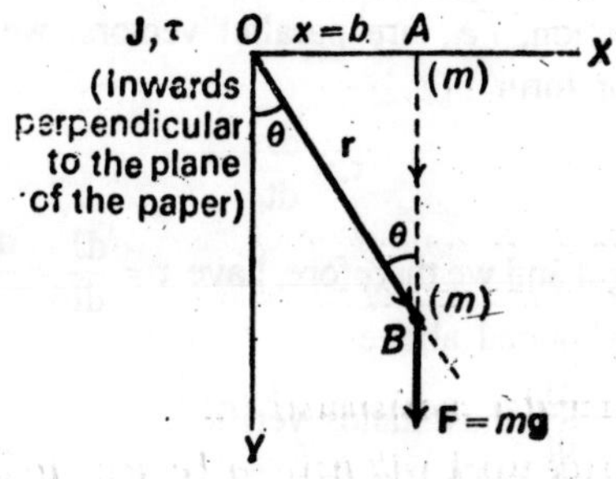

Fig. 1.36

(a) Calculate the angular momentum of the particle and the torque acting on it at any given instant t after its release from the point A.

(b) Verify from your result the relation $\tau = dJ/dt$.

Solution:

(a) Let B be the position of the particle at a given instant t after its release from rest at the point A, so that its position vector with respect to the origin O is r, as shown.

Now, *angular momentum about the origin* O is $J = r \times p$, where p is the *linear momentum*. And the magnitude of J is $J = rp \sin\theta = p\, r \sin\theta = pb$, because $r \sin\theta = b$.

Since the particle is falling under the action of gravity, its velocity at instant is, say, $v = 0 + gt = gt$.

$\therefore$ $p = mv = mgt$ and $J = mgbt$, the angular momentum vector J being, in accordance with the right hand rule, directed *inwards* at O, *perpendicular to the plane of the paper.*

And, clearly, the torque acting on the particle is $\tau = r \times F$, where F is the force mg acting on it vertically downwards, as indicated.

$\therefore$ *magnitude of the torque*, $\tau = r F \sin\theta = F(r \sin\theta) = Fb = mgb$.

Again, in accordance with the right hand rule, the torque vector τ must also be directed inwards at O, perpendicular to the plane of the paper, *i.e., in the same direction as* J.

Let us now see whether we obtain the same value for the magnitude of the torque by the application of the relation $\tau = dJ/dt$.

Since the change in angular momentum dJ and the torque τ both have the same direction, *i.e.,* are parallel vectors, we can write this relation in the scalar form

$$\tau = \frac{dJ}{dt}.$$

(b) Now, $J = mgbt$ and we therefore, have $\tau = \dfrac{dJ}{dt} = \dfrac{d(mgbt)}{dt} = mgb$, the same result as obtained above.

The relation $\tau = \dfrac{dJ}{dt}$ thus stands verified.

Example 5:

A solid cylinder (a) rolls, (b) slides from rest down an inclined plane. Neglect friction and compare the velocities in both cases when the cylinder reaches the bottom of the incline.

Solution:

We know that the acceleration of a body *rolling* down an inclined plane of angle of inclination O is given by $a = \left[\frac{R^2}{(K^2 + R^2)}\right] g \sin\theta$.

For a solid cylinder, $K^2 = \frac{R^2}{2}$ and, therefore,

$$a = \left[\frac{R^2}{\left(\frac{R^2}{2} + R^2\right)}\right] g \sin\theta = \frac{2}{3} g \sin\theta.$$

Hence, if v_1 be the *velocity acquired by the cylinder on reaching the bottom of the incline of length l*, we have

$$v_1^2 - 0 = 2al.$$

Or, $v_1^2 = 2.\frac{2}{3} g \sin\theta.l = \frac{4}{3} g \sin\theta.l$ [∵ u = 0]

And, when the cylinder *slides* down the plane, its acceleration a = g sin θ.

∴ if v_2 be the velocity acquired by the cylinder on reaching the bottom of the incline, we have

$$v_2^2 - 0 = 2al. \text{ Or, } v_2^2 = 2g \sin\theta.l \qquad [\because u = 0].$$

∴ $v_2^2 : v_1^2 :: 2g \sin\theta l : \frac{4}{3} g \sin\theta.l.$

Or, $v_2^2 : v_1^2 :: 2 : \frac{4}{3}.$

$$\frac{v_2}{v_1} = \sqrt{\frac{2 \times 3}{4}}$$

$$= \frac{3}{2} = 1.225.$$

Thus, *final velocity when the cylinder slides down the plane : final velocity when it rolls down the plane* : 1.225 : 1.

Example 6:

Two spheres are identical in mass and volume, but one is hollow and the other solid. How will you identify them experimentally?

Solution:

As we know, the acceleration of a body rolling down an inclined plane is given by $a = [R^2/(K^2 + R^2)]\, g \sin\theta$. So that, *the greater the value of the radius of gyration (K) of the body about the axis of rotation, the smaller its acceleration and hence the longer the time taken by it to rail down the plane.*

Now, *in the case of a solid sphere,* $K^2 = \frac{2}{5}R^2$, where R is its *radius.*

In the case of a hollow sphere, $K^2 = \frac{2}{5}\left(\frac{R^5 - r^5}{R^3 - r^3}\right)$, where R and r are its *external and internal radii* respectively.

$$\text{Clearly, } \frac{2}{5}\left(\frac{R^5 - r^5}{R^3 - r^3}\right) = \frac{2}{5}\cdot\frac{R^5\left(1 - \frac{r^5}{R^5}\right)}{R^3\left(1 - \frac{r^3}{R^3}\right)} = \frac{2}{5}R^2\left(\frac{1 - \frac{r^5}{R^5}}{1 - \frac{r^3}{R^3}}\right)$$

$$\text{Since } \left(\frac{r^5}{R^5}\right) < \left(\frac{r^3}{R^3}\right), \text{ we have } \left(1 - \frac{r^5}{R^5}\right) > \left(1 - \frac{r^3}{R^3}\right)$$

$$\text{And, therefore, } \quad \frac{1 - \frac{r^5}{R^5}}{1 - \frac{r^3}{R^3}} > 1,$$

i.e., for a hollow sphere, $K^2 = \frac{2}{5}R^2 \times$ (*a quantity greater than* 1),

or K^2 for a hollow sphere is greater than $\frac{2}{5}R^2$, which is the value of K^2 for a solid sphere.

It follows, therefore, that *the acceleration of the hollow sphere down the plane will be less than that of the solid sphere. The latter will thus roll down the plane faster than the former and the two may thus be easily distinguished from each other despite their outward identical appearance.*

Example 7:

(a) *A force F equal to 10 kg wt is applied to lhe spoke of a wheel at a distance of 50 cm from the axis of rotation. If the spoke and the applied force make an angle of 45° and 75° respectively with the x-axis calculate the torque acting on the wheel.*

(b) An engine rotating at te rate of 1500 rev/min develops 50 H.P. What is the torque supplied by it?

Solution:

(a) Let OA be the spoke of a wheel W, (Fig. 1.37a), inclined at an angle of 45° to the x-axis and let P be the point distant 50 cm from O where the force is applied in the direction shown, the force vector P making an angle of 75° with the x-axis.

Clearly, then, the angle between the *position vector r* of point P and the force vector F is $\theta = (75° - 45°) = 30°$. If, therefore, τ be the *torque* acting on the wheel, we have $\tau = r \times F$ and its *magnitude* is thus given by $\tau = r\ F \sin\theta = 50(10 \times 1000 \times 980)(\sin 30°) = 50(98 \times 10^5)(1/2) = 2.45 \times 10^8$ dyne-cm and its *direction* is perpendicular to the plane of r and F, *i.e.*, perpendicular to the plane of the paper outwards or along the axis through O.

It may be noted that since $\tau = r\ F \sin\theta$, we could obtain its value by either the relation $\tau = r\ F\perp$, where $F\perp$ is the component of F perpendicular to r or the relation $\tau = F r\perp$, where $r\perp$ is the component of r perpendicular to F, (Fig. 1.37b).

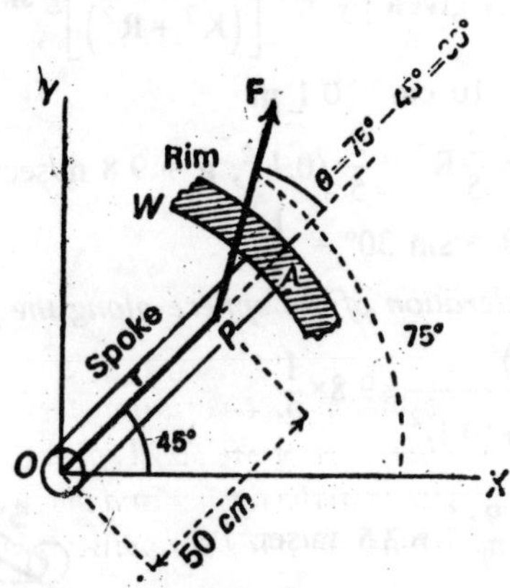

Fig. 1.37(a)

In the former case, $\tau = 50\ (F \sin\theta) = 50\ (98 \times 10^5)(1/2) = 2.45 \times 10^8$ dyne-cm and in the latter case, $\tau = (98 \times 10^5)(r \sin\theta) = (98 \times 10^5)(50)(1/2) = 2.45 \times 10^8$ dyne-cm.

(b) Clearly, *work dune by the engine per second* = 50 × 550 ft lb and the *angle through which it rotates in one second* = (1500/60) (2π) = 25 × 2π = 50π *radian.*

Now, *work done = torque × angle, i.e.,* w = τθ. So that, 50 × 500 = τ × 50π, where τ is the torque supplied by the engine.

Or, $\tau = \frac{50 \times 500}{50\pi} = \frac{500}{\pi} = 175$ft lb.

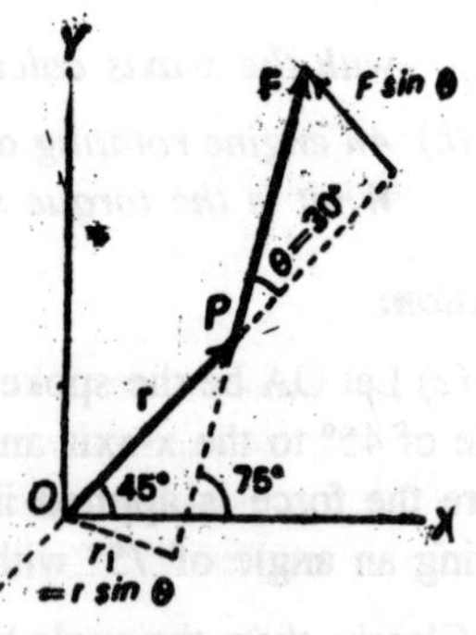

Fig. 1.37(b)

Thus the *torque supplied by the engine* = 175 ft. lb.

Example 8:

A uniform sphere of mass 2 kg and radius 10 cm is released from rest on an inclined plane which makes 30° angle with the horizontal. Deduce (i) its angular acceleration, (ii) linear acceleration along the plane and (iii) kinetic energy as it travels 2 metres along the plane.

Solution:

As we know, [1.10 (ii) (b)], the acceleration of a body rolling down an inclined plane is given by $a = \left[\frac{R^2}{(K^2 + R^2)}\right] g \sin\theta$

Here, $R = 10$ cm = 0.1 m

and $K^2 = \frac{2}{5}R^2 = \frac{2}{5}(0.1)^2$, g = 9.8 m/sec²

and $\sin\theta = \sin 30° = \frac{1}{2}$.

∴ *linear acceleration of the sphere along the plane*

$$= \frac{(0.1)^2}{\frac{2}{5}(0.1)^2 + (0.1)^2} \times 9.8 \times \frac{1}{2}$$

$$= \frac{4.9}{1 + \frac{2}{5}} = \frac{4.9 \times 5}{7} = 3.5 \text{ m/sec}^2.$$

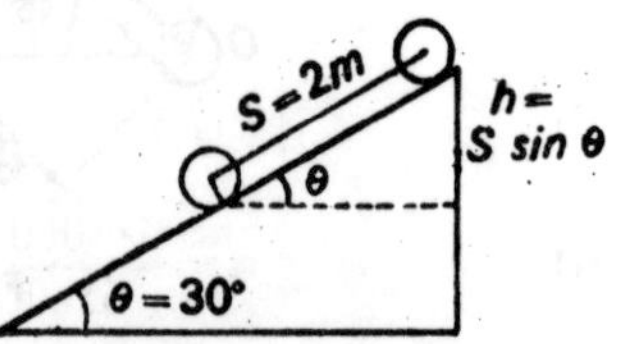

Fig. 1.38

If v be the *linear velocity* of the sphere along the plane at any given instant, we have v = Rω, where ω is its angular velocity.

$\therefore \frac{dv}{dt} = R\frac{d\omega}{dt} = R\alpha = a$, where $\frac{d\omega}{dt}$ is the *angular acceleration* α.

So that, *angular acceleration of the sphere*, $\alpha = \frac{a}{R} = \frac{3.5}{0.1} = 35$ radian/sec^2.

And, *K.E. acquired by the sphere on covering a distance S = 2 metres along the*

plane = P.E. lost by it in falling through a vertical distance h = S sin θ, (Fig. 1.38).

$$= \text{Mg S} \sin\theta = 2(9.8)(2)\left(\frac{1}{2}\right) = 1.96 \text{ joules.}$$

Example 9:

A flywheel of mass 6.4 kg is made in the form of a circular disc of radius 18 cm; it is driven by a belt whose tensions at the points where it runs on and off the rim of the wheel are 2 kg and 5 kg weight respectively. If the wheel is rotating at a certain instant at 60 revolutions per minute, find how long will it be before the speed has reached 210 revolutions per minute. While the flywheel is rotating at this latter speed, ***the belt is slipped off and a brake applied. Find the constant breaking couple required to stop the wheel in 7 revolutions.***

Solution:

Here, obviously, the *resultant tension in the belt* = 5 – 2 = 3 kg wt = 3 × 1000 × 981 *dynes*.

∴ *moment of the couple due to this tension* = 3 × 1000 × 981 × 18 dyne-cm.

But, if $I = \frac{MR^2}{2}$ be the moment of inertia of the flywheel (of the form of a circular disc) about its axis of rotation and α, its *angular acceleration*, the couple acting on the wheel is also Iα. We, therefore, have

$$I\alpha = 3 \times 1000 \times 981 \times 18.$$

Or, $$(65.4 \times 1000 \times 18^2/2)\alpha = 3 \times 1000 \times 981 \times 18,$$

whence, $$\alpha = \frac{3 \times 1000 \times 981 \times 18}{65400 \times 18 \times 9} \mp 5 \text{ radian/sec}^2.$$

Now, we have the relation $\omega_2 = \omega_1 + \alpha t$,

where ω_2 **= 210 rev/min = 210 × 2π/60 = 7π radians/sec,**

$$\omega_1 = 60 \text{ rev/min} = 60 \times 2\pi/60 = 2\pi \text{ radians/sec}$$

and $\alpha = 5$ radians/sec^2.

So that, $7\pi = 2\pi + 5t$.

Or, $5\pi = 5t$.

Or, $t = \pi = 3.142$ sec.

The flywheel will thus acquire the speed of 210 *rev/min in* 3.142 *sec.*

Let the angular retardation produced in the wheel by the braking couple be α'. And, Clearly, angle turned through by the wheel before coming to rest, say, $\theta = 7 \times 2\pi = 14\pi$ *radians*. So that applying the relation $\omega_2^2 - \omega_1^2 = 2\alpha\theta$, where $\omega_2 = 0$ and $\omega_1 = 7\pi$ radian/sec, we have $0 - (7\pi)^2 = 2\alpha' (14\pi)$, because, here,

$$\alpha = \alpha'. \text{ Or, } \alpha' = -\frac{49\pi^2}{28\pi} = -\frac{7\pi}{4} \text{ radian/sec}^2.$$

$\therefore$ *braking couple required to be applied*

$$= I\alpha' = \left(65.4 \times 1000 \times \frac{18^2}{2}\right)\left(\frac{7\pi}{4}\right) \text{gm-wt-cm}$$

$$= (65.4 \times 1000 \times 18 \times 19)\left(\frac{\frac{7\pi}{4}}{1000}\right) = 59.37 \text{ kg wt-cm.}$$

Example 10:

Two masses of 4 gm and 6 gm respectively are attached to the two ends of a light rod, 10 cm long, of negligible mass and the rod rotates anticlockwise at the rate of 2 revolutions per second about an axis passing through its centre of mass and perpendicular to its length. Obtain the values of

(a) the angular momentum of each mass about the centre of mass,

(b) the total angular momentum of the system about the centre of mass,

(c) the angular momentum of the system about the axis of rotation.

How will things change if the axis of rotation were inclined at an angle of 30° to the length of the rod and why?

Solution:

(a) Let m_1 and m_2 be the two masses of 4 gm and 6 gm respectively at ends of the rod AB of negligible mass and of length 10 cm (Fig. 1.39). Then, if x be the distance, of the centre of mass (C) of the system from mass m_1 = 4gm at A, we have 4x = 6(10 − x). Or, 10x = 60. Or, x = 6cm.

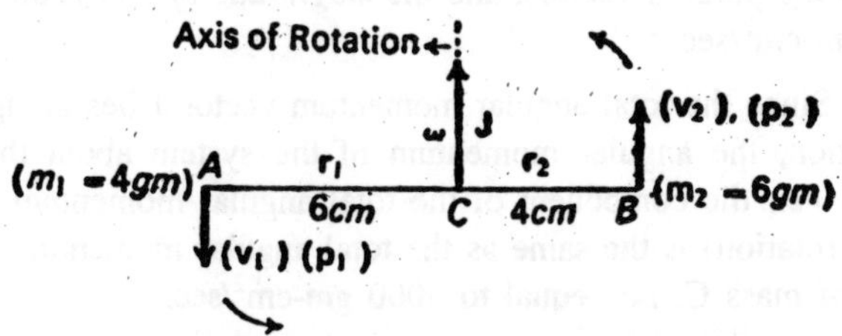

Fig. 1.39

Thus, the centre of mass of the system (C) lies at 6 cm from A and 4 cm from B, with the axis of rotation passing through it, perpendicular to the length of the rod, as shown.

Since the rod is rotating *anticlockwise* about this axis through C with an angular velocity of $2 \times 2\pi \approx 12.5$ radians/sec, the angular velocity vector ω has a magnitude 12.5 rad/sec and lies along the axis of rotation, as indicated in the figure.

Now, since mass m_1 = 4 gm at A rotates about C in a circle of radius CA = 6 cm, its *linear velocity* v_1 = CA × (*angular velocity*) = 6(12.5) = 75.0 cm/sec, the *linear velocity vector* v_1, and hence also the *linear momentum vector* p_1 (= m_1v_1), being perpendicular to CA, in the direction shown.

Similarly, *linear velocity of mass* m_2 = 6 gm at B about C in a circle of radius CB = CB × *(angular velocity)* = 4(12.5) = 50.0 cm/sec, with the *linear velocity vector* v_2, and hence also the *linear momentum vector* p_2 (= m_2v_2) perpendicular to CB, in the direction indicated.

The *angular momentum of mass* m_1 *about* C is thus $r_1 \times p_1 \times (m_1v_1)$ and its *magnitude* = 6 × 4 × 75 = 1800 gm-cm^2/sec.

Similarly, *angular momentum of mass* m_2 *about* C = $r_2 \times p_2 = r_2 \times (m_2v_2)$ and its *magnitude* = 4 × 6 × 50 = 12 gm-cm^2/sec.

(b) Clearly, the linear momentum vectors $p_1 = m_1v_1$ and $p_2 = m_2v_2$ of the two masses respectively are oppositely directed but since their position vectors r_1 and r_2 are also oppositely directed with respect to the centre of mass C, their angular momenta about C are both directed upwards perpendicular to the length of the rod. Hence *the total angular momentum vector about the centre of mass is J, directed upwards at C, perpendicular to the rod AB, or along the axis of rotation, with its direction thus coinciding with that of the angular velocity vector* m, *i.e.*, J and ω are parallel vectors, and the *magnitude of* J is 1800 + 1200 = 3000 gm-cm²/sec.

(c) Since the total angular momentum vector J lies along the axis of rotation, the angular momentum of the system about the axis of rotation *i.e.*, the component of the total angular momentum along the axis of rotation) is the same as the total angular momentum about the centre of mass C, *i.e.*, equal to 3000 gm-cm²/sec.

Alternatively, we could obtain the angular momentum of the system about the axis of rotation directly from the relation J = Iω, where I is the *moment of inertia* of the system about the axis of rotation, equal to $m_1r_1^2 + m_2r_2^2 = 4(6)^2 + 6(4)^2 = 144 + 96 = 240$ gm-cm² and ω, the angular velocity of the system about the axis of rotation, equal to 12.5 *radian/sec.* So that, *angular momentum of the system about the axis of rotation* = 240 × 12.5 = 3000 *cm-gm²/sec.*

Let us now take the case when the axis of rotation of the rod is inclined at an angle of 30° to its length, (Fig. 1.40).

In this case also, the *angular velocity vector* ω naturally lies along the axis of rotation and its magnitude is the same as before, *viz.*, 12.5 *radian/sec.*

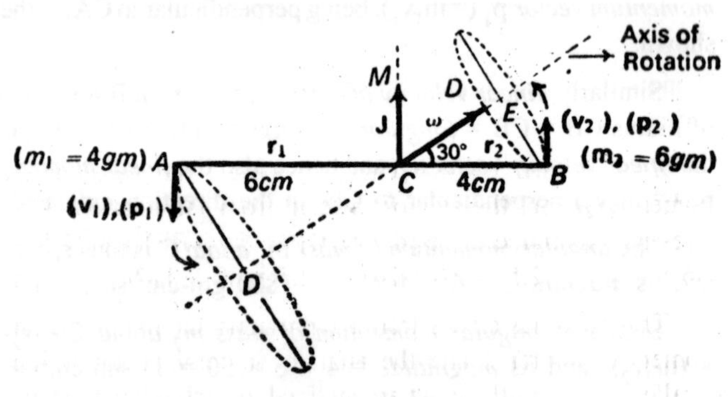

Fig. 1.40

Since mass m_1 = 4gm here rotates about the axis in a circle of radius AD = CA sin 30° = 6 × 1/2 = 3 cm, its *linear velocity* v_1 = AD × ω = 3 × 12.5 = 37.5 cm/sec and the *linear velocity vector* v_1, and hence also the linear momentum *vector* $p_1(= m_1v_1)$ is perpendicular to CA.

Similarly, since mass m_2 = 6 gm rotates about the axis in a circle of radius BE = CB sin 30° = 4 × 1/2 = 2 cm, its *linear velocity* v_2 = BE × ω = 2 × 12.5 = 25.0 cm/sec and the *linear velocity vector* v_2, and hence also the *linear momentum vector* p_2, is perpendicular to CB but directed oppositely to v_1 and p_1.

∴ *angular momentum of* m_1 *about* C = $r_1 \times p_1 = r_1 \times (m_1v_1)$ and its *magnitude* = 4 × 6 × 37.5 = 900 gm-cm²/sec

And, *angular momentum of* m_2 *about* C = $r_2 \times p_2 = r_2 \times (m_2v_2)$ and its *magnitude* = 4 × 6 × 25 = 600 gm-cm²/sec.

Now, although the linear momenta of the two masses about C are oppositely directed, their position vectors too with respect to C being oppositely directed, their angular momenta about C are both directed upwards, perpendicular to the rod at C.

∴ *total angular momentum of the system about C (the centre of mass), i.e.,* J *is directed upwards along CM at C, perpendicular to the rod, as shown,* and *its magnitude* = 900 + 600 = 1500 gm-cm²/sec.

Clearly, therefore, *the total angular momentum vector J here is not in the some direction with the angular velocity vector ω but is inclined to it at an angle of* (90 – 30) = 60°.

∴ *magnitude of the angular momentum of the system about the axis of rotation* (*i.e.,* the component of the total angular momentum (about C) along the axis of rotation) = 1500 cos 60° = 1500 × 1/2 = 750 gm-cm²/sec.

Alternatively, we could directly obtain the value of the angular momentum about the axis of rotation from the relation J = Iω which gives its magnitude to be $[4 \times (AD)^2 + 6 \times (BE)^2] \times 12.5 = [4 \times (3)^2 + 6 \times (2)^2] \times 12.5 = (36 + 24) \times 12.5 = 60 \times 12.5$ = 750 gm-cm²/sec.

This case thus differs from the previous one in that the total angular momentum vector J and the angular velocity vector ω are here not parallel vectors as there but arc inclined to each other at an angle of 60°. This is so because when the axis of rotation of a body is not also its axis of symmetry, not the whole of the applied torque vector but only a component of it lies along the axis and it is this component which produces rotation about this axis. The component at right angles to the axis merely tends to turn the axis from its fixed position and is neutralised by an equal and opposite torque applied by the bearings to the shaft to keep the axis fixed. This component of the torque, in the case shown, is perpendicular to the plane of J and ω, *i.e.*, the plane of the paper, acting outwards at C.

Example 11:

A flywheel of mass 10 kg and radius 20 cm is mounted on an axle of mass 8 kg and radius 5 cm. A rope is wound round the axle and carries a weight of 10 kg. The flywheel and the axle are set into rotation by releasing the weight. Calculate : (i) the angular velocity and the kinetic energy of the wheel and the axle, and (ii) the velocity and kinetic energy of the weight, when the weight has descended 20 cm from its original position.

Solution:

Clearly, P.E. lost by the weight in descending through a distance of 20 cm must be equal to the K.E. gained by the flywheel and axle and the weight itself, *i.e.*, $Mg \times 20 = \frac{1}{2}I\omega^2 + \frac{1}{2}Mv^2$, where M is the *mass of the weight, I, the moment of inertia of the flywheel and the axle*, ω, *its angular velocity* and v, the *linear velocity of the mass M* = rω (r being the radius of the axle).

Now, I = *mass of the flywheel* $\left(\frac{R^2 + r^2}{2}\right) + \frac{1}{2}$ *mass of the axle* (r^2).

where R is the *radius of the flywheel and r that of the axle.*

$$\text{Or,} \qquad I = 80 \times 1000 \frac{(20^2 + 5^2)}{2} + \frac{1}{2} 8 \times 1000 \times 5^2$$
$$= 17 \times 10^6 + 10^5 = 171 \times 10^5 \text{ gm-cm}^2.$$

∴ substituting the values of the various quantities in ihe energy equation above, we have

$$10 \times 1000 \times 981 \times 20 = \frac{1}{2}\left(171 \times 10^5, \omega^2 + \frac{1}{2} \times 10 \times 1000\, 5\omega\right)^2$$

$$\text{Or, } 1962 \times 10^5 = 855 \times 10^4 \times \omega^2$$
$$= 125 \times 10^3 \times \omega^2 = 8675 \times 10^3 \times \omega^2,$$

whence, $\omega^2 = 1962 \times \frac{10^5}{8675} \times 10^3 = 1962 \times \frac{10^2}{8675}$,

Or, $\omega = \sqrt{\frac{1962 \times 10^2}{8674}} = 4.755$ rad/sec.

Thus, *angular velocity of the flywheel and axle* 4.755 radians/sec. and, therefore, *linear velocity of the weight*, v = rω = 5ω = 5 × 4.755 = 23.785 cm/sec.

Clearly, K.E. of the flywheel and the axle

$$= \frac{1}{2}I\omega^2 = 855 \times 10^4 \times (4.755)^2$$

$$= 19.34 \times 10^7 \text{ ergs}$$

and *K.E. of the weight* $\frac{1}{2}Mv^2 = \frac{1}{2} \times 10 \times 1000\ (5 \times 4.755)^2$

$$= 28.27 \times 10^5 \text{ egrs.}$$

Example 12:

A bicycle wheel of mass 2 kg and radius 50 cm is rolling along on a read at 20 km/hr. What torque will have to be applied to the handle to turn it through half a radian in 0.1 sec? (Take the mass of the wheel to be concentrated at the rim).

Solution:

We know that the torque producing precession is given by

$$\tau_1 = \frac{J\,\delta\theta}{\delta t} = J\phi,$$

where J is the *angular momentum* of the body about its axis of rotation and φ, the *rate of precession* of the axis.

Now, J = Iω = mr²ω, where I = mr² is the *moment of inertia* of the wheel about its axis of rotation (its mass being concentrated at the rim) and ω, its *angular velocity.*

Here, I = 2000 × 2500 = 5 × 10⁶ gm-cm² and

$$\omega = \frac{v}{r} = \frac{20 \times 1000 \times 100}{3600 \times 50} = \frac{100}{9}$$

radian/sec, and $\phi = \frac{\delta\theta}{\delta t} = \frac{0.5}{0.1} = 5$ radian/sec.

∴ *torque required to be applied to the handle, i.e.,*

$$\tau_1 = J\phi = I\omega\phi = 5\times10^6 \times \frac{100}{9} \times 5 = \frac{25\times10^8}{9}$$

$$= 2.78 \times 10^8 \text{ gm-cm}^2\text{/sec.}$$

Example 13:

The internuclear distance between the two hydrogen atoms in a hydrogen molecule is 0.71 A and mass of a proton is 1.6 × 10^{-24} gm. Calculate the moment of inertia and energy of the first and third rotational energy levels in ergs, in electron volts and in cm^{-1}. Given h = 6.6 × 10^{-27} erg-sec.

Solution:

Here, *reduced mass of the hydrogen molecule,* $\mu = \frac{m_H m_H}{m_H + m_H} = \frac{m_H}{2}$ where m_H is the mass of a hydrogen atom.

Or, since $m_H \approx m_P$, the *mass of a proton,*

we have $\mu = \frac{m_P}{2} = 1.6 \times 10^{-24}/2 = 0.8 \times 10^{-24}$ gm.

∴ *moment of inertia of the hydrogen molecule about the axis passing through its centre of mass and perpendicular to the line joining the two nuclei (or the two protons),* say $I = \mu r_0^2$ (1.15) = 0.8 × 10^{-24} × $(0.71 \times 10^{-8})^2$ = 4.03 × 10^{-41} gm-cm².

Now, as we know, (1.20), *energy corresponding to the jth energy level is given* by E = Aj (j + 1), where $A = \frac{h^2}{8\pi^2 I}$. So that, *energy of the hydrogen molecule corresponding to the first rotational energy level, i.e.,* for j = 1, is given by $E_1 = 2A = \frac{2h^2}{8\pi^2 I} = 2\frac{(6.6\times10^{-27})^2}{8\pi^2 \times 4.03\times10^{-41}}$

$$= 2.72 \times 10^{-14} \text{ ergs.}$$

Since 1.6 × 10^{-12} *ergs* = 1 *electron volt,* we have

$$E_1 = \frac{2.72\times10^{-14}}{1.6\times10^{-12}} = 1.7 \times 10^{-2} \text{ eV.}$$

And, in cm^{-1}, *i.e.,* in *wave numbers,* E_1 *is given by*

$$v_1 = \frac{E_1}{hc} = \frac{2.72\times10^{-14}}{6.6\times10^{-27}\times3\times10^{10}} = 137.4 \text{ cm}^{-1}.$$

Similarly, *energy of the hydrogen molecule corresponding to the third rotational energy level, i.e.,* for j = 3, is given by

$$E_3 = 12A = \frac{12h^2}{8\pi^2 I} = 12\frac{\left(6.6\times10^{-27}\right)^2}{8\pi^2\times4.03\times10^{-41}} = 16.3 \times 13^{-14} \textit{ ergs.}$$

Or, *in electron volts,* $E_2 = \frac{1.63\times10^{-41}}{1.6\times10^{-12}} = 10.2 \times 10^{-2}$ eV, and *in wave numbers* $\overline{v}_3 = \frac{E_3}{hc} = \frac{16.3\times10^{-14}}{6.6\times10^{-27}\times3\times10^{10}} = 823\ \text{cm}^{-1}$.

Example 14:

The spacing between successive spectral lines in the rotational spectrum of HCl is found to be 20.8 cm⁻¹. Calculate the internuclear distance or the bond length for the HCl molecule. (Atomic weight of Cl = 35.5 amu and 1 amu = 1.6 × 10⁻²⁴ gm; h = 6.6 × 10⁻²⁷ erg-sec).

Solution:

We know that the spacing between the successive spectral lines in the rotational spectrum of a diatomic molecule $= \frac{h}{4\pi^2 Ic} = \frac{h}{4\pi^2\mu r_0^2 c}$,

where μ is *the reduced mass of the molecule and* r_θ *its internuclear distance or bond length.*

We, therefore, have $\frac{h}{4\pi^2\mu r_0^2 c} = 20.8^{-1}$ cm.

Or, $r_0^2 = \frac{h}{4\pi^2\mu(20.8)c}$,

whence, $r_0 = \sqrt{\frac{h}{4\pi^2\mu(20.8)c}}$

Here, $\frac{1}{\mu} = \left(\frac{1}{2} + \frac{1}{35.5}\right) = \frac{36.5}{35.5}$.

Or, $\mu = \frac{35.5}{36.5}$ amu $= \frac{35.5}{36.5} \times 1.6 \times 10^{-24} = 1.56 \times 10^{-24}$ gm.

And, therefore, $r_0 = \sqrt{\frac{6.6\times10^{-27}}{4\pi^2 1.56\times10^{-24}\times20.8\times3\times10^{10}}}$

$$= \sqrt{\frac{6.6\times10^{-13}}{4\pi^2\times1.56\times20.8\times3}} = 1.31\times10^{-8} \text{ cm.}$$

Or, since 10^{-8} cm = 1 *Angstrom unit,* we have *internuclear distance or the bond length for the hydrogen molecule* = 1.31 A.

Example 15:

A flat thin uniform disc D of radius a and centre O has a hole of radius b in it at a distance c from O (Fig. 1.41). If its mass be 90 gm and a, b and c equal to 7 cm, 2 cm and 4 cm respectively, calculate its moment of inertia (i) about an axis through O and perpendicular to its plane, (ii) about an axis through the centre of the hole and perpendicular to its plane and (iii) about an axis passing through O and the centre of the hole.

Solution:

Clearly, *area of the disc with the hole in it = area of the complete disc-area o the hole* = $\pi a^2 - \pi b^2 = \pi(a^2 - b^2) = \pi(49 - 4) = 45\pi$ sq cm.

Since its mass is 90 gm, *mass per unit area of the disc* = $90/45\pi = 2/\pi$ gm/cm^2.

$\therefore$ *mass of the complete disc* = $2/\pi \times$ *area of complete disc* = $2/\pi \times \pi = 49 = 98$ gm and *mass of the small disc of radius* b = 2 cm (which is removed to produce the hole in the disc) $= \left(\frac{2}{\pi}\right) \times$ *area of complete disc* $= \left(\frac{2}{\pi}\right)\left(\frac{\pi}{4}\right) = 8$ gm.

Fig. 1.41

Now, *M.I. of the complete disc about the axis through its centre O and perpendicular to its plane* = *(mass of the disc)* × *(radius)*^{2}/2 = 98 × 49/2 = 49 × 49 = 2401 gm-cm^2.

And, *M.I. of the small disc of radius* b = 2 cm about the axis through O and *perpendicular to its plane* is, in accordance with the principle of parallel axes, equal to *M.I. of the small disc about the axis through its centre and perpendicular to its plane + mass of the small disc* × c^2 = $8(2)^2 = 8 \times (4)^2 = 32 + 128 = 160$ gm-cm^2.

$\therefore$ (i) *M.I. of the disc (D), with the hole, about the axis through O and perpendicular to ifs planes M.I. of the complete disc about this axis–*

M.I. of the small disc about this axis = 2401 – 160 = 2241 gm-cm².

(ii) *M.I. of the disc (D), with the hole, about the axis through the centre C of the hole and perpendicular to its plane = M.I. of the complete disc about the perpendicular axis through C–M. I. of the small disc about the perpendicular axis through C.*

= (M.I. of the complete disc about the perpendicular axis through O = its mass × c^2)

– M.I. of the smaller disc about the perpendicular axis through C
$= 2401 + 98(4)^2 - 8(2)^2 = 2401 + 98 \times 16 - 32 = 3969 - 32 = 3937$ gm-cm²

(iii) *M.I. of the disc (D), with the hole, about the axis passing through O and the centre C of the hole, i.e., about the common diameter of the complete disc and the hole = M I. of the complete disc about its diameter–M.I. of the small disc about its diameter*

$$= 98 \times \frac{a^2}{4} - 8 \times \frac{b^2}{4} = \frac{98 \times 49}{4} - \frac{8 \times 4}{4} = 600 - 8 = 592 \text{ gm-cm}^2.$$

Example 16:

(a) Four solid spheres, A, B, C, and D, each of mass and m and radius a, are placed with their centres on the four corners of a square of side b, (Fig. 1.42). Calculate the moment of inertia of the system about one side of the square.

(b) Also calculate the moment of inertia of the system about a diagonal of the square.

Solution:

(a) Let A. B, C, and D be the four solid spheres, each of mast m and radius a placed at the four corners of the square PQST, as shown, and let it be required to calculate the M.I. of the system about the side PQ of the square.

Clearly, PQ lies along the diameter of each of the spheres A and B; so that, M.I. of each about $PQ = \frac{2}{5}ma^2$ and, therefore.

M.I. of the pair of spheres A and B about PQ $= 2 \times \frac{2}{5}ma^2 = \frac{4}{5}ma^2$

Similarly, *M.I. of the pair of spheres C and D about the side TS of the square*, (which lies along their diameters) $= 2 \times \frac{2}{5}ma^2 = \frac{4}{5}ma^2$

And, therefore, *M.I. of the pair C and D about side* PQ $= \frac{4}{5}ma^2 + 2mb^2$, in accordance with the principle of parallel axes, because PQ is parallel to TS and distant b from it.

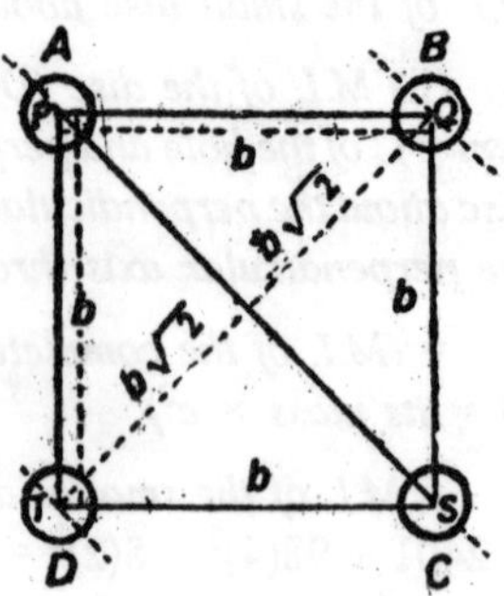

Fig. 1.42

So that *M.I. of the whole system about the side PQ (or, in fact, any side) of the square*

$$= \frac{4}{5}ma^2 + \frac{4}{5}ma^2 + 2mb^2$$

$$= \frac{8}{5}ma^2 + 2mb^2 = \frac{2}{5}m(4a^2 + 5b^2).$$

(b) Here, M.I. of the pair of spheres A and C about their common diameter, *i.e.*, along the diagonal PS $= 2\times\frac{2}{5}ma^2 = \frac{4}{5}ma^2$; *M.I. of sphere B about its diameter* $= \frac{2}{5}ma^2$ and, therefore, *its M.I. about the diagonal* PS, parallel to its diameter and distant $b\sqrt{2}$ from it $= \frac{2}{5}ma^2 + m(b\sqrt{2})^2 = \frac{2}{5}ma^2 + \frac{mb^2}{2}$.

Similarly, M.I. of sphere D about the diagonal PS, parallel to its diameter and distant $b\sqrt{2}$ from it $= \frac{2}{5}\frac{ma^2 + mb^2}{2}$.

∴ *M.I. of the whole system about the diagonal PS (as also along the diagonal QT)*

$$= \frac{4}{5}ma^2 + 2\left(\frac{2}{5}\frac{ma^2 + mb^2}{2}\right) = \frac{8}{5}ma^2 + mb^2 = \frac{2}{5}m\left(4a^2 + \frac{5}{2}b^2\right).$$

Example 17:

The flat surface of a hemisphere of radius r is cemented to one flat surface of a cylinder of the same radius and of the same material. If the length of the cylinder be L and the total mass M, show that the moment of inertia of the combination about the axis of the cylinder is given by

$$Mr^2\frac{\left(\frac{L}{2} + \frac{4}{15}r\right)}{\left(L + \frac{2r}{3}\right)}$$

Solution:

Clearly, *M.I. of the combination (Fig. 1.43) about the axis of the cylinder (i.e., about its axis of symmetry), say, I = M.I. of the cylinder about the same axis,* I_1 *+ M.I. of the hemisphere about the same axis* (*i.e.*, about its own diameter), I_2, *i.e.*, $I = I_1 + I_2$.

Fig. 1.43

As we know, $I_1 = m_1r^2/2$ and $I_2 = 2\,m_2r^2/5$ (*i.e.*, the same as that of a sphere), where m_1 and m_2 are respectively the masses of the cylinder and the hemisphere and r, the radius of either.

$$\therefore \quad I = \frac{m_1r^2}{2} + \frac{2}{5}m_2r^2.$$

Now, if ρ be the density of the material of the cylinder and the hemisphere, we have

$$m_1 = \pi r^2 L\rho$$

and $$m_2 = \frac{2}{3}\pi r^2 \rho.$$

$$\therefore \quad I = \frac{\pi r^2 L\rho r^2}{2} + \frac{2}{5}\left(\frac{2}{3}\pi r^2\rho\right)r^2 = \frac{\pi r^4 L\rho}{2} + \frac{4}{15}\pi r^3\rho = \pi r^4\rho\left(\frac{L}{2} + \frac{4}{15}r\right)$$

Since total mass of the combination, $M = m_1 + m_2 = \pi r^2 L\rho + \frac{2}{3}\pi r^3\rho$

$$= \pi r^2\rho\left(L + \frac{2}{3}r\right),$$

we have $$\rho = \frac{M}{\pi r^2}\left(L + \frac{2}{3}r\right).$$

Substituting this value of ρ in the expression for I above, we have

$$I = \pi r^4 \frac{M}{\pi r^2\left(L + \frac{2}{3}r\right)}\left(\frac{L}{2} + \frac{4}{15}r\right) = \frac{Mr^2\left(\frac{L}{2} + \frac{4}{15}r\right)}{\left(L + \frac{2}{3}r\right)}$$

EXERCISES

1. Define moment of inertia of a body about an axis. Establish the parallel axes theorem. Evaluate the moment of inertia of a uniform circular disc about a diameter.
2. Define moment of inertia of a body and discuss its physical significance. Derive an expression for the moment of inertia of a sphere about an axis tangent to its surface.
3. (a) Find the moment of inertia of a solid cylinder about a line parallel to its axis and touching its surface.

   (b) Derive an expression for the moment of inertia of an annular ring (i) about an axis passing through its centre and perpendicular to its plane (ii) about its diameter.
4. From the results obtained in Question 10 (b) above, deduce the moments of inertia of (i) a plane circular disc about a perpendicular axis through its centre, (ii) a ring or a hoop about its diameter.
5. A thin and uniform metal rod of mass M and length $2l$ is bent sharply at its mid-point so that its two halves are inclined to each other at an angle θ. Show that its moment of inertia about an axis through its mid-point and perpendicular to its plane is $Ml^2/2$ irrespective of the value of θ.
6. Calculate the moment of inertia of a thin spherical shell and hence or otherwise, that of a solid and a hollow sphere, about an axis through its centre.
7. The cross section of five bodies of different geometrical shapes r but the same mass (M) are shown in Fig. 1.44, each cross section having the same height and maximum width (2a).

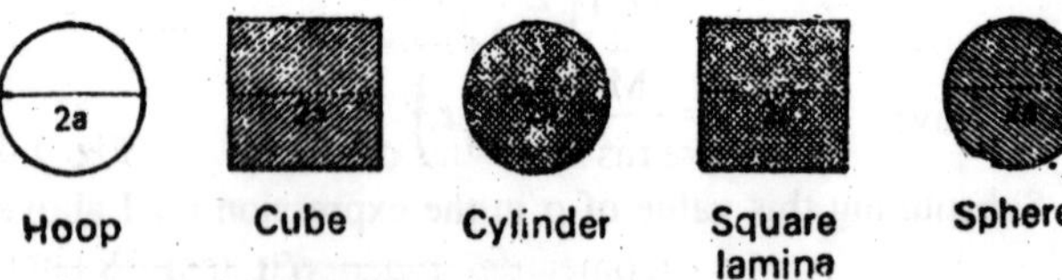

Fig. 1.44

Obtain the values of their moments of inertia about perpendicular axes passing through their respective centres of cross section, pointing out clearly the highest and the lowest values.

8. (a) About which axis, will (i) a plane square lamina, (ii) a cube have the least moment of inertia?

   (b) Will the moment of inertia of two circular discs of the same mass and radius but of materials of different densities be different about the axes through their centres of mass and perpendicular to their cross sections?

9. A uniform thin bar of mass 3 kg and length 0.9 metre is bent to make an equilateral triangle. Calculate the moment of inertia about an axis passing through the centre of mass and perpendicular to the plane of the triangle.

10. A thin uniform disc of radius 25 cm and mass 1 kg has a hole of radius 5 cm in it. If the centre of the hole be at a distance of 10 cm from the centre of the disc, calculate the moment of inertia about an axis perpendicular to the plane and passing through ihe centre of the hole.

    Ans. 4.25×10^5 gm-cm^2.

11. Five masses, each of 2kg, we placed on a horizontal circular disc (of negligible mass,) which can be rotated about a vertical axis passing through its centre. If the masses be equidistant from the axis and at a distance of 10 cm from it, what is the moment of inertia of the whole system? **Ans.** 10^6 gm-cm^2.

12. Calculate the moment of inertia of a sphere about its diameter and also about a tangent.

13. A sharp impulse is given to a stationary billiard ball of mass m by means of a cue held horizontally at a height A above the centre line, as shown in Fig. 1.45. As a result, the ball starts moving away from the cue with velocity v but soon acquires a final velocity (9/7)v. Show that h = (4/5)r, where r is the radius of the ball.

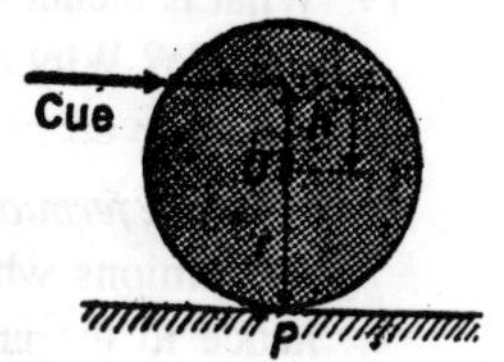

Fig. 1.45

[**Hint:** Angular momentum imparted to the ball about the point P = mv(r + h). This must be equal to Iω, where I is the M.I. of the ball about the axis through P and parallel to the one through O, *i.e.*, $= \frac{2}{5}mr^2 + mr^2 = \frac{7}{5}mr^2$, and ω, the angular velocity of the ball. So that,

$$\frac{7}{5}mr^2\omega = mv(r+h),$$

whence, $\omega = \frac{5v(r+h)}{7r^2}$.

And therefore, $\frac{9}{7}v = \omega \times r = \frac{5v(r+h)}{7r^2} \times r,$

which gives $h = \frac{4}{5}r.$

14. Obtain an expression for the moment of inertia of a uniform right circular cone of mass M about its axis, than radius of the base being r and deduce from it the moment of inertia of the cone about an axis through its vertex and parallel to its base.

15. State and illustrate *Routh's rule* for the moments of inertia of bodies of geometrical shapes about any one of their axes of symmetry.

16. What is a gyrostat? Explain is working and mention some of its important applications.

17. Explain the theory underlying a gyrastatic pendulum and obtain an expression for its time-period.

18. What is meant by the quantisation of the angular momenta of electrons around the nucleus? How does it lead to the conclusion that the radii of the permissible orbits around the nucleus are proportional to n^2, where n = 1, 2, 3 etc?

19. What is meant by the spin angular momentum of a fundamental particle? Why is it refined to as intrinsic angular momentum of the particle?

20. What are *fermions and bosons*? What is the characteristic property of fermions which has led to *Pauli's exclusion principle* and hence to the arrangement of electrons in an atom?

21. Two point-masses m_1 and m_2 separated by a massless a link of length r rotate about an axis passing through the centre of mass of the system and perpendicular to the link. If the angular momentum is $\sqrt{j(j+1)}h/2\pi$, find the rotational energy.

Ans. $E = h^2 j(j + 1)/8\pi^2 I$, where $I = \mu r^2$ and $\mu = \frac{m_1 m_2}{m_1 + m_2}$

22. Deduce an expression for the wavelength of light absorbed by a molecule of moment of inertia I in rotational transition j = 0 to j = 1 assuming that the angular momentum J is quantised according to the relation $J^2 - \left(\frac{h}{2\pi}\right)^2 j(j+1)$.

23. Obtain the values of the three longest wavelengths in the rotational spectrum of W, given that the bond length of the HF molecule is 0.92×10^{-8} cm and the atomic weight of F is 19.

Ans. 0.024 cm; 0.012 cm; 0.008 cm.

24. A hoop of radius 0.5 m and mass 16 kg is rolling along a horizontal floor, such that its centre of mass has a velocity of 0.25 m/sec. What work will have to be done to stop it?

Ans. 1 joule.

25. A thin hollow cylinder, open at both ends and of mass M, (a) slides with a velocity v *without rotating*, (b) rolls *without slipping*, with the same speed. Compare the kinetic energies it possesses in the two cases. Ans. 1 : 2.

26. (a) A solid spherical ball rolls on a horizontal table. What fraction of its total kinetic energy is rotational?

(b) A circular disc of mass m and radius r is set rolling on a table. If ω is its angular velocity, show that its total energy E is given by $E = \frac{3}{4} mr^2\omega$. Ans. (a) $\frac{2}{7th}$

27. Asymmetrical diatomic molecule of a gas, of mass 5.30×10^{-28} kg and moment of inertia about the axis through the centre of its bond length and perpendicular to it, 1.94×10^{-46} kg-m² is moving with a speed of 500 m/sec. If its kinetic energy of rotation be two thirds of its kinetic energy of translation, calculate its mean angular velocity.

Ans. 6.747×10^{12} radians/sec.

28. A body of mass M and radius R, rolling on a horizontal surface, without slipping, with a velocity v, rises up an incline to a maximum height h = 7Mv²/10. What is the body?

Ans. *A solid sphere.*

29. A string is wrapped round a cylinder of mass M and radius R. The string is pulled vertically upward to prevent the centre of

mass of the cylinder falling as the string round it sets unwound. (i) What is the tension in the string? (ii) What is the work done on the cylinder when it has attained an angular velocity ω? (iii) What is the length of the string unwound during this time?

Ans. (i) Mg, (ii) $Mr^2\omega^2/4$, (iii) $R^2\omega^2/4g$.

30. A wheel of radius 6 cm is mounted so as to rotate about a horizontal axis through its centre. A string of negligible mass, wrapped round its circumference, carries a mass of 200 gm attached to its free end. When let fall, the mass descends through 100 cm in the first 5 seconds. Calculate the angular acceleration of the wheel and its moment of inertia. .

Ans. 4/3 radian/sec^2; 8.75×10^5 gm-cm^2.

31. A small sphere of radius r and mass m. rolls without slipping on the inner surface of a large hemisphere of radius R, whose axis of symmetry it vertical (Fig. 1.46), where R >> r. If the sphere starts from rest at the top off the hemispherical surface, calculate (i) the kinetic energy of the sphere at the bottom of the surface, (ii) the rotational and translational fraction of this energy and (iii) the normal force' exerted by the sphere on the bottom of the hemispherical surface.

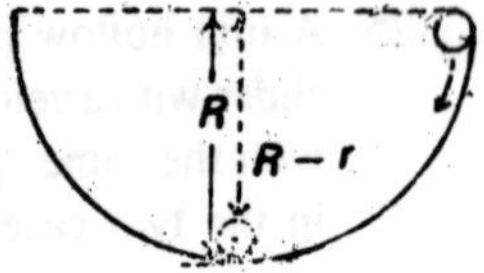

Fig. 1.46

32. The free end of a string wrapped round the axle of a flywheel of moment of inertia 27.61×10^5 gm-cm^2 carries a weight of 5 kg which is allowed 10 fall. What is the number of revolutions made by the wheel when the weight has fallen through 1 metre. The kinetic energy of the weight may be neglected.

33. A pair of rails is supported in a horizontal position and the axle of a wheel tests on the rails. A thread is wrapped round the axle and a weight hung on the end of the thread. As the weight falls, the wheel moves along the rails. How would you determine the moment of inertia of the wheel with this arrangement?

33. Obtain an expression for the acceleration of a body rolling down an inclined plane.

35. Show that for a sphere to be able to roll along an inclined plane of inclination θ, the coefficient of static friction must not be less than 2/7 tan θ.

36. Starting from rest at the top of a zigzag path, with its right hand end horizontal, as shown in (Fig. 1.47), a small solid sphere rolls without slipping until it finally rolls off the horizontal end. If the top of the path be 195 cm and its horizontal end 20 cm above the ground level, how far from the latter will the sphere hit the, ground?

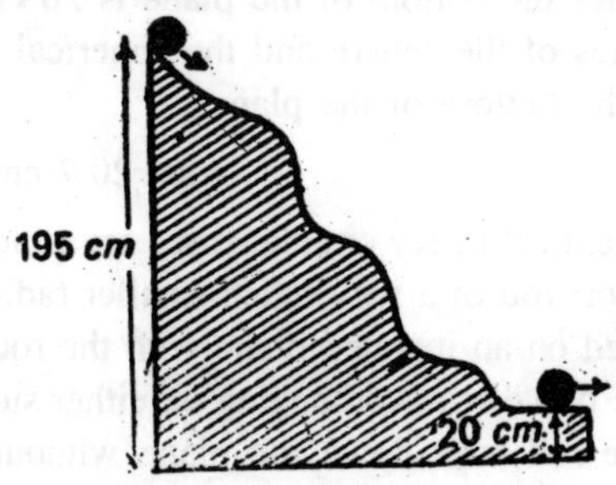

Fig. 1.47

37. A circular disc, starting from rest, rolls (without slipping) down an inclined plane of 1 in 8 and covers a distance of 5.32ft m 2 sec. Calculate the value of g. **Ans.** 31.92 ft/sec^2.

38. A sphere rolls up an incline of 1 in 2. If its linear speed at the bottom of the incline be 20 ft/sec, how far will it go up the incline and how long will it be before it comes back to the bottom? **Ans.** 17.5 ft; 3.50 sec.

39. A solid sphere rolls down two different inclined planes right from their tops. The heights of the two inclined planes are the same but their angles of inclination are 60° and 30° respectively. Show that if the sphere takes time t to reach the bottom of the plane in the first case, the time taken by it to do so in the second case is $\sqrt{3}t$.

40. Two spheres are identical in mass and volume, but one is hollow and the other, solid. How will you identify them experimentally? Explain the theory underlying the experiment.

41. A solid cylinder (a) rolls, (b) slides from rest down an inclined plane. Neglect friction and compare the velocities in both cases when the cylinder reaches the bottom of the incline.

42. Show that if a sphere, a disc, a cylinder, a spherical shell and a hoop roll down the same inclined plane, the sphere, the disc, the cylinder, the spherical shell, the hoop reach the bottom of the incline in that order.

43. Starting from rest, a solid sphere, a circular disc and a spherical shell roll down an inclined plane. The velocity of the disc as it reaches the bottom of the plane is 20 cm/sec. Calculate the velocities of the sphere and the spherical shell when they too reach the bottom of the plane.

Ans. 20.7 cm/sec; 18.97 cm/sec.

44. Two identical heavy circular discs are attached to the two ends of a short rod of a very much smaller radius and the assembly is placed on an inclined plane with the rod touching the plane and the two discs overhanging on either side of the plane, such that the rod rolls down the plane without slipping. Near the bottom of the plane as the discs touch the horizontal surface of the table, the assembly starts moving with a very much higher linear velocity. Explain why?

45. Define moment of inertia. Describe how you would determine experimentally, the moment of inertia of a flywheel about its usual axis of rotation. Discuss briefly the part played by a flywheel in a machine.

46. A thin string is slipped on to a small peg on the axle of a flywheel and wound round it with a mass of 2 kg suspended from its lower free end which is initially held in position at a height of 100 cm from the floor. The mass is then allowed to fall. When it just touches the floor and the string slips off the peg, the flywheel comes to rest after making 20 revolutions in 8 sec. If the radius of the axle be 1.0 cm, obtain the value of the moment of inertia of the wheel about its axle and the kinetic energy of the 2 kg mass as it just touches the floor.

Ans. 2.20×10^5 gm-cm^2; 9.87×10^5 *ergs.*

47. What is meant by the term '*precession*'? Show that if the axis of the torque applied to a body be perpendicular to its axis of rotation, the body processes about an axis perpendicular to either of the first two axes.

48. The internuclear distance between the two protons in a hydrogen molecule is 0.74 A. Calculate its moment of inertia and the first

two rotational energy, levels. (Given that angular momentum I is quantised by the rule $J^2 = j(j + 1)\ h^2/4\pi^2$, with j = 0, 1, 2, 3 etc.). **Ans.** 9.6 × 10^{-4} gm-cm^2; 0.153 eV and 0.046 eV.

49. It is found that the wave numbers of the successive spectral lines in the rotational spectrum of CO differ by 3.82 cm^{-1}. Obtain the value of energy for the third rotational energy level. What is the maximum wavelength of the radiation emitted? What is the bond length of the molecule and its moment of inertia about the right bisector of the bond length?

50. Deduce the fundamental equation of motion of a rigid body about a fixed axis and show how it is the rotational analogue of the equation F = dp/dt in the case of translational motion.

51. Show that the dimensions of torque are the same as those of work or energy and those of angular momentum the same as those of *energy* × *time*.

52. Show that the torque on a rigid body about given axis is equal to the product of moment of inertia of the body and its angular acceleration about that axis and that its angular momentum about the given axis is equal to the product of its moment of inertia and its angular velocity about that axis.

53. Describe the vector representing the angular velocity of the earth (i) about its own axis, (ii) about the sun.

54. Show that the angular momentum about any point of a single particle, moving with constant velocity remains the same throughout its motion.

55. A heavy disc rotating about the axis through its centre and perpendicular to its plane is slowed down due to friction at the bearings, such that its initial rate of rotation of 100 rotations per second is reduced to 90 rotations per second at the end of the first minute. If the frictional resistance be assumed to be constant, what will be its rate of rotation at ihe end of the second minute?

Ans. 80 rotations per second.

56. Define Moment of inertia and Radius of gyration. Explain their physical significance. State the laws of (i) parallel and (ii) perpendicular axes and prove any one of them.

57. A flywheel in the form of a solid circular disc of mass 500 kg and radius 1 *metre* is rotating, making 120 rev/min. Compute

the kinetic energy and the angular speed if the wheel is brought to rest in 2 seconds; friction is to be neglected.

Ans. 1.974×10^{12} *ergs*; 3.142×10^{11} radian/sec.

58. The moment of inertia of a reel of thread about its axis is MK^2. If the loose end of the thread is held in the hand and the reel is allowed to unroll itself while falling under the action of gravity, show that it falls down with an acceleration $ga^2/(a^2 + k^2)$, where a is the radius of the reel.

59. A metre stick is first held vertically with its lower end resting on a horizontal surface and is then allowed to fall. Assuming that its lower end does not slip, calculate the velocity of the upper end as it hits the horizontal surface.

60. A thin uniform disc of radius 25 cm and mass 1 kg has a hole of radius 5 cm in it. If the centre of the hole be at a distance of 10 cm from the centre of the disc, calculate the moment of inertia about an axis perpendicular to the plane and passing through ihe centre of the hole.

2

Gravitation Fields and Potentials

(Inverse Square Law Forces—Fundamental Lengths and Numbers)

INTRODUCTION

We are aware from early child wood that fruits from a tree fall down, a stone thrown upwares falles back to earth, the motion of planets and stars take place following some well-known laws. All these and many other activities are due to what is known as "Gravitational Effect of Earth". The gravity of earth was studied systematically first by Galileo. It was however, the genius of Issac Newton that result in the discovery that gravitation is a universal property. To the best of our knowledge even todey, we believe in the truth of the Universal Gravitational Law over very small and very large distance.

There are only four known ways for the interaction of matter leading to four fundamental types of forces, viz;

(i) *gravitational interaction*, leading to gravitational forces,

(ii) *electromagnetic interaction, leading* to electromagnetic forces

(iii) *strong interaction, and*

(iv) *weak interaction* leading to what are referred to as nuclear forces.

Of these, the *gravitational interaction is the weakest* and yet it is the gravitational force which is responsible for :

(a) holding the earth together and retaining on it the atmosphere, as we know it, with its life-giving constituent oxygen;

(b) binding the earth and the other planets to that perennial source of energy, the sun, into a well-knit solar system, and

(c) similarly, binding the stars together into what are called galaxies, etc.

The *electromagnetic interaction* is equally important in that the electromagnetic force binds the electrons in an atom to the nucleus, the atoms into molecules and the molecules into crystals. It further accounts for such properties of matter as elasticity, viscosity, surface tension, refractive index, conductivity, specific heat and latent heat etc. The reason why it is so called is that it operates not only between electric charges but also between a moving charge and a magnetic field (since a moving charge too, as we know, develops a magnetic field about itself). In Chemistry and Biology, this alone is the dominating force.

Among the short-range nuclear forces, the *strong interaction* provides the strong force which binds together the nucleons, *i.e., protons, carrying* positive charges and *neutrons* (carrying no charge) into a stable nucleus despite the force of repulsion between the similar charges carried by the protons. This force is, in fact, the strongest of all the four fundamental forces and its importance is obvious from the fact that all material bodies in the universe are actually made up of atoms and these atoms are, in a majority of cases, stable in consequence of this force.

Finally, the *weak Interaction* gives rise to comparatively *feebler forces* which operate between lighter fundamental particles (like *electrons, leptons, muons* and *neutrinos*) or between a light fundamental particle and a heavier particle. They cannot, however, form any such stable systems as the solar system formed by gravitational forces.

In this chapter, we shall for the most part, concern ourselves only with gravitational forces.

NEWTON'S LAW OF GRAVITATION

This law was first announced by *Newton* in the year 1687 in his monumental work *Principal*, hailed as the greatest production *of the human mind.*

The law states that *every particle of matter in the universe attracts every other particle with a force which is directly proportional to the product of their masses and inversely proportional to the square of the distance between them.*

Thus, if the masses of two particles, distant *r* from each other be *m* and *m'*, the force of gravitational attraction between them, say, $F \alpha\ mm'\ hr^2$.

or $$F = \frac{mm'}{r^2} G,$$

where *G* is a *universal constant*, called the *universal gravitational constant*, or usually simply the *Gravitational constant.*

Clearly, if $m = m' = 1$ *gm*

and $r = 1$ *cm*,

we have $F = G$,

i.e., the *Gravitational constant is the force of gravitational attraction between two unit masses unit distance apart.*

Its dimensions are $M^{-1} L^3 T^{-2}$ and its accurate value (determined by *Heyl* in 1930) is taken to be 6.669×10^{-8} dynes-cm^2/gm^2

The reason why the law is said to be a universal law is that it operates from enormous interplanetary distances down to the small terrestrial ones. The minimum distance up to which it holds good is probably not yet known with absolute certainty but it appears to break down at molecular distances, of the order of 10^{-7}cm.

EXPERIMENTAL DETERMINATION OF THE GRAVITATIONAL CONSTANT (G)

Cavendish's Method

Cavendish was the first to have made use of a *torsion balance* to devise a laboratory method for determining the value of *G* in the year 1798.

Apparatus : The apparatus used by Cavendish, shown in Fig. 2.1, consisted of a cross bar PQ about 6 ft (or 180 cm) long, suspended from the ceiling of a chamber, so as to be free to rotate about a vertical axis by means of a wheel and string arrangement, manipulated from outside the chamber. Two *metal rods* attached to the two ends of the cross bar carried two *equal lead spheres* C and D at their lower free ends, 20 to 25 cm in diameter and weighing about 160 kg each.

Directly below the mid-point of the cross bar, was a *torsion head* M, also manipulated from outside from which was suspended a lighter, deal rod RS (of the same length as PQ) by means of a fine *torsion wire* W of silver-plated copper. To increase the strength of the rod without increasing its moment of inertia, two *wires* w, w were fastened to its two ends R and S and to a *small vertical rod* r fixed at its midpoint and attached to the suspension wire. Two *smaller of balls* A and B, about 5 cm in diameter and weighing about 740 gm each were suspended from the two ends of the deal rod RS, such that their centres, along with those of the lead spheres C and D lay in the same horizontal plane, roughly in a horizontal circle of about 3 ft (or 90 cm) radius. A small vernier carried by each end of the torsion rod (RS) moved over a fine ivory scale fixed to vertical stands, with each division equal to 0.05".

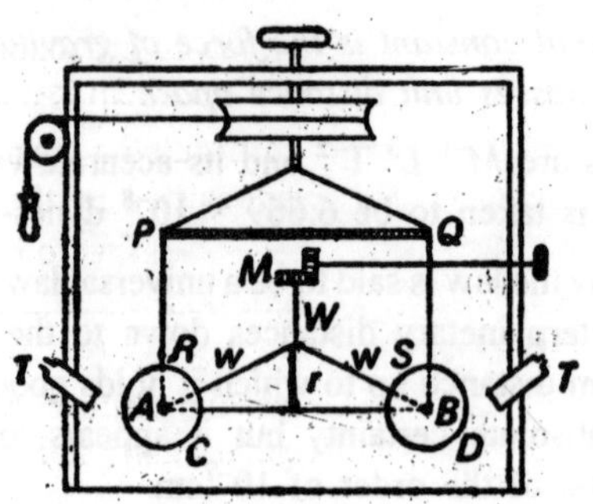

Fig. 2.1

As a safeguard against any changes of temperature and consequent setting up of air draughts (or convection currents) which might mask the rather feeble gravitational effect, the chamber was kept completely closed and the observations taken with the help to telescopes T, T fixed into the walls of the chamber, as shown. Further, to shield the apparatus from the effect of any outside electric charges, it was enclosed in a gilded glass case, supported on four levelling screws.

Working : The cross bar PQ wap first rotated until the line joining the centres of the lead spheres C and D, carried by it, was at right angles to the torsion rod RS carrying the smaller lead balls *A* and *B* at its two ends, as shown in Fig. 11.2 (a), it being so arranged that *in this position there was no twist in the suspension wire W*, and the readings on the verniers at either end of the torsion rod taken.

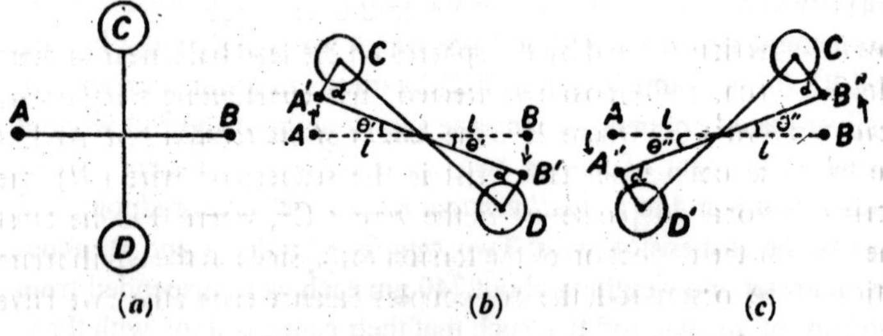

Fig. 2.2

The cross bar PQ was then rotated until the lead spheres C and D lay on opposite sides of the lead balls A and B respectively as shown in Fig. 2.2 (b), such that the lines joining the centres of A and C and of B and D were *equal in length and perpendicular to the* torsion rod. Obviously, then, the gravitational force of attraction on ball A due to sphere C was equal and opposite to that on ball B due to sphere D. This pair of equal and opposite forces constituted a couple tending to rotate the torsion rod clockwise. This was resisted by the restoring torsional couple set up in the suspension wire W, tending to bring the torsion rod back into its original position. Equilibrium was naturally attained when the two couples balanced each other, when the torsion rod had been deflected, say through an angle θ into the position A' B', from its initial position AB (shown dotted), this angle being noted on the verniers at the two ends of the torsion rod by the method of oscillations, as in the case of a physical balance.

The cross bar was now rotated the other way about, such that the lead spheres C and D now lay opposite the lead balls B and A respectively, in the positions C' and D', as shown in Fig. 2.2 (c), and the same adjustment was made as before, *viz.*, that the lines joining the centres of B and C' and of A and D' were of the same length as before and perpendicular to the torsion rod. The torsion rod was thus deflected in the opposite direction to that in the first case, *i.e.*, anticlockwise, occupying the position A" B" when the deflecting couple just balanced the restoring torsional couple. The deflection of the torsion rod, θ", was again read on the verniers at its two ends. The mean of these two values of deflection, θ' and θ", thus obtained, was taken as the true deflection θ of the torsion rod.

Calculations : Let M and m be the masses of each lead sphere and lead ball respectively and d the distance between their centres in the deflected position of the torsion rod [Fig 2.2(b) or (c)]. Then, clearly, forces of attraction exerted by the spheres on the lead balls near to them $= MmG/d^2$ each, but oppositely directed, thus constituting a *deflecting couple* $(MmG/d^2).\,2l$, where $2l$ is the *length of the torsion rod*. And, if C be torsional couple per unit twist in the suspension wire (W), the restoring torsional couple set up in the wire = Cθ, where θ is the twist in the wire (or the deflection of the torsion rod). Since in the equilibrium position of the torsion rod, the two couples balance each other, we have

$$\frac{Mm}{d^2}G.2l = C\theta, \quad \text{whence,} \quad G = \frac{Cd^2}{Mm.2l}\theta$$

To determine the value of C, the torsion rod (RS), together with the lead balls A and B suspended from its two ends was set into torsional vibration about the suspension wire (W) and its time-period t noted, (found to be 28 minutes in Cavendish's own experiment). Then, if I be the moment of inertia of the torsion rod (along with the lead balls) about the wire W as axis, we have $t = 2\pi\sqrt{I/C}$, whence, $C = 4\pi^2 I/t$.

Substituting this value of C in the expression above, we have

$$G = \frac{2\pi^2 \; Id^2}{Mml\,t^2}\,\theta.$$

And, if δ be the *diplacement* of each end of the torsion rod, clearly, $\theta = \delta/l$. So that,

$$G = \frac{2\pi^2 \; Id^2}{Mml\,t^2}\,\delta.$$

Or, if we ignore the mass of the torsion rod compared with that of the balls A and B, we have $I = 2m\,(2l/2)^2 = 2ml^2$. And, therefore,

$$G = \frac{4\pi^2 l d^2}{Mt^2}\theta = \frac{4\pi^2 d^2}{Mt^2}\delta$$

Corrections Applied : Corrections were applied *for the force of attraction between (i) a lead sphere and the distant small ball, (ii)* the two lead spheres and the torsion rod and (*iii*) *the rods carrying the lead spheres and the smaller lead balls.*

Sources of Error : The following are the chief sources of error in the experiment:

(i) The force of attraction between each pair of lead sphere and lead ball being small, the torsion rod had to be a long one to increase the deflecting couple. This necessitated a large chamber in which convection currents could hardly be avoided.

(ii) The suspension wire (W) required a large torque per unit angular twist and hence for a given deflecting couple; the deflection of the torsion rod was rather small.

(iii) The suspension wire was not perfectly elastic, so that the torque (or the couple) required was not strictly proportional to the deflection of the torsion rod.

(iv) The lead spheres tended to decrease the deflection of the torsion rod due to their force of attraction on the distant small balls, whereas the rods carrying the lead spheres tended to increase its deflection due to their force of attraction on the small balls.

(v) The method of measuring the angle of deflection was far from sensitive.

Result : The value of G obtained by *Cavendish*, as the mean of twenty-nine observations, was 6.754×10^{-8} dyne-cm^2/gm^2 in the C.G.S. units or 6.754×10^{-11} *newton-metre*^{2}/kg^2 in the M.K.S. (or SI) units.

Boys' Method

Sir Charles Vernon Boys devised, in the year 1895, a far more accurate method for the determination of the value of G. He removed all the sources of error of Cavendish's method, greatly reduced the size of the apparatus and yet increased its sensitiveness (by reducing the dimensions of different parts of it in different proportions).

Apparatus : The apparatus used by *Boys*, shown diagrammatically in Fig. 2.3, consists of *two coaxial glass tubes*, with the bore of the inner one about 4 cm. This inner tube is fixed and the outer one can be rotated about their common axis, with the whole assembly mounted on a suitable platform provided with levelling screws. A large chamber is thus done away with.

The moving system here consists of two *gold balls*, A and B, about 0.5 cm in diameter and weighing 2.65 gm each, suspended by fine gold wires from the two ends of a small *beam* PQ which is itself suspended by a fine *quartz fibre f* from a torsion head T, inside the inner tube. This serves to protect the system from any outside disturbances. Also, if suspended *centrally* inside the tube, the system remains unaffected by the effects due to attraction by the tube itself.

Immediately below the beam PQ is a *plane mirror strip* m, 2.5 cm long and rigidly attached to the beam, with grooves in its two vertical edges inside which slide the wires carrying the gold balls. This ensures that (i) *the two wires are kept a constant horizontal distance apart* and (ii) *the effective points of suspension of the gold balls lie on the two edges of the mirror strip*, which thus replaces the long torsion rod of Cavendish's apparatus.

Two large lead spheres C and D, about 11.0 cm in diameter and weighing 74 kg each, are suspended in the outer tube as indicated, such that the centre of C is in a level with that of ball A and that of D in a level with that of ball B (to ensure greater precision in the measurement of the distance between each pair), the *distance between each pair being the same.* And, to minimise of effect of a large sphere on the distant small ball (*i.e.*, of C on B and of D on A), one of the pairs, say, BD, is arranged to lie some 15 cm above the level of the other (AC), the centres of both, the sphere and the ball, lying in the same vertical plane, with no twist in the suspension fibre.

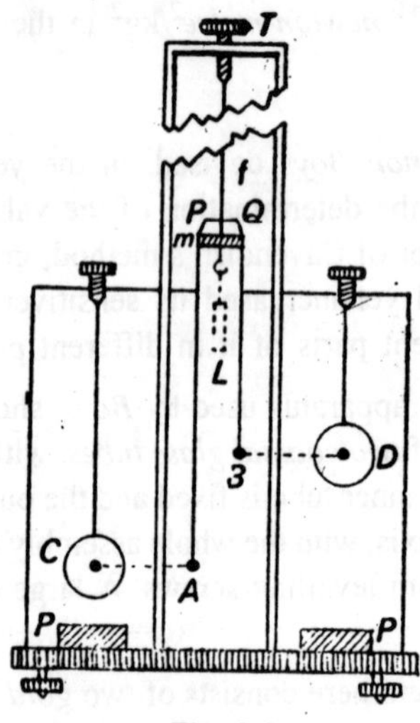

Fig. 2.3

The deflection of the torsion rod (*i.e.*, the mirror strip) m is measured accurately by a scale and telescope arrangement and a half-millimeter scale is placed at a distance of about seven meters from it for the purpose.

Finally, as a safeguard against damage to the tube by an accidental fall of the lead spheres (C and D), rubber pads P and P are placed directly below them, as shown.

Working : The outer tube is rotated until *the lead spheres C and D lie on opposite sides of the two gold balls* A *and* B *respectively (but not in a line with the mirror strip)*, so as to exert the maximum torque on the suspended system and its deflection is, therefore, the largest. This deflection is noted on the half-millimeter scale.

The outer tube is next rotated until the lead spheres now lie *on the other sides* of the gold balls *in a similar position to the previous one*, again exerting the maximum torque on the suspended system and hence producing the maximum deflection of the mirror strip. The mean of the two deflections is then taken. Let it be θ.

Calculations : Let A,B and C,D (Fig. 2.4) be the positions of the gold balls and the lead spheres respectively when they are in equilibrium in the position of maximum deflection θ, with their centres in the same vertical plane again, as to start with.

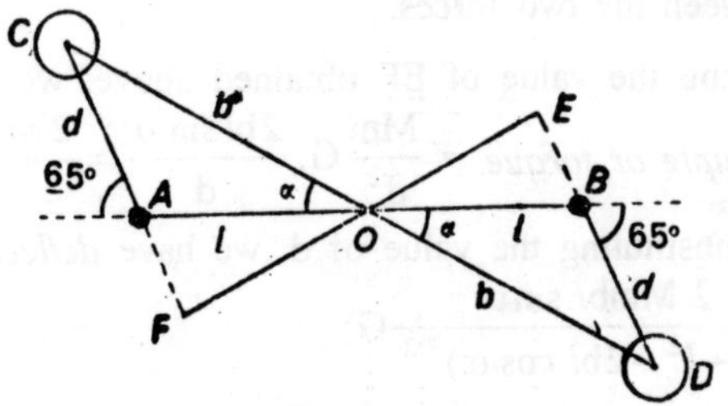

Fig. 2.4

It may be noted that the gold balls A and B are here shown in their initial positions, corresponding to $\theta = 0$ and the lead spheres C and D in their final positions into which they have been rotated to exert the maximum torque on the suspended system and to produce the maximum deflection θ. In this position, therefore, the deflecting gravitational couple exerted by the lead spheres on the suspended system is just balanced by the restoring torsional couple set up in the suspension fibre.

Now, let O be the mid-point of the mirror strip and l, its half-length (*i.e.*, OA = OB = l) and let OC = OD = b, AC = BD = d, $\angle$ AOC = $\angle$ BOD = α and OE = OF, where EF is the perpendicular through O on tc DB and CA produced.

Obviously, in Δ OBD, we have

$$BD = \sqrt{OD^2 + OB^2 - 2OD.OB \cos \alpha}, \text{ i.e.,}$$

$$d = (b^2 + l^2 = 2bl \cos \alpha)^{1/2}$$

Again, in the same triangle, $\dfrac{\sin \alpha}{\sin BDO} = \dfrac{BD}{OB} = \dfrac{d}{l}$

Or, $\sin BDO = \dfrac{l \sin \alpha}{d}$.

And, in the right-angled triangle OED, we have OE = OD sin EDO = b sin BDO or, substituting the value of sin BDO, we have

$$OE = bl \sin \alpha/d \text{ and } \therefore EF = 2OE = 2bl \sin \alpha/d$$

Now, if M be the mass of each lead sphere and m, that of each gold ball, we have

force of attraction between each pair of lead sphere and gold hall $= MmG/d^2$. The two forces being *equal, opposite* and *parallel*, constitute a *couple or torque* $= (MmG/d^2)$. EF, where EF is the perpendicular distance between the two forces.

Substituting the value of EF obtained above, we therefore have

deflecting couple or torque $= \dfrac{Mm}{d^2}G.\dfrac{2bl \sin \alpha}{d} = \dfrac{2\,Mmbl \sin \alpha}{d^3}G.$

Again, substituting the value of d, we have *deflection couple or torque* $= \dfrac{2\,Mmbl \sin\alpha}{(b^2 + l^2 - 2bl \cos \alpha)^{3/2}}G.$

And, if C be the torsional couple per unit twist of the suspension fibre, the restoring torsional couple or torque $= C\theta$.

Since the two couples just balance each other in the position of equilibrium of the system, we have

$$\frac{2\,Mmbl \sin\alpha}{(b^2 + l^2 - 2bl \cos\alpha)^{3/2}}G = C\theta,$$

whence, $G = \dfrac{(b^2 + l^2 - 2bl \cos \alpha)^{3/2}}{2\,Mmbl \sin \alpha}C\theta.$

Since a *quartz fibre* is nearly perfectly elastic, the value of α maybe taken to be the same as that of θ, and was therefore have

$$G = \frac{(b^2 + l^2 - 2bl \cos\theta)^{3/2}}{2Mmbl \sin\theta}C\theta,$$

whence, the value of G may be easily obtained. Its value, as obtained by *Boys*, was 6.6576×10^{-8} dyne-cm^2/gm^2 in CGS *units* and 6.6576×10^{-11} *newton-metre*^{2}/kg^2 in M.K.S. (or SI) units.

Advantages over Cavendish's Method: The following are the obvious advantages of the method over Cavendish's method.

(i) The size of the chamber was greatly reduced, thus almost completely eliminating convection currents and enabling its temperature to be controlled.

(ii) The two pairs of lead spheres and gold balls being arranged at different levels, the gravitational force of a lead sphere on the distant gold ball is almost negligible.

(iii) The deflection of the mirror strip is measured more accurately by the scale and telescope method.

(iv) A quartz fibre is used as the suspension wire which, besides being fine and strong, (a) is almost perfectly elastic and (b) requires a small torque per unit twist. The angular deflection of the mirror strip in thus quite appreciably large as well as proportional to the applied torque.

Heyl's Method

The method adopted by *P. R. Heyl*, in the year 1930, for the determination of the value of G is taken to be the most accurate to date. It is a modification of *Braun's Torsion balance experiment* which itself was a revised version of Boys' method, dealt with under (2) above.

*Apparatus : He*yl arranged his apparatus in a *constant temperature enclosure* (in fact, the constant temperature room of the American Bureau of Standards, 12 metres below ground level) with the pressure inside it reduced to 2 mm of mercury column to minimise convection currents.

The attracting large *masses* used by him were massive *cylinders of steel* (with 0.9% of carbon), each weighing about 66.3 kg and suspended from a system free to rotate about a vertical axis midway between the two.

The reason why he took his large masses in the form of cylinders rather than that of the spheres (which obviously make the calculations easier) was that it was difficult to machine a sphere of such a huge mass. As some one well remarked, *the burden was thus 'shifted to the broad shoulders of the mathematician.'*

The *smaller masses*, each weighing 2.44 gm, were *balls of gold, platinum* and *optical glass* in three different sets of experiments respectively, and were suspended from the two ends of a *light aluminium torsion rod* R, 28.6 cm long (Fig. 2.5), supported by a *tungsten thread*

T.W. (1 metre long and 0.25 mm in diameter) and two inclined copper wires (w,w), so that almost the whole moment of inertia of the assembly remains in the balls themselves. *Heyl* preferred a tungsten thread to a quartz fibre as suspension wire because the latter is sometimes found to break quite unexpectedly and for no apparent reason.

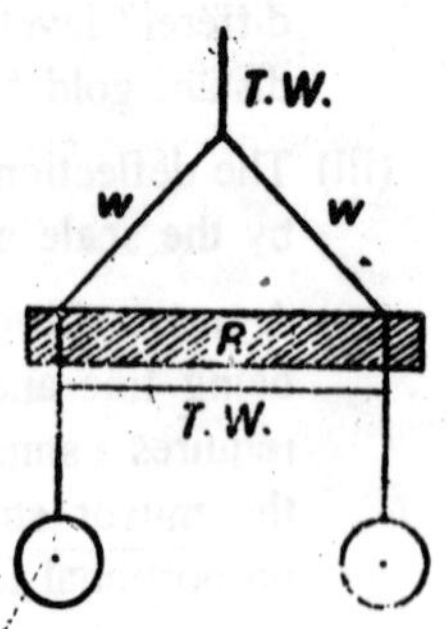

Fig. 2.5

It may be mentioned that the gold balls, first used by Heyl, were discarded by him when he found that in the near complete vacuum in the chamber, they absorbed mercury from the mercury gauges used to measure the pressure there. He next tried platinum balls, coated with lacquer, and finally preferred balls of optical glass in which any internal cavities could be easily detected visually.

Working : The suspension system (*i.e.*, the torsion rod together with the two small masses) was made to oscillate in the gravitational field of the two large masses which were first arranged with their centres in the same horizontal line with those of the smaller masses, as shown in Fig. 2.6 (a) and then with the line joining their centres along the right bisector of the torsion rod, as shown in Fig. 2.6 (b), the two positions being referred to as the *near* and the *distant* positions respectively. The gravitational attraction obviously accelerates the oscillations in the first case and retards them in the second.

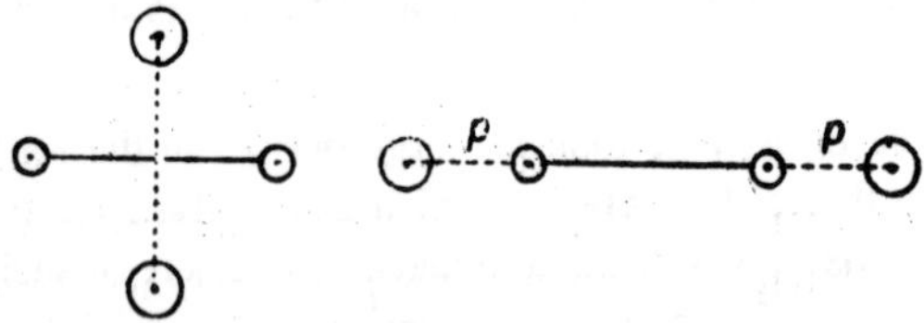

Fig. 2.6

Or, as a variation of this, the time-period of the suspended system was first noted with no other masses in the neighbourhood and then with the large masses in the *near position*, shown in Fig. 2.6 (a).

The system was set oscillating by bringing bottles of mercury near to the small masses for a while and then removing them. It was found that for an angular displacement of 4° ,the system continued oscillating for about 20 hours. The usual scale and telescope method was used to observe the oscillations, the passage of the lines of the image of a scale across a vertical cross wire of the telescope being recorded automatically by a pen on a chronograph, with another pen marking down on it the second signals from a standard clock.

Calculation : From the time-periods of the suspended system in the near and distant positions, the value of G was obtained as indicated in brief outline below.

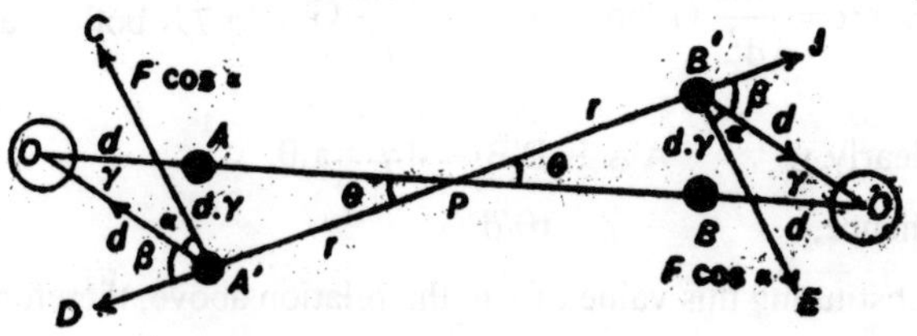

Fig. 2.7

Let T_1 be the time-period of the suspended system when the large masses are not yet brought in its neighbourhood. Then, if C be the couple or torque per unit twist of the suspension wire and *I*, the moment of inertia of the system about it, we have

$$T_1 = 2\pi\sqrt{\frac{I}{C}}.$$

With the large masses brought into the *near position*, as shown in Fig. 2.7, such that the distance between each pair of large and small masses is the same, say, d, the gravitational pull due to each large mass on the small mass near to it will clearly be $F = MmG/d^2$, where M and m are the magnitudes of each large mass and small mass respectively. The value of F remains unaffected due to each small mass being deflected through a small angle θ from its initial position A or B into the position A' or B'.

This force F will naturally be directed in either case from the centre of the small mass towards the centre of the neighbouring large mass, *i.e.,* along A'O and B'O' respectively. Resolving F into two rectangular components, along A'B' (the line joining the centres of the two small masses in their displaced positions) and perpendicular to it, we have *component perpendicular to A'B' (in either case)* = F cos α = $(MmG/d^2)\cos\alpha$, represented by A'C and B'E in the two cases respectively, where, $\angle$ OA'C = $\angle$ O'B'E = α.

Since $\angle$ OA'D = $\angle$ O'B'J = (90-α) = β, we have

$$F\cos\alpha = (MmG/d^2)\sin\beta$$

Now, since $\angle$ A'OP = $\angle$ B'O'P' = γ

we have $\beta = (\theta + \gamma)$. And, therefore,

$$F\cos\alpha = \frac{Mm}{d^2}G.\sin(\theta+\gamma) = \frac{Mm}{d^2}G.(\theta+\gamma),$$ both θ and γ being small

Clearly, A'A = B'B = d.γ = r.θ,

whence, $\gamma = r\theta/d$.

Substituting this value of γ in the relation above, therefore, we have

$$F\cos\alpha = \frac{MmG}{d^2}\left(\theta+\frac{r\theta}{d}\right) = \frac{MmG}{d^2}\left(1+\frac{r}{d}\right)\theta.$$

These two *equal, opposite* and *parallel* forces acting at A' and B' obviously form a couple, tending to bring the small balls back into their original positions A and B respectively and, clearly,

Moment of the couple $= \dfrac{MmG}{d^2}\left(1+\dfrac{r}{d}\right)\theta.\,2r$ [$\because$ A'B' = 2r]

And, if C be the torsional couple per unit twist of the suspension wire, the restoring couple also, tending to bring the small balls back into their original positions after being deflected through angle θ, is equal to $C\theta$. So that,

total couple acting on the suspended system

$$= C\theta + \frac{MmG}{d^2}\left(1+\frac{r}{d}\right)\theta.\,2r = \left[C+\frac{2MmG}{d^2}\left(\frac{rd+r^2}{d}\right)\right]\theta.$$

If, therefore, T_2 be the time-period of the suspended system *now* we have

$$T_2 = 2\pi\sqrt{\frac{I}{C+\frac{2MmG}{d^2}\left(\frac{rd+r^2}{d}\right)}}.$$

And thus, $$\frac{T_1^2}{T_2^2}=\frac{\left[C+\frac{2MmG\left(rd+r^2\right)}{d^3}\right]}{C}=1+\frac{2MmG\left(rd+r^2\right)}{Cd^3}.$$

Or, $$\frac{2MmG\left(rd+r^2\right)}{Cd^3}=\frac{T_1^2}{T_2^2}-1=\frac{T_1^2-T_2^2}{T_2^2},$$

$$G=\frac{T_1^2-T_2^2}{T_2^2}\cdot\frac{Cd^3}{2Mm\left(rd+r^2\right)}.$$

The mean results (for the different masses, mentioned earlier) gave the value of G to be 6.670×10^{-8} dyne-cm^2/gm^2 in C.G.S. units and 6.670×10^{-11} newton-metre^2/kg^2 in the M.K.S. (or SI) units.

Birge later estimated the probable error in the result to be 0.005. So that, the *true value of* G (obtained to date) = $(6.67 \pm 0.005) \times 10^{-8}$ dyne-cm^2/gm^2 or $(6.67 \pm 0.005) \times 10^{-11}$ *newton-metre^2/kg^2*

DENSITY OF THE EARTH

The weight of a body on the surface of the earth (or the gravitational force of attraction on it due to the earth) is, as we know, given by mg, where m is the *mass of the body* and g, the acceleration due to gravity at the place. If M be the *mass of the earth* and R, its *radius*, this gravitational force of attraction, in accordance with Newton's law of gravitation is equal to MmG/R^2. We, therefore, have

$$-mg = -MmG/R^2$$

or $$g = MG/R^2$$

or $$M = gR^2/G$$

Taking the earth to be a homogeneous sphere and hence of volume $V = 4\pi R^3/3$ and it density to be Δ, we have $M = V\Delta = 4\pi R^3\Delta/3 = g R^2/G$, whence,

density of the earth, $\Delta = 3g/4pRG$ and may be easily evaluated.

The radius of the earth (R) may be easily estimated by choosing two points P_1 and P_2 on the earth's surface, a known distance x apart and in the same latitude (Fig. 2.8) and measuring the altitudes of the sun at both the points *simultaneously* at about 12.0 noon. If these be ϕ_1

and ϕ_2 at P_1 and P_2 respectively, the angle subtended by them at the centre of the earth is clearly $\phi_1 \sim \phi_2 = \phi$, say. So that, $\tan \phi = x/R$. Or, since $\phi = (\phi_1 \sim \phi_2)$ is small, we have $\phi = x/R$, whence, $R = x/\phi = x/(\phi_1 \sim \phi_2)$.

Taking the value of G to be 6.6576×10^{-8} dynes-cm^2/gm^2, (as found by *Boys*), the *density of the earth.* Δ, works out to be 5.5270 gm/c.c. and with Heyl's value of G (*i.e.*, 6.670×10^{-8} dyne-cm^2/gm^2) it comes to (5.515 ± 0.004)gm/c.c.

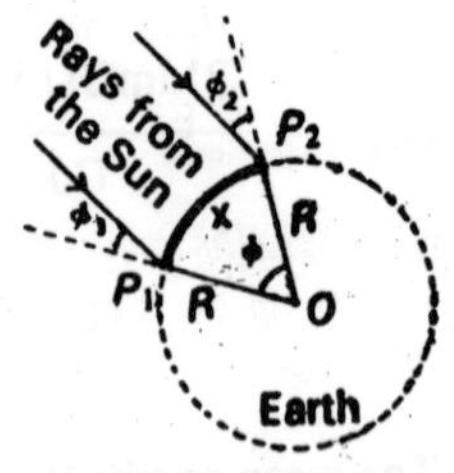

Fig. 2.8

The most probable value of Δ is, however, taken to be 5.5247 gm/c.c. and since the density of the outer layers of the earth is found to be only 2.7 gm/c.c., obviously, the density of the inner layers must be very much greater than 5.5247 gm/c.c.

Interestingly enough, Newton had intuitively guessed the probable density of the earth to lie between 5 and 6 gm/c.c.

GRAVITATIONAL FIELD—INTENSITY OF THE FIELD

The area round about a body within which its force of gravitational attraction is perceptible (no other body being near about it) is called its *gravitational field.*

The *intensity* (or *strength*), E, of the gravitational field of a particle of mass m at a point distant r from it is *the force experienced by a unit mass placed at that point in the field* (it being assumed that the presence of the unit mass does not in any way affect the gravitational field of mass m). Thus, $E = -m/r^2$ Gr, where r is the unit vector along the direction of r.

or $\quad E = -mG/r^2.$

It is thus clear that the intensity of the field is directed towards the particle, *i.e.*, opposite to the vector r.

The intensity of the field at a point is quite often referred to simply as the *field at the point,* and may also be defined as the *space rate of change of gravitational* potential (or the *potential gradient*) at the point, *i.e.*, $E = dV/dr$, where dV is the small change of gravitational potential for a small distance dr.

GRAVITATIONAL POTENTIAL AND GRAVITATIONAL POTENTIAL ENERGY

The gravitational potential V at a point distant r from a body of mass m is equal to *the amount of work done in moving a unit mass from infinity (where the gravitational force and potential are zero) to that point*. Thus,

$$v = -\int_{\infty}^{r} E dr = -\int_{\infty}^{r} \frac{m}{r^2} G - \frac{m}{r} G.$$

Clearly, *this is also the potential energy of unit mass at the point distant r from the body of mass m*. So that, *the gravitational potential at a point is equal to the potential energy of unit mass at that point.*

It follows, therefore, that the potential energy of a mass m' at the point in question will be $U = m'V = -mm'G/r$.

It will be noted that the *gravitational potential (V) and the potential energy (U) are always negative in sign, their highest value being zero at infinity*. This is clearly a consequence of the fact that in the case of gravitation we come across only forces of attraction and never those of repulsion.

MASS OF THE EARTH AND THE SUN

Knowing the mean density of the earth, it is but an easy step to obtain its *mass* equal to *volume* × *density.*

The earth, however, is not actually a sphere but a spheroid of revolution about its polar axis; so that, if its equatorial and polar radii be R_1 and R_2 respectively, we have

its volume $= \frac{4}{3}\pi R_1^2 R_2$ and hence *its mass* $= \frac{4}{3}\pi R_1^2 R_2 \Delta$.

Now, taking the orbit of rotation of the earth around the sun to be circular (though, in actual fact, it is elliptical), of radius r, it is clear that the *centripetal force* acting on the earth towards the centre of its orbit (*i.e.*, the centre of the sun) is $M_e r \omega^2$, where M_e is the mass of the earth and ω, its angular velocity = $(2\pi/365 \times 24 \times 60 \times 60)$ *radians/sec.*

This centripetal force is obviously supplied by the *force of gravitational attraction exerted by the sun on the earth* = $M_e M_s G/r^2$, where M_s is the mass of the sun. We, therefore, have

$$M_e M_s G/r^2 = M_e r\omega^2, \text{ whence,}$$

mass of the sun, $M_s = r^3\omega^2/G$ and can be easily evaluated.

This, incidentally, enables us to obtain an expression for the time-period of the earth's revolution (T, say) around the sun in terms of the mass of the sun and the radius of the earth's orbit. For $\omega^2 = M_s G/r_3$ and, therefore, $\omega = \sqrt{M_s G / r^3}$.

Hence, $T = 2\pi/\omega = 2\pi\sqrt{r^3 / M_s G}$.

VELOCITY OF ESCAPE FROM THE SOLAR SYSTEM

If a body of mass m be situated on the earth in the gravitational field of the sun, its potential energy, say, $U = \frac{M_s m}{R_s} G$, where M_s is the mass of the sun = 1.33×10^{33} gm and R_s, the distance between the sun and the earth = 1.49×10^{13} cm.

In order that the body may escape from the solar system, it should be projected with a velocity v (*i.e.*, its *escape velocity* should be v) such that its kinetic energy $\frac{1}{2}mv^2 = U = M_s m\frac{G}{R_s}$, whence, its escape *velocity*

$$v = \sqrt{\frac{2M_3}{R_3}G} = \sqrt{\frac{2\times1.33\times10^{33}}{1.49\times10^{13}}\times6.67\times10^{-3}} = 4.2\times10^6 \text{ cm/sec or}$$

42 km/sec.

If however, we take into account the rotation of the earth around the sun, *i.e.*, the fact that the earth has a relative velocity of 3×10^6 cm or 30km/sec with respect to the sun, the *escape velocity* v = 42-30 = 12 km/sec.

AN INTERESTING CONSEQUENCE OF ESCAPE VELOCITY

An obvious, interesting and important consequence of escape velocity is that it enables us to form an idea as to the probable nature of the atmosphere on other planets as also, of course, on our own. For, clearly, the nearer the average velocity of the molecules of a gas to the velocity of escape, the greater the chances of their escaping away from the upper regions of the atmosphere which may thus get completely denuded of them in course of time.

Thus, for example, on the *moon*, a landing on which has now been effected, the escape velocity is only 2.4×10^5 cm/sec or 2.4 km/sec as against 11.2 km/sec on the earth. No wonder, then, that all the constituents of the atmosphere (as we know it here on earth), like *oxygen, nitrogen, carbon dioxide and water vapour,* the *root mean square velocities* of which (at 0°C) lie between 0.4×10^5 to 0.8×10^5 cm/sec not to speak of *hydrogen* and *helium,* whose r.m.s. velocities are very much greater, about 2×10^5 cm/sec—should all be absent from its surface and that the latter two lighter gases (*hydrogen* and *helium*) should be so rare even in our own atmosphere on the earth. Again, the presence of these two latter gases in relative abundance in the atmosphere of the sun should not surprise us in view of the much stronger gravitational field of the sun and consequently a much higher escape velocity of 618 km/sec there.

LAW OF GRAVITATION AND THE THEORY OF RELATIVITY

As mentioned already, it was only when *Einstein* came forward with his *theory of relativity* that the infallibility of Newton's law of gravitation began to be seriously questioned and it came to be realised that it was only an approximation, although an extremely close one, to the true or a more fundamental law of gravitation.

A detailed discussion of this new law is beyond our present scope but we shall consider here how Newton's law lays itself open to criticism in view of some of the salient points of Einstein's theory of relativity.

(i) One starting consequence of Einstein's theory, the concept of what may be called the *inertia of energy, viz.,* that *whenever a change in the energy of a body is brought about, it is accompanied by a corresponding change in its mass.* In other words, *energy and mass are mutually convertible,* the relation between the two being given by the famous mass-energy relation $\Delta mc^2 = \Delta E$,

where Δm is the change in the mass of the body corresponding to a change ΔE in its energy and c, the velocity of light in free space (equal to 3×10^{10} cm/sec).

Further, in accordance with this theory $m = \dfrac{m_0}{\sqrt{1 - u^2/c^2}}$,

where m is the mass of a body when moving with velocity u, called its *moving mass* and m_0, its mass when at rest, called its *rest mass.*

Thus, the mass of a body is different when in motion from that when at rest, *i.e.,* it changes with the velocity of the body, and *Newton* has not specified which one of them is to be used in his formula for gravitational attraction. This is, indeed, a noticeable omission.

Another consequence of the theory is that the numerical value of the distance between two points varies according to the system of space-time coordinates chosen. Thus, the distance between them varies with the situation of the observer making the measurement.

The effect of these two discrepancies is, of course, only very slight but it is there all the same. *Einstein* has taken both these factors into account in the formulation of his law of gravitation which satisfactorily explains the deviations from the Newtonian law.

Thus, the correct law of gravitation is the one due to *Einstein*, which is true for both strong and weak gravitational fields. However, although Newton's law is true only for weak gravitational fields, it is a most satisfactorily close approximation to the correct law for most of our ordinary purposes (engineering and others), any slight discrepancies becoming noticeable only at extremely small distances.

VELOCITY OF ESCAPE FROM THE EARTH

Ordinarily, as we know, when a body, say, a rifle bullet, is projected upwards, it comes down to the earth due to the gravitational pull of the earth on it. If, however, it can be given a velocity which can take it beyond the gravitational field of the earth, it will never come back and escape into space. This velocity of the body is, therefore, called the *velocity of escape.* Let us try to obtain its value.

If m be the mass of the body, M, that of the earth and R, its radius, clearly, the gravitational force acting on the body at a distance x from the centre of the earth is mMG/x^2.

Therefore work done by the body against the gravitational field of the earth in moving upwards through distance dx = $(mMG/x^2)dx$.

Therefore, total work done by the body in escaping away from surface of the earth, *i.e.*, in moving away to an infinite distance from it $= \int_R^{\infty} \frac{mMG}{x^2}\, dx = \frac{mM}{R} G = m \times$ *the gravitational potential on the surface of the earth.*

If v_e be the initial or the *escape velocity* that thus takes the body away from its surface, into space, the *initial kinetic energy of the body*

$= 1/2\ mv_e^2$. This must obviously be equal to the work done by the body in escaping away from the earth. We, therefore, have

$$1/2\ mv_e^2 = mMG/R$$

or $$v_e^2 = 2\ MG/R$$

or $$V_e = \sqrt{2MG/R}. \qquad ...(i)$$

or, since $MG/R^2 = g$, we also have

$$v_e = \sqrt{\left(\frac{2MG}{R^2}\right)R} = \sqrt{2gR}. \qquad ...(ii)$$

Thus, velocity of escape from the surface of the earth,

$$v_e = \sqrt{\frac{2MG}{R}} = \sqrt{2gR}.$$

Substituting the values of M, G and R in relation (i) or of g and R in relation (ii), we have *velocity of escape,*

$$v_e = 11.19 \times 10^5 \text{cm/sec} \approx 11.2 \text{ km/sec}$$

or, approximately 25000m/hr.

The smallest value of the velocity of escape (v_e) is for the planet *Mercury*, being only 4.2 km/sec and the highest, viz., 61 km/sec, for the planet *Jupiter*, not to speak of the *sun* for which it is as high as 618 km/sec but from which *only atoms*, and no other bodies or rockets, can escape.

Velocity to become a satellite of the earth. Let us now calculate the velocity with which the body should be projected from the surface of the earth so as to revolve *close* around it, *i.e.*, to become its *satellite*. Let it be v_0. Then, clearly, the *centrifugal force* on the body, tending to take it away from the surface of the earth will be mv_0^2/R, where R the radius its orbit around the earth, is practically the same as the radius of the earth, (it being close to the earth). This must obviously be just balanced by the gravitational pull mMG/R^2 on it due to the earth. So that, we have

$$mv_0^2/R = mMG/R^2. \text{ Or, } v_0^2 = MG/R. \text{ Or, } v_0 = \sqrt{MG/R}.$$

Or, since $MG/R^2 g$, we also have $v_0 = \sqrt{\left(\frac{MG}{R^2}\right)} = \sqrt{gR}.$

Thus, *velocity of projection ρf a body to become a satellite of the earth,*

$$v_o \sqrt{\frac{MG}{R}} = \sqrt{gR.}$$

Clearly, therefore, $\dfrac{v_o}{v_e} = \sqrt{\dfrac{gR}{2gR}} = 1\sqrt{2}.$

Or, $$v_o = \frac{v_e}{\sqrt{2}} = 0.7073\ v_e$$

The following table shows at a glance how the path of a satellite of the earth depends upon its velocity of projection v in relation to v_o and v_e.

Velocity of satellite	Nature of path
(i) $v = v_o$	*Circular path around the earth*
(ii) $v < v_o$	*Elliptical path—Return to earth*
(iii) $v > v_o$ but $< v_e$	*Elliptical path around the earth*
(iv) $v = v_e$	*Parabolic path— Escape from the earth*
(v) $v > v$	*Hyperbolic path—Escape from the earth*

When the orbit of the satellite is elliptical, the *closest point* to the earth is referred to as the *perigee* and the farthest point as the *apogee,* the speed of the satellite being the highest, *i.e.,*

$\sqrt{\left(\dfrac{GM}{a}\right)\left(\dfrac{1+e}{1-e}\right)}$ at the former, and lowest, *i.e.,* $\sqrt{\left(\dfrac{GM}{a}\right)\left(\dfrac{1-e}{1+e}\right)}$, at the latter point, where a and e are respectively the *semi-major axis* and the *eccentricity* of the ellipse.

EQUIPOTENTIAL SURFACE

A surface, *at all the points of which the gravitational potential is the same,* is called an *equipotential surface.*

Thus, as we shall presently see, the gravitational potential at all points on the surface of a spherical shell is the same, viz., $-\dfrac{MG}{R}$, where M is the mass of the shell and R, its radius. The surface of the shell is thus an *equipotential surface.*

A consequence of this is that the difference of potential between any two points on such a surface being *zero, no work is done against the gravitational force in moving a unit, or any other, mass along it.*

This means, in other words, that there is *no component of the gravitational field along an equipotential surface* or that *the field is at every point perpendicular to it.*

Thus, consider two points P and Q, a small distance δr apart, on an equipotential surface AB, (Fig. 2.9), and let the intensity of the gravitational field at P be E, directed along PR at an angle θ with PQ. Then, clearly, component of the field along PQ = $E\cos\theta$ and, therefore, work done in moving a unit mass from P to Q = $E\cos\theta.\delta r$.

Fig. 2.9

Since P and Q lie on an equipotential surface, the work done, *i.e.*, $E\cos\theta.\delta r = 0$ and since neither E nor δr is *zero*, we have $\cos\theta = 0$ or $\theta = 90°$, *i.e.*, the field is directed along the perpendicular to the surface at P. And, obviously, what is true of the point P, is also true of all other points on the surface AB. Thus, *the direction of the field at every point on an equipotential surface is perpendicular to the surface at that point.*

We shall now proceed to calculate gravitational fields and potentials in particular cases.

GRAVITATIONAL POTENTIAL AND FIELD DUE TO A SPHERICAL SHELL

Gravitational Potential

(a) *At a point outside the shell*: Let P be a point, distant r from the centre 0 of a spherical shell of *radius* R (Fig. 2.10) and *surface density* (*i.e., mass per unit area of surface*) σ.

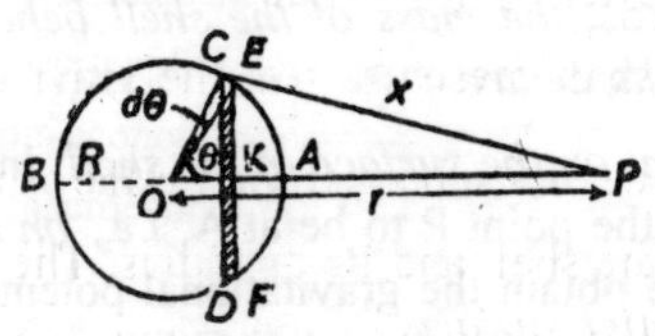

Fig. 2.10

Join OP and cut out a slice CEFD in the form of a ring by two planes CD and EF close to each other and perpendicular to the radius OA of the shell, and let angle EOF be θ and the small angle COE = $d\theta$.

clearly, *radius of the ring*, EK = OE sin θ = R sin θ, so that its *circumference* = $2\pi R \sin\theta$ and its width = CE = $Rd\theta$.

Therefore, *surface area of the ring* = *circumference* × *width* = $2\pi R \sin\theta . R\, d\theta$ and hence its *mass* = $2\pi R \sin\theta . Rd\theta\, \sigma = 2\pi R^2 \sin\theta d\theta\sigma$.

If EP = x, every point of the slice or the ring is at a distance x from P and, therefore, *potential* at P *due to the ring*, say,

$$dV = -\frac{\text{mass of slice}}{x}G = -\frac{2\pi R^2 \sin\theta d\theta\sigma}{x}G. \qquad ...(i)$$

Now, in Δ OEP, $EP^2 = OE^2 + OP^2 - 2OE.OP \cos\theta$. Or, $x^2 = R^2 + r^2 - 2Rr\cos\theta$ which, on differentiation, gives $2xdx = 0 + 0 + 2Rr \sin\theta d\theta$, whence, $x = Rr \sin\theta d\theta/dx$. [R and r being constants]

Substituting this value of x in expression (i) above, we have

$$dV = -\frac{2\pi R^2 \sin\theta d\theta\sigma}{Rr\sin\theta d\theta}Gdx = -\frac{2\pi R\sigma G}{r}dx.$$

The integral of this between the limits x = AP = (r – R) and x = BP = (r + R) then clearly gives the potential ν at P due to the whole shell. Thus,

$$V = \int_{(r-R)}^{(r+R)} -\frac{2\pi R\sigma G}{r}dx = -\frac{2\pi R\sigma G}{r}\int_{(r-R)}^{(r+R)} dx$$

$$= -\frac{2\pi R\sigma G}{r}\left[x\right]_{(r-R)}^{(r+R)} = -\frac{2\pi R\sigma G}{r}2R = -\frac{4\pi R^2\sigma G}{r}.$$

Clearly, $4\pi R^2$ is the surface area of the shell and, therefore, $4\pi R^2 \sigma$, its mass M. We, therefore, have *gravitational potential at P due to the whole shell*, $V = -\frac{M}{r}G$, *i.e.*, the same as due to a mass. M at O.

In other words, *the mass of the shell behaves as though it were concentrated at its centre.*

(b) *At a point on the surface of the shell*: In Fig. 2.10 above, if we imagine the point P to be at A, *i.e.*, *on the surface of the shell itself*, we obtain the gravitational potential there by integrating the expression for dV (obtained above) between the limits x = PA = 0 and x = PB = 2R. So that, *gravitational potential at a point on the surface of the shell,*

i.e., $$V = \int_0^{2R} -\frac{2\pi R\sigma G}{r}dx = -\frac{2\pi R\sigma G}{r}[x]_O^{2r}$$

$$= -\frac{4\pi R^2 \sigma G}{r} = -\frac{M}{r}G = -\frac{M}{R}G \qquad [\because \hbar\odot\_\odot\ r = R]$$

Again, therefore, the mass of the shell behaves as though it were concentrated at its centre.

(c) *At a point inside the shell* : Let the point P now lie anywhere inside the spherical shell (Fig. 2.11), such that OP = r. Then, proceeding as in case (a) above, we have *potential at P due to the ring CEFD, i.e.,*

$$dV = -2\pi\frac{R\sigma G}{r}dx.$$

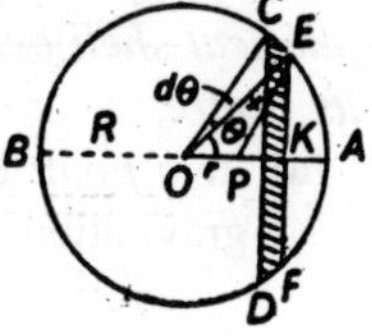

Fig. 2.11

Differentiating this expression between the limits x = PA = (R – r) and x = PB = (R + r), we therefore have *potential at P due to the whole shell, i.e.,*

$$V = \int_{(R-r)}^{(R+r)} -\frac{2\pi R\sigma G}{r}dx = -\frac{2\pi R\sigma G}{r}[x]_{(R-r)}^{(R+r)}$$

$$= -\frac{2\pi R\sigma G}{r}.\,2r = -4\pi R\sigma G.$$

Or, multiplying and dividing by R, we have

$$V = -\frac{4\pi R^2\sigma G}{R} = -\frac{M}{R}G. \qquad [\because 4\pi R^2\sigma = M.$$

the same as at a point on the surface of the shell.

Since the point P has been taken *anywhere* inside the shell, it follows that the *gravitational potential at all points inside a spherical shell is the same and is numerically equal to the value of the gravitational potential on the surface of the shell itself, which is also its maximum (negative) value.*

Gravitational Field

(a) *At a point outside the shell.* Since the potential at a point *outside the shell*, distant r from its centre (Fig. 2.10), *i.e.*, when r > R, is given by $V = -\frac{MG}{r}$, we have *intensity of the gravitational field at the point,*

$$E = -\frac{dV}{dr} = -\frac{d}{dr}\left(-\frac{M}{r}G\right) = -\frac{M}{r^2}Gr.$$

Or $$E = -\frac{M}{R^2}G.$$

i.e., the same as though the whole mass of the shell were concentrated at its centre.

Force on a point mass m : If instead of a unit mass at P (Fig. 2.10), we have a point mass m there, its potential energy will be given by $U = mV = -\frac{mMG}{r}$ and the force acting on it by $F = -\frac{dU}{dr} = -\left(\frac{mM}{r^2}\right)G$, the same as due to a mass M at O.

Thus, *for all points lying outside it (i.e., for all values of* r > R), *a spherical shell behaves as though its whole mass were concentrated at its centre.*

(a) *At a point on the outer surface of the shell*: As we know, the gravitational potential at a point on the outer surface of the shell is given by $V = -\frac{MG}{R}$.

Therefore, *intensity of the gravitational field at the point, i.e.,*

$$E = -\frac{d}{dr}\left(-\frac{MG}{R}\right) = -\frac{M}{R^2}G.$$

again, as though the mass of the shell were concentrated at its centre O.

(b) *At a point inside the shell*: We have seen under 1 (c) above that the gravitational potential at all points inside a spherical shell is the same.

Now, the gravitational field at a point is given by the space rate of change of potential (*i.e.,* the potential gradient) there. Therefore, *field at the point is* given by $E = -\frac{dV}{dr}$.

Since V is constant for all points inside the shell, $\frac{dV}{dr} = 0$, *i.e.,* the field in the interior of the shell is zero at all points. In other words, *there is no gravitational field inside a spherical shell.*

Force acting on a mass m at a point inside the shell. Since the gravitational field inside a spherical shell is *zero*, the force acting on a unit mass or any mass m at any point inside the shell is zero. This may also be seen from the following:

Considering a mass m at P, we have its P.E. $= U = mV = -\frac{mMG}{R}$, where M is the mass of the shell and R, its radius. And, therefore, *force acting on mass* m, *i.e.*, $F = -\frac{dU}{dR} = 0$.

Further, it will also be readily seen that whereas the gravitational field at a point on the inner surface of the shell is *zero*, that at a point on its outer surface is $-\frac{MG}{R^2}$, *i.e.*, there is a sudden change in the value of the field from $-\frac{MG}{R^2}$, on its outer surface to *zero* on its inner surface. *There is thus a definite, sharp discontinuity in the gravitational intensity at the surface of a spherical shell*

COMPARISON OF GRAVITATIONAL POTENTIAL AND FIELD DUE TO A SPHERICAL SHELL ELECTROSTATIC POTENTIAL AND FIELD DUE TO A CHARGE SPHERICAL SHELL

We have seen under That the electrostatic potential due to a charge Q at a distance r from it is given by V = Q/kr, where k is the dielectric constant of the intervening medium. And, therefore, intensity of the electrostatic field there is given by

$$E = \frac{dV}{dr} = \frac{Q}{Kr^2} . r \text{ . Or, } E = \frac{Q}{Kr^2}.$$

For air, k = 1 and, therefore, in air, $V = \frac{Q}{r}$ and $E = \frac{Q}{r^2}$.

Considering the case of a spherical shell carrying a charge Q, and proceeding exactly as in deducing an expression for gravitational potential due to a spherical shell, and taking σ as the *surface density of the charge* on the shell, (equal to $Q/4\pi R^2$), the *electrostatic potential at a point distant r from the centre of the shell,* where r > R (*i.e.*, the point lies outside the shell) is given by V = Q/kr and the *intensity of the electrostatic field* at the point, given by

$$E = \frac{Q}{Kr^2} r \text{ or, } E = \frac{Q}{Kr^2}.$$

Thus, *M* and *G* in the case of gravitational potential and field are here replaced by Q and 1/k respectively and the negative sign by a positive one.

In case the intervening medium between the spherical shell and the point in question be *air*, k = 1 and, therefore, V = Q/r and E = Q/r^2.

And, at a point *inside the shell, i.e* when r < R, the potential due to the shell is given by V = Q/R, *the same as on the surface of the shell* and the intensity of the electrostatic field there is, therefore, E = dV/dr = 0, because all points inside the shell are at the same potential.

If now we plot gravitational potential (V) due to a spherical shell of mass *M* and radius *R*, against distance (*r*) from its centre, we obtain a curve of the form shown in Fig. 2.12 (a), indicating that the gravitational potential *inside the shell* (*i.e.,* for r < R) has the maximum negative value, – MG/R but *outside the shell* (*i.e.,* for r > R), the –ve value goes on decreasing, or the gravitational potential goes on increasing with distance *r* from the centre, tending to its highest value, zero, as r → ∞.

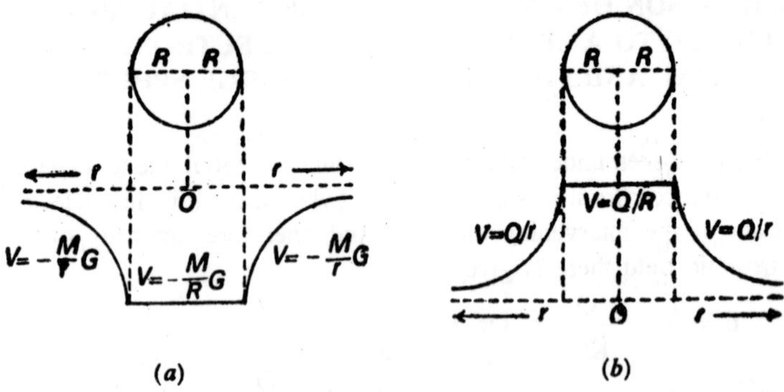

Fig. 2.12

On the other hand, if we plot electrostatic potential (*V*) due to a spherical shell, of radius *R* and carrying a charge *Q*, against distance (*r*) from the centre of the shell, we obtain a curve of the form shown in Fig. 2.12 (b), which is the *reciprocal* of that in Fig. 2.12 (*a*), indicating that *inside the shell* (*i.e.,* for r < R) the potential has its maximum value, Q/R, but *outside the shell*, (*i.e.,* for r > R) it goes on progressively decreasing, tending to its lowest value, zero, as r → ∞.

Similarly, if we plot intensities of gravitational field and electrostatic field against distance *r* from the centre of the shell, we obtain curves such as those shown in Figs. 2.13 (*a*) and (*b*) respectively.

Curve (*a*) shows that the intensity of the *gravitational field inside the shell* (*i.e.,* for r < R) is *zero* at all points; just outside the shell, it falls to its highest negative value,— MG/R^2 and then becomes —MG/r^2, *i.e.,* its negative value goes on decreasing, or the intensity goes on increasing with distance r (where r > R), tending to its *highest value 0 as r→∞*

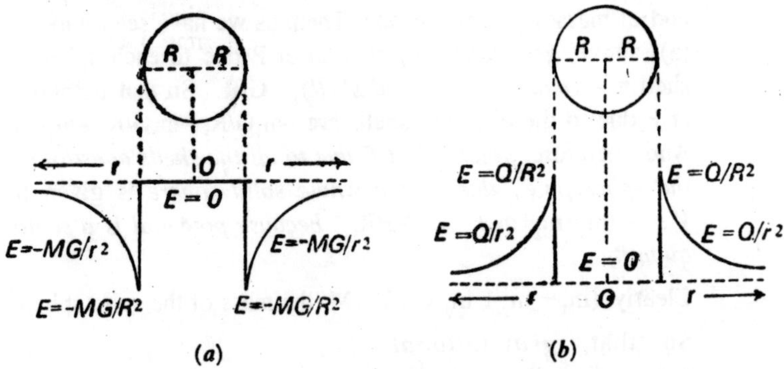

Fig. 2.13

Curve (b), on the other hand, shows that the intensity of the *electrostatic field inside the shell* (*i.e.,* for r < R) is *zero* at all points but at the surface of the shell itself it suddenly rises to its highest value Q/R^2 and thereafter becomes Q/r^2, *i.e.,* goes on falling with distance r (where r > R), tending to its lowest value, 0, as r→∞.

Again, if we plot gravitational potential energy (*U*) of a mass m at a point P against its distance r from the centre of the shell, we obtain a curve identical in' form with the one in Fig. 2.12 (a). And, if we plot P.E. of a charge Q' at a point P against its distance r from the centre of the shell, we obtain a curve identical in form with that in Fig. 2.12(b).

And, similarly, the graph between gravitational force on a mass m and its distance r from the centre of the shell is of the same form as that in Fig. 2.13(a) and that between electrostatic force on a charge *Q'* and its distance r from the centre of the shell, of the same form as that in Fig. 2.13(b).

GRAVITATIONAL POTENTIAL AND FIELD DUE TO A SOLID SPHERE

Gravitational Potential

(a) *At a point outside the solid sphere*: Let P be a point distant r from the centre *0* of a solid sphere, of mass M and radius R, outside the sphere, *i.e.,* with $r > R$, (Fig. 2.14), where the gravitational potential due to the sphere is to be determined. Imagine the sphere to consist of a number of spherical shells (shown dotted), one inside the other, concentric with the sphere, and of masses m_1, m_2, m_3 etc. Then, as we have seen under, 1 (a), above, *gravitational potential* at P due to each spherical shell = – *(mass of spherical shell)* × G/R. So that potentials at P due to the different shells are $-m_1G/R$, $-m_2G/R$, $-m_3G/R$ And, therefore, *potential at P due to all the shells constituting the sphere, i.e., due to the whole solid sphere* is given by $V = -(m_1+m_2+m_3+\ \ldots)G/R$, *because potential is a scalar quantity.*

Clearly, $(m_1 + m_2 + m_3 + \ldots) = M$. the mass of the solid sphere.

So that, *gravitational potential at P due to the solid sphere, i.e.,*

$$V = -\frac{M}{r}G.$$

Again, *therefore, the sphere behaves as though its whole mass is concentrated at its centre.*

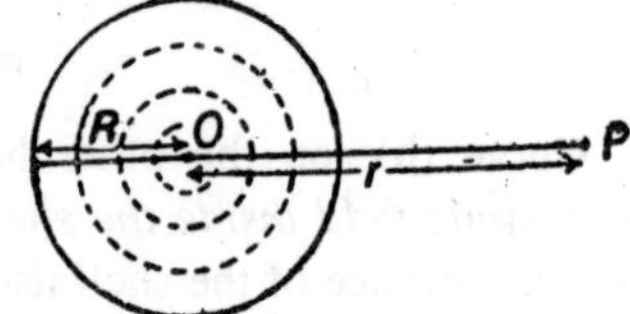

Fig. 2.14

(b) *At a point on the surface of the solid sphere*: Clearly, if the point lies on the surface of the solid sphere, we have $r = R$, the radius of the sphere.

So that, gravitational potential at a point on the surface of a solid sphere $= -\dfrac{M}{R}G.$

(c) *At a point inside the solid sphere* : Let the point P now lie inside the solid sphere at a distance r from the centre *O* of the sphere, (Fig. 2.15) *i.e.,* now $r < R$.

The solid sphere may be imagined to be made up of *an inner solid sphere of radius* r surrounded by a number of spherical shells, concentric with it and with their radii ranging from r to R. The potential at P due to the whole solid sphere is then clearly equal to the sum of the potentials at P due to the inner solid sphere and all the spherical shells outside it.

Clearly, point P lies on the surface of the *inner* solid sphere of radius r and inside all the spherical shells of radii greater than r. So that, *potential at P due to the inner solid sphere of radius r*

$$= -\frac{\textit{mass of the sphere}}{r}G = -\frac{4}{3}\pi r^3\rho\frac{G}{r} = -\frac{4}{3}\pi r^2\rho G,$$

because *mass of the inner solid sphere*

$= \frac{4}{3}\pi r^3\rho$, where ρ is the *volume density* of the sphere.

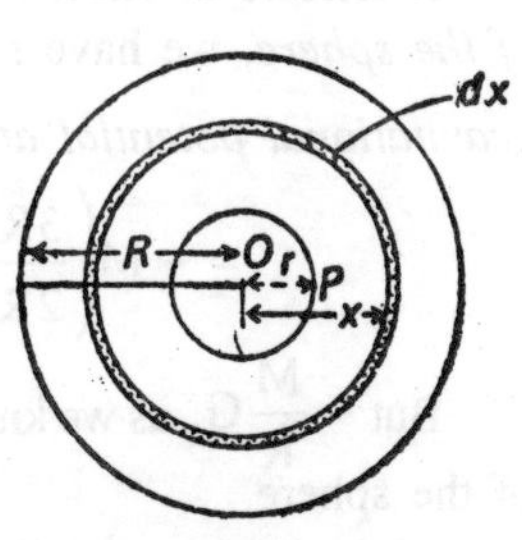

Fig. 2.15

To determine the potential at P due to all the outer shells, let us consider one such shell of radius x and thickness dx, *i.e.,* of *volume* = *area* × *thickness* = $4\pi x^2 dx$ and hence of mass = $4\pi x^2 dx\rho$.

Since potential at a point inside a shell is the same as that at a point on its surface we have

$$\textit{potential at P due to this shell} = -\frac{4\pi x^2 dx\rho}{x}G = -4\pi x dx\rho G.$$

∴ *potential at P due to all the shells*

$$= \int_r^R -4\pi\rho G x dx = -4\pi\rho G\int_r^R x dx = -4\pi\rho G\left[\frac{x^2}{2}\right]_r^R$$

$$= -4\pi\rho G\left(\frac{R^2 - r^2}{2}\right) = -\frac{4}{3}\pi\rho G.\frac{3(R^2 - r^2)}{2} = -\frac{4}{3}\pi\rho G\frac{(3R^2 - 3r^2)}{2}$$

∴ *potential at P due to the whole solid sphere = potential at P due to inner solid sphere + potential at P due all the outer spherical shells*

$$= -\frac{4}{3}\pi r^2\rho G - \frac{4}{3}\pi\rho G.\left(\frac{3R^2 - 3r^2}{2}\right) = -\frac{4}{3}\pi\rho G\left(r^2 + \frac{3R^2}{2} - \frac{3r^2}{2}\right)$$

$$= -\frac{4}{3}\pi\rho G.\left(\frac{3R^2 - r^2}{2}\right) = -\frac{4}{3}\pi R^3\rho G\left(\frac{3R^2 - r^2}{2R^3}\right).$$

[Multiplying and dividing by R^3.

Clearly, $\frac{4}{3}\pi R^3\rho$ is the mass of the whole solid sphere, *i.e.,* M.

∴ *gravitational potential at P' due to the solid sphere, i.e.,*

$$V = -\frac{M\left(3R^2 - r^2\right)}{2R^3}G.$$

It follows at once, therefore, that if the point *P lies at the centre of the sphere*, we have r = 0. So that,

gravitational potential at the centre of the solid sphere

$$= -M\left(\frac{3R^2}{2R^3}\right)G = -\frac{3}{2}.\frac{M}{R}G.$$

But $-\frac{M}{R}G$, as we know, is the gravitational potential on the *surface* of the sphere.

We thus have *gravitational potential at the centre of solid sphere* = $\frac{3}{2}$ *times the gravitational potential on its surface.* Or,

gravitational potential at the centre of the solid sphere : gravitational potential on the surface of the sphere : : 3 : 2.

This means, in other words, that *the gravitational potential due to a solid sphere has its maximum (negative) value at its centre.*

Gravitational Field

(a) *At a point outside the solid sphere*: We know that the gravitational potential at a point P outside a solid sphere distant r from its centre (*i.e.*, with r > R) is given by V = – MG/r,

And, Since intensity of the gravitational field at a point is equal to the potential gradient there, we have

gravitational field due to a solid sphere at a point P distant r from its centre (r > R), *i.e.*, E = — dV/dr $= -\frac{d}{dr}\left[-\frac{MG}{r}\right] = -\frac{MG}{r^2}$,

the same as though the whole mass (M) of the sphere were concentrated at its centre.

Alternatively, we may imagine the sphere to consist of a number of concentric spherical shells one inside the other, with their radii ranging from 0 to R. Then, considering one such spherical shell of radius x and thickness dx and hence of mass $4\pi x^2 dx\rho$, we have *gravitational field at a point P distant r from the centre of the shells* $= -4\pi x^2 dx\rho G/r^2$

And, therefore *gravitational field at P due to the whole solid sphere*

$$= -\frac{4\pi\rho G}{r^2}\int_0^R x^{2dx} = -\frac{4}{3}\pi R^3\rho\frac{G}{r^2}$$

Clearly, $\frac{4}{3}\pi R^3\rho = M$, the mass of the sphere. So that,

gravitational field at P due to the solid sphere, i.e.,

$$E = -MG/r^2$$

Force on a point-mass m : If we replace the unit mass at P by a point-mass m, we have P.E. *of the mass,*

$$U = mV = -mMG/R.$$

And, *therefore force acting on the point-mass m, i.e.,*

$$F = -dU/dr = -mMG/r^2.$$

(b) *At a point on the surface of the solid sphere*: For a point on the surface of the solid sphere, obviously, r = R, the radius of the sphere. We, therefore, have

gravitational field at a point on the surface of the solid sphere, i.e., $E = -MG/R^2$.

(c) *At a point inside the solid sphere*: As we know, the gravitational potential at a point inside a solid sphere distant r from its centre (*i.e.*, with $r < R$) is given by $V = -MG(3R^2 - r^2)/2R^3$. Since the gravitational field at a point is given by the potential gradient there, we have

gravitational field due to a solid sphere at a point P inside it, distant r from its centre, i.e.,

$$E = -\frac{dV}{dr} = -\frac{d}{dr}\left[-MG\frac{(3R^2 - r^2)}{2R^3}\right]$$

$$= -MG\frac{2r}{2R^3} - \frac{MG}{R^3}r, \text{ showing that } E \propto r.$$

Thus, the intensity of the gravitational field at a point inside a solid sphere is directly proportional to the distance of the point from the centre of the sphere.

Obviously, therefore, *at the centre of the sphere*, since r = 0, *the intensity of the gravitational field is zero, whereas the gravitational* potential there, as we have seen under 1 (c) above, has its maximum (negative) value.

It will also be readily seen that in the expressions for the intensities of the gravitational field (E) outside and inside a solid sphere, deduced above under (a) and (c), if we put r = R, the radius of the sphere, we have $E = MG/R^2$ as the field on the surface of the sphere in either case, *i.e.*, whether the point P lies on the outer or the inner side of the surface of the sphere.

This shows clearly that there is a continuity in the intensity of the gravitational field at the surface of a solid sphere , unlike the case of a spherical shell, where there is a sharp discontinuity in the intensity of the gravitational field at the surface of the shell

Force on a point-mass m : As usual, the potential energy of a point-mass m at a point P inside the solid sphere at a distance r from its centre is given by

$$U = mV = -mMG\left(\frac{3R^2 - r^2}{2R^3}\right) \text{ and, therefore,}$$

force acting on the mass, i.e., $F = -\dfrac{dU}{dr} = -\dfrac{mMG}{R^3}r,$

i.e., the force acting on a mass m placed at a point inside a solid sphere is also directly proportional to its distance from the centre of the sphere and is, therefore, zero at the centre of the sphere.

COMPARISON OF GRAVITATIONAL POTENTIAL AND FIELD DUE TO A SOLID SPHERE WITH ELECTROSTATIC POTENTIAL AND FIELD DUE TO A CHARGED SOLID SPHERE

Here *two cases* arise, *viz.*, (i) the sphere may be *of a conducting material*, a metallic sphere, for instance, (ii) the sphere may be a homogeneous one *of a non-conducting material.*

Since in the case of a conducting sphere, the charge on it is not uniformly distributed over the whole of its volume but resides only on its surface, it *effectively* functions as a charged spherical shell which we have considered already under

We shall, therefore, consider here only the case of a *homogeneous non-conducting sphere*, in which the charge is uniformly distributed over its entire volume.

Proceeding in the same manner as for gravitational potential and field due to a solid sphere and taking charge Q on the sphere in place of its mass M, with ρ now representing the volume density of its charge $\left(\text{i.e., } \rho = Q/\frac{4}{3}\pi R^3\right)$, instead of the volume density of its mas$\left(M/\frac{4}{3}\pi R^3\right)$, we obtain expressions similar to those for gravitational potential and field, *viz., for points outside the charged sphere, (i.e., for* $r > R$), *electrostatic potential,* $V = Q/kr$ and *intensity of the electrostatic field,* $E = \frac{Q}{kr^2}$ r Or, $E = \frac{Q}{kr^2}$ And, *on the surface of the sphere,* $V = Q/kR$ end $E = Q/kR^2$.

Thus, M in the case of gravitational potential field is here replaced by Q and G by 1/k, with, of course, the negative sign replaced by a positive one.

If the charged sphere and the point in question be in air, $k = 1$, and we have $V = Q/r$ and $E = Qr^2$. And, *on the surface of the sphere,* $V = Q/R$ and $E = QR^2$, *i.e., the sphere behaves as though the whole of the charge on it is concentrated at its centre.*

And, *for points inside the sphere,* $V = \frac{Q}{k}\left(\frac{3R^3 - r^2}{2R^3}\right)$

and $E = \frac{Qr}{kR^3}$ r Or, $E = \frac{Q}{kR^3}r$.

At the centre of the sphere, (i.e., for $r = 0$). clearly, $V = \frac{3}{2}\frac{Q}{kR}$ and $E = 0$.

If, therefore, we plot gravitational and electrostatic potentials against distances from the centres of a uniform solid sphere and a uniformly charged solid sphere respectively, we obtain curves such as those shown in Figs. 2.16 (a) and (b).

Curve (a) shows that the gravitational potential (negative) due to a uniform solid sphere is the *highest,* $-\frac{3}{2}\frac{M}{R}G$ *at its centre,* $-MG\left(\frac{3R^2 - r^2}{2R^3}\right)$ for points between the centre and the surface of the sphere, $-\frac{M}{R}G$ at the surface of the sphere and $-\frac{M}{r}G$ thereafter, tending to zero as $r \to \infty$.

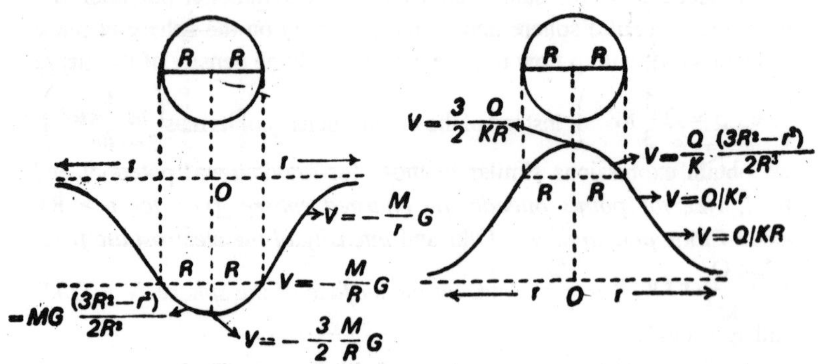

(a) Gravitational potential due to a uniform solid sphere, **(b) Electrostatic potential due to a uniformly charged solid sphere.**

Fig. 2.16

Curve (b), as will be readily seen, is the exact reciprocal of curve (a), because although the electrostatic potential due to the charged sphere varies with distance from its centre precisely in the manner of the gravitational potential in case (a), its sign is reversed and is now *positive*. So that, here, electrostatic potential is $\frac{3}{2}\frac{Q}{kR}$ at the centre of the charged sphere, $\frac{Q}{k}\left(\frac{3R^2-r^2}{2R^3}\right)$ for points between the centre and the surface of the sphere Q/kR at the surface of the sphere, and Q/kr thereafter, tending to zero as $r \to \infty$.

Similarly, if we plot intensities of the gravitational and electrostatic fields due to a uniform solid sphere and a uniformly charged solid sphere respectively against distances from their centres, we obtain curves of the form shown in Figs. 2.17 (a) and (b).

Clearly, Fig. 2.17 (a) shows that the gravitational field E is *zero* at the centre of the uniform solid sphere and its (negative) value increases linearly with distance r from the centre in accordance with the relation $E = -(MG/R^3)\,r$, so that at the surface of the sphere it attains its maximum (negative) value — MG/R^2 ($\because$ here, $r = R$). Thereafter, its (negative) value goes on decreasing with distance r in accordance with the relation $E = -MG/r^2$, tending to zero as $r \to \infty$.

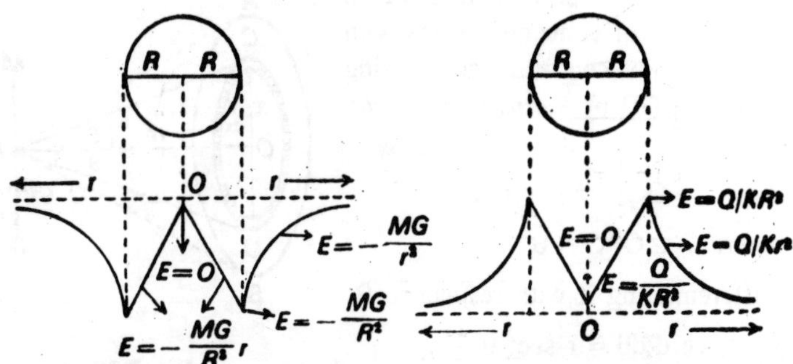

(a) Gravitational field due to a uniform solid sphere **(b) Electrostatic field due to a uniformly charged solid sphere.**

Fig. 2.17

Similarly, Fig. (b) shows that the electrostatic field due to a uniformly charged solid sphere is zero at its centre and its value goes on increasing linearly with distance r from the centre in accordance with the relation $E = (Q/kR^3)r$, so that at the surface of the sphere it attains its maximum value $E = Q/kR^2$ ($\because$ here, r = R). Thereafter, its value goes on decreasing with distance r in accordance with the relation $E = Q/kr^2$, tending to zero as $r \to \infty$.

Again, if we plot a graph between gravitational potential energy of a mass (m) and distance (r) from the centre of the sphere, we obtain a curve of the form shown in Fig. 2.16 (a). And, if we plot a graph between electrostatic potential energy of a charge (Q') and distance (r) from the centre of the charged sphere, we obtain a curve of the form shown in Fig. 2.16 (b).

And the graphs between gravitational force on a mass m and the electrostatic force on a charge Q' at distance r from the centres of the uniform solid sphere and the uniformly charged solid sphere respectively, will be of the respective forms shown in Figs. 2.17 (a) and (6).

INTENSITY AND POTENTIAL OF THE GRAVITATIONAL FIELD AT A POINT DUE TO A CIRCULAR DISC

Let AB, (Fig. 2.18), represent a circular disc of radius R with its plane perpendicular to the plane of the paper and let P be a point on its axis, distant r from its centre O where the intensity and potential due to its gravitational field are to be determined.

Imagining the disc to consist of an infinite number of concentric rings with O as their common centre and considering one such ring CD of *radius* x and *width* dx, we have

$$OC = x = r \tan \theta,$$

where $\angle OPC = \theta$.

Differentiating it with respect to θ, we have $d/d\theta = r \sec^2 \theta$.

or $dx = r \sec^2 \theta d\theta$ and $CP = r \sec \theta$.

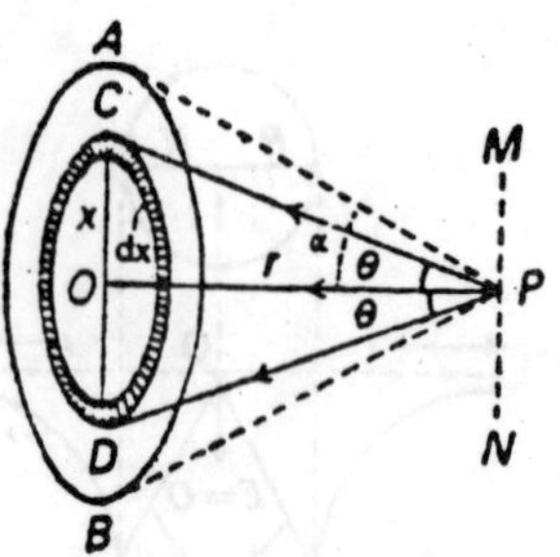

Fig. 2.18

Now, *area of the ring = circumference × width* = 2πxdx.

∴ *mass of the ring* = 2πxdxσ, where σ is the *surface density* (*i.e., mass per unit area*) of the disc.

Considering a small element of mass δm of the ring at C, we have *gravitational intensity at P due to the element* = δ mG/CP² *along the direction PC.*

Resolving it into its rectangular components along and perpendicular to PO, we have *component along* $PO = \frac{\delta m}{CP^2} G \cos\theta$ and component at right angles to it, *vertically upwards* $= -\frac{\delta m}{CP^2} G \sin\theta$, along PM.

Similarly, for an equal element at D, diametrically opposite to C, we may resolve the gravitational intensity at P due to it (*i.e.*, $-\delta m\ G/DP^2$) into similar components, *viz.*,

component $-\frac{\delta m}{DP^2} G \cos\theta = -\frac{\delta m}{CP^2} G \cos\theta$ along PO

and component $-\frac{\delta m}{DP^2} G \sin\theta = -\frac{\delta m}{CP^2} G \sin\theta$, perpendicular to it, vertically downwards, along PN. [∵ DP = CP.

The two vertical components (along PM and PN) being equal and opposite along the same line of action, cancel out and the components along PO alone being *effective* are added up. The same will obviously be the case with all other elements into which the ring may be supposed to be broken up. So that, the gravitational intensity at P due to the whole ring is equal to the sum of the components along PO due to the different elements, *i.e.*,

$$= -\frac{\sum \delta m}{CP^2} G \cos\theta = -\frac{\text{mass of the ring}}{CP^2} G \cos\theta$$

$$= -\frac{2\pi x dx \sigma}{CP^2} G \cos\theta$$

Or, substituting the values of x, dx and CP obtained above, we have *gravitational intensity* at P *due to the ring*

$$= -\frac{2\pi\sigma . r \tan\theta . r \sec^2\theta d\theta}{r^2 \sec^2\theta} G\cos\theta = -2\pi\sigma G \sin\theta d\theta, \text{ along PO.}$$

Clearly, therefore, the intensity at P due to all the rings into which the disc is supposed to be divided up, *i.e., due to the whole disc*, can be easily obtained by integrating the above expression between the limits $x = 0$ and $x = R$ or $\theta = 0$ and $\theta = \alpha$, where α is the angle that each extremity of the diameter AB of the disc makes with PO.

Thus, *gravitational intensity* at *P due to the whole disc, i.e.*,

$$E = -2\pi\sigma G \int_0^\alpha \sin\theta d\theta.$$

$$\text{or } E = -2\pi\sigma G \left[-\cos\theta\right]_0^\alpha = -2\pi\sigma G \left[-\cos\alpha - (-\cos\theta)\right]$$

$$= -2\pi\sigma G (1 - \cos\alpha) \quad \text{...(i)}$$

or since $\cos\alpha = \frac{OP}{AP} = \frac{r}{\sqrt{r^2 + R^2}}$, we also have

$$E = -2\pi\sigma G \left(1 - \frac{r}{\sqrt{r^2 + R^2}}\right) \quad \text{...(ii)}$$

Again, because $2\pi(1 - \cos\alpha)$ is the *solid angle* ω subtended by the disc at the point P, we have, from relation (i) above,

$$E = -\sigma G\omega. \quad \text{...(iii)}$$

Now, *potential at the point P due to the ring of radius* x = intensity at the *point P due to the ring* $\times r = -2\pi\sigma G \sin\theta d\theta \times r$.

[∵ *intensity of the field at P = potential gradient there.*

Hence, *gravitational potential at P due to the whole disc, i.e.*,

$$V = -2\pi\sigma G r \int_0^\alpha \sin\theta d\theta = -2\pi\sigma G r (1 - \cos\alpha) \quad \text{...(iv)}$$

or, since $\cos\alpha = \frac{r}{\sqrt{r^2+R^2}},$ we have

$$V = -2\pi\sigma G r\left(1-\frac{r}{\sqrt{r^2+R^2}}\right) \quad ...(v)$$

or, again, since $2\pi(1-\cos\alpha) = \omega$, we have $V = -G\omega r$. ...(vi)

FLUX OF GRAVITATIONAL INTENSITY—GAUSS'S THEOREM

Let P be a point in the gravitational field of a mass m at O, (Fig. 2.19) distant r from it. Then, clearly, *gravitational intensity or field at P due to mass m, i.e.,* $E = -mG/r^2$, directed along OP. Let it be represented by PQ.

Draw any closed and finite surface through P, enclosing mass m and consider a surface element ab of it, of area ds, with PN as the outward drawn normal to it at point P.

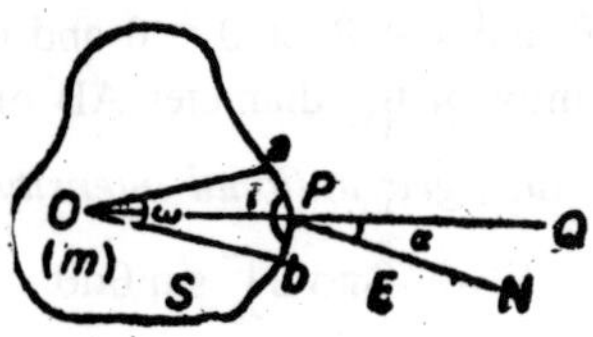

Fig. 2.19

If < QPN, that the intensity E makes with the normal PN at P, be α, we have

component of gravitational intensity E *along the normal PN*

$$= E_n = E\cos\alpha.$$

The product $E_n ds$, of the normal component E_n of intensity E and the small area ds of the surface element ab, is called the *flux of gravitational intensity or field* across the area ds. And, therefore,

flux of gravitational intensity over the whole closed surface S

$= \int E_n.\ ds = \int E\cos\alpha.ds$, ds, where the integration extends over the entire closed surface.

Gauss's Theorem : This theorem states that *the total flux of gravitational intensity over a closed surface (having a unique outward drawn normal to it at every point) in a gravitational field is* — 4πG *times the total mass enclosed by the surface, i.e.,* $\int E_n\ ds = -4\pi G\sum m$, where Σm *is the total mass enclosed by the surface*. This may be easily proved as follows:

We have just seen above (Fig. 2.19) that the flux of gravitational intensity over the surface element ab of area ds is $E_n ds = E\cos\alpha\ ds = -(mG/r^{2)\ \cos\alpha}$ ds.

Now ds. $\cos\alpha/r^2 = d\omega$, the solid angle subtended by area ds at O. So that, *flux of gravitational intensity over area ds* = $-mG.d\omega$.

And, therefore, *flux of gravitational intensity over the whole of the closed surface* $= -\int mG.d\omega. \ = -G\omega\sum m \ = -4\pi G\sum m$, where ω is the solid angle subtended by the whole of the closed surface, at the point O = 4π and Σm is the total mass enclosed by the surface.

Thus, if there be a number of masses m_1, m_2, m_3 etc. enclosed by the surface, $\Sigma m = m_1 + m_2 + m_3 + ...$and the *total flux of gravitational intensity over the closed surface* $= -4\pi\ G\ (m_1 + m_2 + m_3 + ...)$

And, if there be no mass enclosed by the surface, $\Sigma m = 0$ and, there fore, *total flux of gravitational intensity over the closed surface* = 0.

Some Simple Applications of Gauss's Theorem

Gravitational Intensity due to a Solid Sphere

(a) *At a point outside the sphere*: Consider a solid sphere of mass M and radius R, the gravitational intensity E due to which is to be determined at a point P, distant r from its centre O, (Fig. 2.20).

Imagine a sphere to be drawn with centre O, and radius r as shown dotted in the figure, so as to completely enclose within it the given sphere, with point P lying on its surface.

By sheer symmetry, the gravitational intensity will be the *same* at every point on the surface of *this* sphere in a direction normal to the surface at that point. We, therefore, have $E_n = E$.

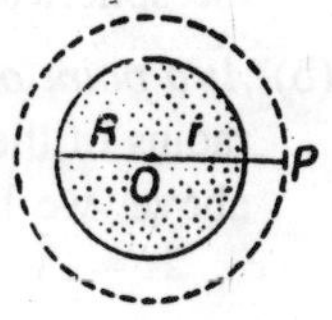

Fig. 2.20

Since the surface area of the sphere is $4\pi r^2$, we have *total flux of gravitational intensity across its surface* = $4\pi r^2 E$.

This should, in accordance with *Gauss's theorem*, be equal to $-4\pi GM$. So that, we have $4\pi r^2 E = -4\pi GM$, whence, $E = -MG/r^2$, the same result as obtained earlier.

(b) *At a point on the surface of the sphere*. In this case, obviously r = R, So that, proceeding as above, we have $E = -MG/r^2$.

(c) *At a point inside ihe sphere.* Let the point P now lie inside the sphere at a distance r from the centre O of the sphere, (Fig. 2.21).

Again, draw a sphere with centre O and radius r, as shown dotted in the figure, with point P lying on its surface.

As before, due to symmetry, the gravitational intensity (E) will be the *same* at every point on *this* sphere, in a direction normal to its surface at that point, *i.e.,* $E_n = E$.

Therefore total flux of gravitational intensity across the surface of this sphere $= 4\pi r^2 E$.

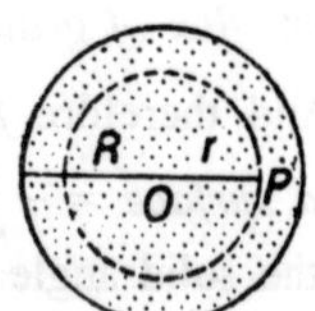

Fig. 2.21

By *Gauss's theorem*, this should be equal to $-4\pi G$ times the mass enclosed by the sphere. Thus, $4\pi r^2 E = -4\pi G\left(\frac{4}{3}\pi r^3 \rho\right)$, where ρ is the density (*i.e.,* volume density) of the material of the sphere.

$$\text{Or, } E = -\frac{4}{3}\pi G r \rho = -\frac{MG}{R^3}r, \quad [\because \frac{4}{3}\pi R^3 \rho = M. \text{ Or, } \frac{M}{R^3} = \frac{4}{3}\pi\rho.$$

i.e., the intensity varies directly as the distance (r) from the centre of the sphere.

At the centre of the sphere, since r = 0, E = 0, the same result as obtained before.

Gravitational Intensity due to a Spherical Shell

(a) *At a point outside the shell.* Proceeding exactly as in the case (i) (a), we have here too, $E = -MG/r^2$, where M is the mass of the spherical shell.

(b) *At a point on the surface of the shell.* Here, clearly, r = R, the radius of the spherical shell. So that, proceeding as in case (a) above, we have

$$E = -MG/r^2.$$

(c) *At a point inside the shell.* Here, we describe a sphere with O as centre and r, as radius, such that the point P (inside the shell) lies on its surface, (Fig. 2.22)

Then, by symmetry $E_n = E$ And, therefore,

$4\pi r^2$ E = $-4\pi G$ (*mass enclosed by the sphere*).

Since the mass enclosed by the sphere is clearly *zero*, we

have $4\pi r^2 E = -4\pi G \times 0$

or $E = 0$

i.e., there is no gravitational field or intensity Inside a spherical shell.

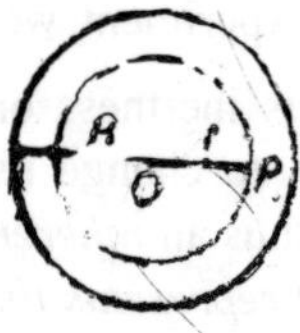

Fig. 2.22

INTENSITY AND POTENTIAL OF THE GRAVITATIONAL FIELD AT A POINT DUE TO AN INFINITE PLANE

In the case of the disc above if the radius R becomes infinite, the disc becomes an *infinite plane*, when, obviously, $\alpha = \pi/2$ and, therefore, $\cos\alpha = 0$, and $\omega = 2\pi$.

Thus, if we put $R = \infty$ in expression :

(i), $\cos\alpha = 0$ in expression

(ii) or $\omega = 2\pi$ in expression

(iii), we have *gravitational intensity at P due to an infinite plane* $= -2\pi\sigma G$, which is, clearly, quite independent of the distance r from it.

Similarly, putting these values of R, $\cos\alpha$, and ω in relations (v), (iv) and (vi) respectively, we have

gravitational potential at a point P distant r from as infinite plane

$$= -2\pi\sigma Gr.$$

INERTIAL AND GRAVITATIONAL MASS

Inertial Mass

We are aware of the property of inertia possessed by a material body, *i.e.,* its property of resisting any change in its position of rest or uniform motion along a straight line. Thus, we know that some effort or force has to be applied to put into motion a block, of whatever material, lying at rest on a smooth, frictionless horizontal surface, or to stop or even to show it down or decrease its acceleration if it be in motion, and also that the bulkier the block, *i.e.,* the greater its mass, the greater the effort required for the purpose. Obviously, gravity does not come into the picture in this case (the motion being perpendicular to the direction in which the gravity acts). In fact, the effort or the force required to set the block in motion or to stop it would still be the same if the experiment were performed in gravity-free space.

This inertness or reluctance of the block, or a material body, in general, to change its state of rest or uniform motion along a straight line is thus an *inherent property of the body in virtue of its mass which, since it represents the inertia of the body,* is called its *inertial mass.* It can obviously be measured in terms of the ratio between the force (F) applied and the acceleration (a) produced in the body. For, in accordance with Newton's second law of motion, we have

$$F = ma \text{ and, therefore, } m = F/a.$$

Here, m represents the *inertial mass* of the body and may, therefore, more appropriately, be denoted by the symbol m_i.

If we apply the same force F to two bodies of inertial masses m_i and m_i', and if the Bending of Beams—Columnss produced in them be a and a' respectively, we have

$$F = m_i a = m'_i a', \text{ Whence, } m_i/m'_i = a'/a.$$

Thus, *the inertial masses of two bodies are inversely proportional to the Bending of Beams—Columnss produced in them by a given force.* This holds good irrespective of the magnitude of the force applied.

We can, therefore, measure the inertial mass (m_i) of a body by comparing the acceleration produced with that produced in *standard mass* (M), such as the *standard kilogramme* (preserved at *Sevres*) by the application of the same force F to both. If the Bending of Beams—Columnss produced in the two masses be a and A respectively, we have

$$F = m_i a = MA,$$

whence, $m_i/M = A/a.$

Or, $m_i = (A/a) M.$

So that, m_i can be easily evaluated.

Gravitational Mass

We can also measure the mass of a body in terms of the gravitational force of attraction exerted on it by another body, such as the earth. For, as we know, we have to exert an upward force to prevent a body from falling under the gravitational force of the earth and again, the bulkier the body, the greater the force required to hold it. Here, clearly, the property of inertia plays no part but only the action of gravity. *This mass of a body, measured in terms of the gravitational force of attraction on it due to another body, usually the earth, is called its gravitational mass.* Thus, if F and F' be the forces of attraction exerted by a given body A, of mass M, on two bodies B and C, of masses m and m' respectively, at the same distance R from it, we have

$$F = mMG/R^2$$

and $$F' = m'MG/r^2,$$

Whence, $F/F' = m/m'$.

Here, m and m' are the *gravitational masses* of the two bodies and may be denoted by m_g and m'_g respectively to distinguish them from their inertial masses m_i and m'_i. So that, $F/F' = m_{g},/m'_g$.

Thus, we see that *the gravitational masses of the two bodies B and C, are directly proportional to the forces of attraction exerted on them by the body A.*

If the body A happens to be the earth, clearly F = w and F' = w', where w and w' are the *weights* of bodies B and C respectively at the given place. We, therefore, have

$$w = \frac{m_g M}{R^2}G \text{ and } w' = \frac{m'_g M}{R^2}G, \text{ whence}, \frac{w}{w'} = \frac{m_g}{m_g'}.$$

Or, *the ratio between the gravitational masses of two bodies is equal to the rátio between their weights at a given place.*

It follows, therefore, that the mass of a body, as measured by a spring balance or by balancing it in one pan against a standard mass in the other, of a physical balance, is its *gravitational mass.*

Equality of inertial and gravitational masses : The question naturally arises whether the inertial mass of a body is any different from, or the same as, its gravitational mass. As a first step to enquire into the problem, let us try to see whether the inertial masses of two bodies too are proportional to their weights at a given place. For this, we first determine the inertial masses m_i and m'_i of the two bodies in the manner discussed under (i) above. Then, since in free space (or vacuum) all bodies fall with the same acceleration (g) at a given place, under the action of gravity, we have forces acting on the two bodies due to attraction by the earth, *i.e.*, their weights w and w', equal to $m_i g$ and $m'_i g$ respectively. And, if m_g and m'_g be the gravitational masses of the two bodies, their weights w and w' are also equal to $m_g MG/R^2$ and $m'_g MG/R^2$ respectively, as we have just seen above.

So that, $m_i g = m_g MG/R^2$

and $m'_i g = m'_g MG/R^2$,

whence, $m_i m'_i = m_g/m'_g$,

showing that *the ratio of the inertial masses of two bodies is the same as that of their gravitational masses.*

The inertial and gravitational masses of a body are thus clearly proportional to each other. Newton, Eotvos and others have conclusively shown that inertial and gravitational masses agree with each other to within 1 part in 10^{10} We can thus safely assert that the two are equal.

INVERSE SQUARE LAW FORCES AND STABLE EQUILIBRIUM

If we have a group of point-charges, or masses, subject to inverse law forces, they cannot *all* remain at rest.

This follows at once from a consideration of the equipotential lines or surfaces due to the charges or masses constituting the group. For, as we know, in the case of a point-charge, the equipotential lines are all concentric circles of increasing radii and decreasing values of potential about the point-charge as centre. And so also for a charged sphere, where the charge behaves as though it were concentrated at its centre, the equipotential surfaces are all concentric spheres of increasing values of potential about the charged sphere. Now, for a test charge to be in stable equilibrium at any point in an electrostatic field, the potential at all other neighbouring points must be higher or lower than at the point in question according as the test charge is positive or negative. And this is obviously impossible in the field either of the point-charge or of the charged sphere.

Similarly, if we have two equal positive (or negative) charges or charged spheres, in fixed positions, the equipotential lines due to them will be somewhat like those shown in Fig. 2.23. So that, if a test charge be placed at any point in the electrostatic field of the two charges, the potential there will again be neither lower nor higher than that at all other neighbouring points and the test charge will not therefore remain in static equilibrium at that or at any other point in the electrostatic field of the two charges. And what is true of two charges is equally true for any number of them.

A charged particle, positive or negative, may of course remain in equilibrium in an electrostatic field in the sense that the electrostatic force on it is *zero,* as for example, at a point somewhere between two similar charges (positive or negative) where the intensity of the field due to them is zero. But then, it will not be in stable equilibrium, for its slightest displacement from the equilibrium position will alter the very configuration of the field and it will never be able to come back to its previous position.

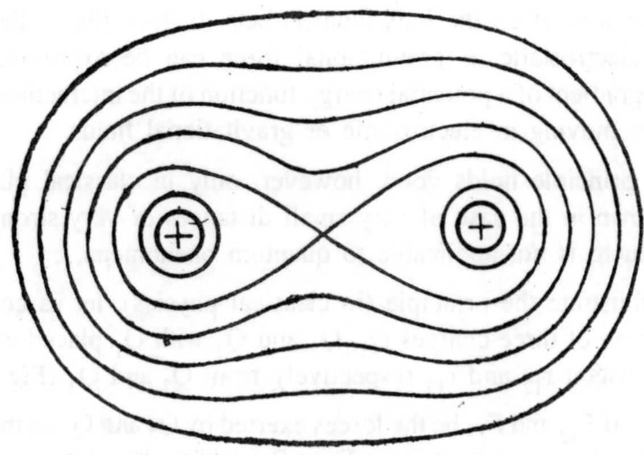

Fig. 2.23

The same would be the case with a small mass-particle if we place it in the gravitational field of either a single large mass or that of two or more large masses.

We can, therefore, safely state that *no static equilibrium is possible in an inverse square law force field.* This means that in any system of bodies in which inverse square law forces are operative, no state in which the kinetic energy of the system becomes zero can possibly be stable. This is true whether the system of bodies be the electrons in an atom, the atoms in a star or the stars in a galaxy. It follows, therefore, that *the entire known universe must consist of moving bodies,* in *which we may have a steady state in the sense that no great changes may occur but never a static state in which all bodies are motionless.*

THE SUPERPOSITION PRINCIPLE

This principle states that the resultant force on a charge or a mass placed at a point, due to a group or a combination of charges or masses is the vector sum of the forces exerted on it by the individual charges or masses comprising the group or the combination.

The principle of superposition thus asserts, in effect, the physical independence of the interaction of each pair of charges or masses that the individual charges or masses form with the given charge or mass at the point. It means, in other words, that each such pair of charges or masses interacts in the same manner as it would in the absence of the other charges or masses and that the total effect is thus *additive, i.e.,* the vector sum of all the individual effects. In fact, this is the reason why an electrostatic or gravitational force can be expressed as the negative gradient of a potential energy function of the interacting charges or masses moving in electrostatic or gravitational fields.

The principle holds good, however, only in classical physics. It breaks down in the case of very small distances or very strong forces and, as such, is not applicable to quantum phenomena.

To illustrate the principle (in classical physics), let us consider a combination of three charges Q_1, Q_2 and Q_3 with Q_1 placed at a point P in air distant r_{12} and r_{13} respectively from Q_2 and Q_3 (Fig. 2.24).

Then, if F_{12} and F_{13} be the forces exerted by Q_2 and Q_3 on the charge Q_1 at P, we have resultant force F_1 at P equal to the vector sum of F_{12} and F_{13}, *i.e.,*

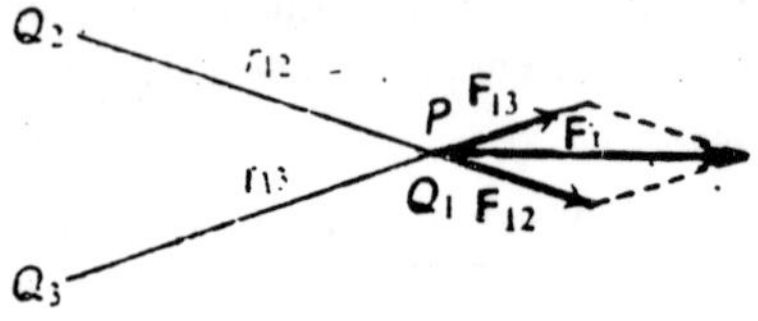

Fig. 2.24

$$F_1 = F_{12} + F_{13}.$$

Or, since $F_{12} = (Q_1Q_2/r^2_{12})\, r_{12}$ and $F_{13} = (Q_1Q_3/r^2_{13})\, r_{13}$, where r_{12} and r_{13} are the unit vectors directed along r_{12} and r_{13}, respectively, we have

$$F_1 = \frac{Q_1Q_2}{r^2_{12}}\, r_{12} + \frac{Q_1Q_3}{r^2_{13}}\, r_{13}.$$

The same naturally holds good for any number of charges and, obviously, similar considerations apply to a combination of masses.

POTENTIAL ENERGY OR SELF ENERGY OF A MULTI-PARTICLE (OR MULTI-CHARGE) SYSTEM

The gravitational potential energy of two masses m_1 and m_2, a distance r_{12} apart, is defined, as we know, as the amount of work done in bringing them from their initial infinite separation to their present distance apart.

Similarly, the gravitational potential energy of a system of three masses, placed as shown in Fig. 2.25, is equal to the amount of work done in assembling them from their initial infinite separation into their present configuration.

This may, therefore, equally well be called the *self energy* of this configuration of the system.

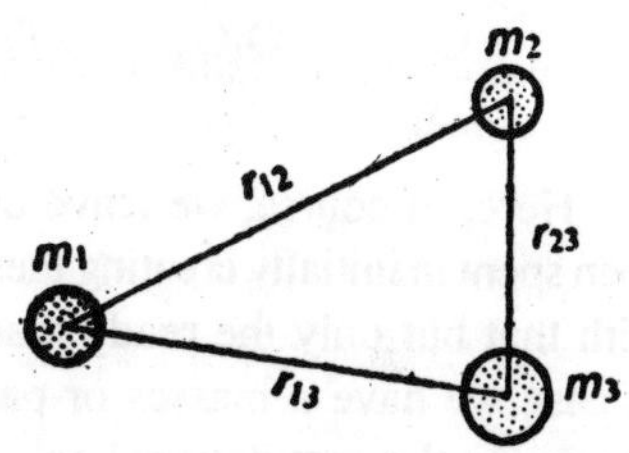

Fig. 2.25

To calculate the gravitational P.E. (or the self energy) of the system, all we need do is to calculate the work done in arranging them in this particular manner.

Clearly, the *potential energy* U_{12} of masses m_1 and m_2 or the *work done* W_{12} *in bringing them from infinity to their present distance* r_{12} apart is given by the relation $U_{12}\ (\text{or } W_{12}) = \int_{\infty}^{r_{12}} F_{12}.ds_{12}$

$$= \int_{\infty}^{r^{12}} -\frac{m_1 m_2 G}{r^2_{12}} dr.ds_{12} = \int_{\infty}^{r^{12}} -\frac{m_1 m_2 G}{r^2_{12}}.dr_{12} = -\frac{m_1 m_2}{r_{12}} G.$$

Similarly, work done in bringing mass m_3 from infinity to distance r_{13} from m_1 and r_{23} from m_2 is given by U_{13} (or W_{13}) + U_{13} (or W_{23})

$$= \int_{\infty}^{r^{13}} F_{13}\, dr_{13} + \int_{\infty}^{r^{23}} F_{23}\, dr_{23} = \int_{\infty}^{r^{13}} -\frac{m_1 m_3 G}{r^2_{13}} dr_{13}$$

$$+ \int_{\infty}^{r^{23}} -\frac{m_2 m_3 G}{r^2_{23}} dr_{23} = -\frac{m_1 m_3}{r_{13}} G - \frac{m_2 m_3}{r_{23}} G.$$

Hence, *potential energy or gravitational self energy* U_s *of the system* is equal to the total work done in bringing the three masses into their present configuration, *i.e.*,

$$U_s = U_{12} + U_{13} + U_{23} = W_{12} + W_{13} + W_{23}.$$

or $$U_3 = -\left(\frac{m_1m_2}{r_{12}} + \frac{m_1m_3}{r_{13}} + \frac{m_2m_3}{r_{23}}\right)G.$$

If we had three charges Q_1 Q_2 and Q_3 (in air) in place of the three masses m_1, m_2, and m_3, the *potential energy* or the electrostatic self energy of the system would similarly be given by $U_3 = U_{12} + U_{13} + U_{23}$

$$= \int_{\infty}^{r_{12}} \frac{Q_1Q_2}{r^2_{12}}\, dr_{12} + \int_{\infty}^{r_{13}} \frac{Q_1Q_3}{r^2_{13}} dr_{13} + \int_{\infty}^{r_{23}} \frac{Q_2Q_3}{r^2_{23}} dr_{23}$$

$$= \frac{Q_1Q_2}{r_{12}} + \frac{Q_1Q_3}{r_{13}} + \frac{Q_2Q_3}{r_{23}}.$$

Here, of course, we leave out of account any energy that may have been spent in initially creating these charges. We are actually not concerned with that but only the ready made charges, so to speak, as given to us, in case we have n masses or particles in the system, the total potential energy or the gravitational self energy of the n-particle system will be given by

$$U_s = -G\sum \frac{m_i m_j}{r_{ij}}$$

all pairs

$i \neq j$

where the summation, extends over all pairs of masses i and j excluding the particular case $i = j$, which is not a pair at all but a case of self-interaction which contributes nothing to the potential energy of the system.

Or, to indicate more explicitly that *each pair of masses i, j is to be counted only once* (*e.g.*, either 3, 2 or 2, 3), we may put the above expression in the form

$U_s = -G\sum_{i>j}^{n}\sum_{j=1}^{n} \frac{m_i m_j}{r_{ij}}$. And, similarly, the expression for the

System of charges in the form, $U_s = -\sum_{i>j}^{n}\sum_{j=i}^{n} \frac{Q_i Q_j}{r_{ij}}$,

where the first summation refers to all values of i greater than j, *ensuring that the same pair is not counted twice.*

Alternatively, we may count *all possible pairs*, which obviously means counting each pair twice, first as i, j and then as j, i (*e.g.*, 2. 3 and 3, 2), omitting, however, the case $i = j$ for the reason explained.

Thus, ***the number counted is twice the actual number of pairs***. To correct for it we put 1/2 before the expression for U_s obtained. So that. now it takes the form

$$U_s = -\frac{1}{2}G\sum_{i=1}^{n}\sum_{\substack{j=1\\ \neq i}}^{n}\frac{m_i m_j}{r_{ij}}$$

Similarly, for a system of n charges, we have *potential energy or electrostatic*

given by
$$U_s = -\frac{1}{2}\sum_{i=1}^{n}\sum_{\substack{j=1\\ \neq i}}^{n}\frac{Q_i Q_j}{r_{ij}}.$$

A positive sign would indicate that the charges are all similar, *i.e.*, either all positive or all negative.

It must be emphasised very clearly here that the forces involved being conservative, it does not matter which of the masses or charges we move first or last and by what path We bring them into their assigned positions in the system. *The potential energy or the self energy of the system or the work done in assembling it will always be the same so long as its configuration remains the same.*

It follows, therefore, that the potential energy (U_s), given by the expressions above, is the potential energy of the system, as a whole, and cannot be assigned proportionately or otherwise to the individual masses or charges, comprising the system. It is, in fact, a sort of *binding energy* which keeps the masses or the charges bound into a system of this particular configuration.

So that, just as we associate the elastic potential energy of a spring with its compressed or stretched configuration and consider it to be stored up in the spring itself, so also we must associate the potential energy or the self energy of a system of masses or charges with the configuration of the system and consider it to be stored up in their gravitational or electrostatic field respectively. Any change in the configuration of the system (of the spring, the masses or the charges) will thus inevitably result in a change in its potential energy.

Further, the negative sign of the expression for the P.E. or self energy (U_s) of the system indicates that energy is liberated when the masses are brought into their respective assigned positions in the system. And the positive sign of the expression for U_s in the case of a system of similar charges indicates that energy has to be supplied to the charges to be brought into their respective assigned positions in the system.

GRAVITATIONAL SELF ENERGY OF A BODY

Just as the gravitational self energy of a multi-particle system is its potential energy or the work done in assembling the particles into a system of a given configuration, the gravitational self energy of any material body may be defined as its potential energy or the amount of work done in assembling the body from its infinitesimal particles initially supposed to lie infinite distance apart from each other. It *is thus the energy of the total mass content of the body.*

If there be n such infinitesimal particles which, assembled together, comprise the body, we have, as explained under.

gravitational self energy of the body,

$$U_s = -\frac{1}{2} G \sum_{i=1}^{n} \sum_{\substack{j=1 \\ \neq i}}^{n} \frac{m_i m_j}{r_{ij}},$$

the negative sign again indicating that as the particles assemble together to form the body, this much energy is released which may get converted into kinetic energy of the particles and ultimately into heat energy and get dissipated in the form of radiation.

PRINCIPLE OF EQUIVALENCE

It has already been mentioned in the article above how the mass of a body, when *subjected to a gravitational attraction but no acceleration,* (*i.e., its gravitational mass*) comes out to be the same as when it is *subjected to an acceleration but no gravitational attraction* (*i.e., its inertial mass*). *This gave Einstein the idea that a gravitational field can be imitated by a field of acceleration and this,* ultimately, led to the formulation of his *general theory of relativity,* wherein he showed that a *non-accelerating or inertial frame of reference in which there is a gravitational field is physically equivalent to a reference frame accelerating uniformly with reference to the inertial frame but in which there is no gravitational field.* This means, in other words, that experiments carried out in the two frames, under the *same* condition, will yield identical results. This is called the *principle of equivalence.*

Thus, if we have two reference frames S and S', the former an *inertial* or *non-accelerating* frame on the surface of the earth, so as to have a uniform gravitational field in it, and the latter, a frame of reference moving away from the earth with an acceleration –g with respect to S into far space where there is no gravitational field in it, all physical phenomena occurring in the two frames of reference will be identical. For instance, just as an object dropped in the inertial frame S will fall freely with an acceleration g with respect to it, so also will an object dropped in frame S' fall freely with an acceleration g with respect to it; or, just as a person sitting in frame S will experience an upward force equal to his weight mg by way of reaction of the floor, so will a person sitting in frame S' experience an identical reaction equal to his weight. So that, if the frame S be a closed ship, a person inside it would not be able to say from any observations he makes inside it whether he is in a uniformly accelerating frame without a gravitational field or in a non-accelerating frame with a uniform gravitational field.

Einstein thus showed that just as, in accordance with his special theory of relativity, we can only talk of *relative*, and *not absolute*, velocity of a frame of reference, so also in accordance with the *principle of equivalence*, we can only talk of relative and *not absolute* acceleration of a frame of reference.

The principle of equivalence thus demands the equality of inertial and gravitational mass. *Indeed, this equality of inertial and gravitational mass is itself sometimes referred to as the principle of equivalence.*

It thus follows from this principle that in a reference frame accelerating towards the inertial frame of the earth, with an acceleration g (*e.g.*, a *falling lift*), all particles, originally subject to its gravitational field, will become free because of the gravitational effect getting cancelled by the effect due to acceleration. This explains the *weightlessness* of a person inside an earth satellite.

CENTRAL FORCES—INVERSE SQUARE LAW FORCES

A *central force between two particles is one which is directed along the line joining the two particles and whose magnitude is a function of the distance r between them.*

Thus, if r be the distance between the two particles, the *central force* operating between them is given by $F = F_{(r)}\, r = F_{(r)} r/r$, where $F_{(r)}$ is a *function of distance* r and r or r/r, the *unit vector along* r.

This central force is –ve or + *ve, i.e.,* of attraction or repulsion, according as $F_{(r)} < 0$ or >0.

Now, if one of the particles be fixed in its position, the central force acting on the other is $F = F_{(r)}\, r/r$. If, therefore, J be the *angular momentum* of the second particle about the fixed particle or the origin, we have *torque acting on it,* $\pi = dJ/dt = r \times F = r \times F_{(r)}\, r/r = 0$.

Since the torque $\pi = 0$, it is clear that J = constant.

In other words, *if a particle be moving under the influence of a central force, its angular momentum* (J) *remains conserved,* (meaning, obviously, that there is no change in its magnitude and direction).

Further, *the constancy of* J *also implies that the motion of the particle remains confined to a plane.* For if ρ be the linear momentum of the moving particle, we have $J = r \times p$. So that, $r \cdot J = r \cdot (r \times p)$.

Or, since in a triple vector product, we may interchange the *dots* and *crosses*, we have $r \cdot J = (r \times r) \cdot p = 0$.

[∵ the cross product $r \times r = 0$.

This shows clearly that *vectors* r *and* J *are perpendicular to each other or that the motion of the particle is confined of a plane.*

The general form of a central force is represented by what may be called the *inverse* n^{th} *power law,* viz., $F = \frac{C}{r^n} r$, where C is a *constant.*

The forces is –ve or attractive or + ve or *repulsive* according as $C < 0$ or > 0.

Since a central force is a *conservative* force, we have

$F_{(r)} = -\text{grad } U = -\frac{du}{dr} = \frac{C}{r^n}$, where U is the P.E. of the particle.

So that, $dU = -\frac{C}{r^n} dr$, whence, on integration, we have

$U_{(r)} = \frac{C}{(n-1)\, r^{(n-1)}} + C_1$, where C_1 is a constant of integration.

Since at $r = \infty$ P.E. = 0, we have $C_1 = 0$. And, therefore,

$$U_{(r)} = \frac{C}{(n-1)\, r^{(n-1)}}.$$

Now, if n = –l and C negative, we have $F_{(r)} = -Cr$ and $U_{(r)} = \frac{C}{2}r^2$, which as we know, is the case of a *simple harmonic oscillator*, the force being always directed towards the fixed point or the origin.

And, if n = 2, we have $F_{(r)} = C/r^2$ and $U = C/r$, where C may be positive or negative if the force be an electrostatic one, depending upon the same or opposite signs of the two point-charges, and *always* negative if the force be a gravitational one.

Clearly, in this case, the force between the two particles (or point-charges) is *inversely proportional to the square of the distance between them.* This force is, therefore, known as the *central inverse square law force,* or more commonly, simply as the *Inverse square law force* and the law pertaining to it, the *Inverse square law,* viz., that *the force between two particles (or two point-charges) is inversely proportional to the square of the distance between them.*

This is the one law that occurs most frequently in the entire field of Physics. One notable example of it is the force of gravitational attraction between two masses m_1 and m_2, viz., $F = -\frac{m_1 m_2}{r^2} G\,r$, which is the familiar *Newton's law of gravitation*, and where the constant C $= m_1 m_2 G$, *the force being always one of attraction.*

Another equally familiar example is *Coulomb's law of electrostatic attraction or repulsion* (in *C.G,S. Gaussian units*) *between two point-charges q_1 and q_2* (in air or vacuum), viz., $F = \frac{q_1 q_2}{r^2} r$, where the constant C is obviously equal to $q_1 q_2$, its sign being +ve or –ve according as the force is one of repulsion or attraction, *i.e.,* according as the two charges are of the same or opposite signs.

The inverse square law of force may also be expressed as the *inverse first power law of potential energy.* For, the inverse square law force (F) being a *central, and hence a conservative, force,* we have $F_{(1)} = -dU/dr = C/r^2$, whence, $dU = -(C/r^2)$, which, on integration, gives $U = \frac{C}{r} + C_1$, where, C_1 is a *constant of integration.*

Since we assume P.E. to be *zero* for an infinite distance between the two particles, the constant $C_1 = 0$ and we, therefore, have $U_{(r)} = C/r$, *i.e.,* the *potential energy* of system of two particles at a separation of r is C/r.

Thus, if $C = -m_1m_2G$, we have *gravitational potential energy of a system of two particles of masses* m_1 and m_2 given by $U = -m_1m_2 G/r$ And, if $C = q_1q_2$, we have *electrostatic potential energy of a system of two charges* q_1 and q_2 *given by* $U = q_1q_2r$ *in C.G.S. Gaussian units.*

GRAVITATIONAL SELF ENERGY OF A UNIFORM SOLID SPHERE

The gravitational self energy (U_s) of a uniform solid sphere is obviously equal to the amount of work done in assembling together its infinitesimal particles initially lying infinite distance apart, so that it may be obtained from the expression for U_s above .This will, however, be somewhat radius to first convert the two summations into integrals and then perform multiple integrations.

We shall, therefore, adopt a simpler method and imagine the sphere to be formed by continuous deposition of mass particles (brought, of course, from ∞) in the form of successive spherical shells around an inner spherical core of radius r until it becomes a full-fledged solid sphere of radius R, (Fig. 2.26).

If ρ be the density of the material of the sphere, and hence of the inner spherical core, we have

mass of the inner spherical cores = *its volume* $\times$ *density* $= \frac{4}{3}\pi r^3\rho$.

And, if the *thickness of the spherical shell* deposited on it be dr, we have

mass of this shell = *its surface area* $\times$ *its thickness* $\times$ *density*

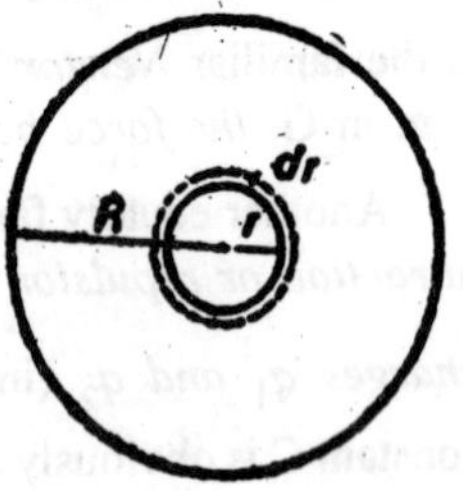

Fig. 2.26

$$= 4\pi r^2 dr\rho.$$

$\therefore$ *change in P.E. or self energy of the core due to this additional mass deposited on it,* say, dU_3 = *potential of the core (i.e., its P.E. per unit mass)* $\times$ *additional mass.*

$$= -\frac{\text{mass of the core}}{\text{radius of the core}} G \times 4\pi r^2 dr\,\rho = \frac{-\frac{4}{3}\pi r^3\rho\,(4\pi r^2 dr\,\rho)\,G}{r}$$

$$= -\frac{16}{3}\pi^2\rho^2 G r^4 dr.$$

This, then, is the *increase in the self energy of the core* as its radius increases, from r to (r + dr).

The integral of this expression for dU_s between the limits r = 0 and r = R, then gives the self energy of the whole sphere on being built from the very start, *i.e.*,

$$U_s = \int_0^R \frac{16}{3}\pi^2\rho^2\, Gr^4 dr = -\frac{16}{3}\pi^2\rho^2 G\left[\frac{r^5}{5}\right]_0^R = -\frac{16\pi^2\rho^2 R^5 G}{15}$$

$$= -\frac{3}{5}\left(\frac{16}{9}\frac{\pi^2 R^6\rho^2}{R}\right)G = -\frac{3}{5}\frac{\left(\frac{4}{3}\pi R^3\rho\right)^2}{R}G.$$

Since $\frac{4}{3}\pi R^3 P = M$, the *mass of the solid sphere,* we have

gravitational self energy of the solid sphere, $U_s = -\frac{3}{5}\frac{M^2}{R}G.$

As mentioned earlier, under the –ve sign of the expression for U_s indicates that this much energy is actually evolved or released during the process of assembling the solid sphere.

TWO-BODY PROBLEM REDUCED TO ONE-BODY PROBLEM

A two-body problem, involving central forces, can always be reduced to the form of a one-body problem, thereby greatly simplifying calculations. This will be seen from the following.

A central force between two point-particles, or two uniform spherical bodies, as we know, is one whose magnitude depends only upon their separation, *i.e.*; is a function of the distance between the two particles (or the centres of the two spherical bodies) and which is directed along the line joining them, the force being reckoned *negative or positive*, according as it is one of *attraction* or *repulsion.*

Further, a central force may, in general, be *any function* of the distance between the two particles (or spherical bodies) and not necessarily the familiar inverse square law force, and may, of course, be gravitational or electrostatic.

We shall now proceed to consider the motion of a system of two particles, subjected to such a force,—an obviously two-body problem,—and see how it may be reduced to a one-body problem.

Suppose we have two particles of masses m_1 and m_2, whose instantaneous position vectors in an inertial frame are r_1 and r_2, with respect to the origin O, (Fig. 11.27)

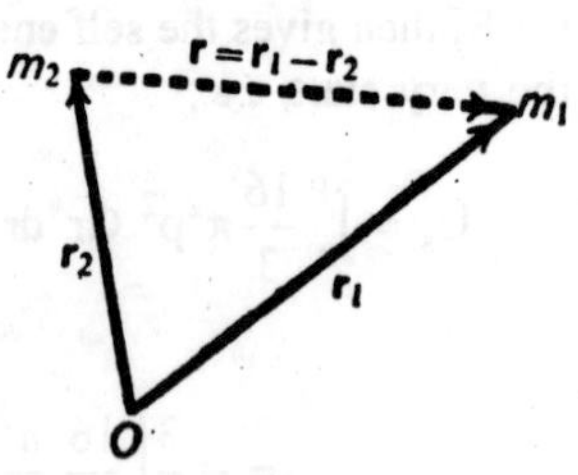

Fig. 2.27

Then, clearly, vector distance of particle m_1 from particle m_2 is $r = r_1 - r_2$.

So that, *central force exerted by the second particle on the first, i.e.,* F_{12} is given by the relation

$$F_{12} = m_1(d^2r_1/dt^2), \quad ...(i)$$

And, *central force exerted by the first particle on the second. i.e.,* Figs. given by the relation

$$F_{12} = m_2(d^2r_2/dt^2), \quad ...(ii)$$

both the forces acting along the line joining m_1 and m_2.

Since, there is no external force acting on the system, we have, in accordance with Newton's third law of motion, $F_{12} = -F_{21} = F = Fr$, (r being the unit vector along r). The magnitude F of the force F may be any function of distance r between m_1, and m_2.

We may, therefore, express relations (i) and (ii) in the forms

$$\frac{d^2r_1}{dt^2} = \frac{F}{m_1} = \frac{Fr}{m_1} \quad ...(iii)$$

and
$$\frac{d^2r_2}{dt^2} = -\frac{F}{m_2} = -\frac{Fr}{m_2} \quad ...(iv)$$

Subtracting relation (iv) from relation (iii), we have

$$\frac{d^2r_1}{dt^2} - \frac{d^2r_2}{dt^2} = \frac{F}{m_1} + \frac{F}{m_2}.$$

Or
$$\frac{d^2}{dt^2}(r_1 - r_2) = \left(\frac{1}{m_1} + \frac{1}{m_2}\right) F$$

$$= \left(\frac{1}{m_1} + \frac{1}{m_2}\right) Fr. \qquad ...(v)$$

Or, putting $\left(\frac{1}{m_1} + \frac{1}{m_2}\right) = \frac{1}{\mu}$,

where, $\mu = \frac{m_1 m_2}{m_1 + m_2}$, called the *reduced* mass of m_1 and m_2 and $r_1 - r_2 =$ we have

$$\frac{d^2r}{dt^2} = \frac{F}{\mu} = \frac{Fr}{\mu} \qquad ...(vi)$$

which may also be put as $F = Fr = \mu \frac{d^2r}{dt^2}$. ...(vii)

This equation (vii), it will readily be seen, is the equation of motion of a particle of mass μ at a vector distance r from one of the particles relative to the other (in our case m_2, because r is the vector distance from m_2 to m_1), *i.e.*, as though the latter (m_2) were fixed at the origin 0 of the inertial frame and exerting a force of attraction F = Fr on the former (m_1). In other words, *considering* m_2 *as a fixed centre which exerts a force* F = Fr *on* m_1 *at vector distance r from it, we can obtain the relative motion of m_1* (*with respect to* m_2) *by using* μ *in place of* m_1 *as the mass.*

Similarly, *considering m_1 as a fixed centre which exerts a force* F = –Fr on m_2 *at vector distance* –r *from it (because the distance from* m_1 to is –r), *we can obtain the relative motion of* m_2 (*with respect to* m_1 *by using μ, in place of m_2 as the mass.*

We have thus been ablè to reduce a two-body problem (of the motion of two particles or bodies of masses m_1 and m_2), involving a central force, to a one-body problem of the motion of a single body of mass μ, *i.e.*, the problem of calculating two position vectors r_1 and r_2 of the two particles or bodies (m_1 and m_2) has been reduced to the calculation of a single vector distance r as a function of time.

It may be noted, however, that if we solve relation (vi) above for r, we can also obtain the values of r_1 and r_2. For, at any given instant, $r_1 - r_2 = r$ and $m_1r_1 + m_2r_2 = (m_1 + m_2) R_{c.m.}$, where $R_{c.m.}$, is the position vector of the centre of mass of the system. So that, solving these two equations, we have and $$r_1 = R_{c.m.} + \frac{m_2 r}{m_1 + m_2} = R_{c.m.} + \frac{\mu}{m_1} r$$

and $$r_2 = R_{c.m.} - \frac{m_1 r}{m_1 + m_2}$$

$$= R_{c.m.} - \frac{\mu}{m_2} r \text{ in the inertial frame,}$$

$$r_1 = \frac{\mu}{m_1} r$$

$$r_2 = \frac{\mu}{m_2} r \text{ in the ceatre-of-mass frame}$$

[$\because$ $R_{c.m.}$ is then *zero*].

Further, as we know, in the absence of any external force acting on the system and, therefore, its total linear momentum remaining conserved, the velocity of the centre of mass remains constant in the inertial frame and, in the centre-of-mass frame, *zero*. The two particles or bodies, therefore, move around the centre of mass in such a way that the vector distance r between them changes in the same manner as the vector distance of a single particle or body of mass μ would do when moving around a fixed centre. The actual paths of two particles or bodies will, of course, depend on the law of force operating (*i.e.*, inverse square law or any other) and on initial positions and velocities.

In the case of rotational motion around the centre of mass, the two particles or bodies will move with identical angular velocities, such that the distance r between them remains intact Fig. 2.28 throughout, as shown in

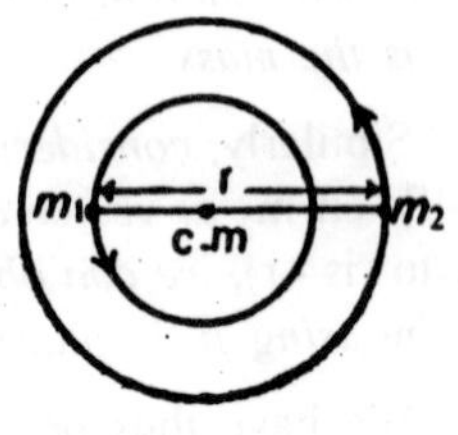

Fig. 2.28

REDUCED MASS

We have mentioned under that $(m_1 m_2)/(m_1 + m_2)$ is referred to as the *reduced mass* μ of the two particles or bodies of masses m_1 and m_2. This is obviously so because the introduction of $1/\mu$ in place of $(1/m_1 + 1/m_2)$ in relation (v) there reduces a two-body problem into a simpler one-body problem. And perhaps also because the reduced mass of two particles is always less than the mass of either particle, *i.e.*, $\mu < m_1$ or m_2. This may be easily seen from the fact that if $m_1 = m_2 = m$, their reduced mass $\mu = m \times m/(m + m) = m/2$, *i.e.*, *equal to half the mass of either particle.*

And, if one of the masses, say, m_1, be very much smaller than the other (m_2), we have $\mu = \dfrac{m_1 m_2}{m_1 + m_2} = m_1\left[\dfrac{1}{(m_1/m_2)+1}\right]$

$$= m_1\left(1+\frac{m_1}{m_2}\right)^{-1} = m_1\left(1-\frac{m_1}{m_2}\right).$$

Or, since $m_1/m_2 \approx O$ (m_1 being $<< m_2$), we have $\mu \approx m_1$, showing that *the value of the reduced mass tends to be nearest to the value of the smaller mass.*

This may perhaps best be seen if we calculate the *reduced mass of a hydrogen atom* which consists, as we know, of a single proton in the nucleus with a single electron going round it, where a proton is 1836 times heavier than an electron. If, therefore, m_e, be the *mass of the electron* and m_ρ, *that of the proton,* we have

reduced mass of the hydrogen atom, $\mu\,H \approx m_e\left(1-\dfrac{m_e}{m_\rho}\right)$

$$\approx m_e\left(1-\frac{1}{1836}\right) \approx m_e,$$

i.e., nearly the same as that of the electron

On the other hand, the *reduced mass of the positronium atom,* a short-lived combination of a positron (having the *same mass as an electron* but a positive charge) and an electron is given by

$$\mu_\rho = \frac{m_e m_e}{m_e + m_e} = \frac{1}{2} m_e,$$

i.e., equal to *half the mass of an electron* and, therefore, *nearly half the reduced of a hydrogen atom.*

Thus, in calculating the time-period of an electron or its energy state etc. as it moves in its orbit around the nucleus, we must, *in the case of hydrogen,* imagine a mass μ_H revolving round the fixed proton and, *in the case of positronium,* a mass μ_ρ revolving round the fixed positron, rather than a mass m_e.

Since, however, $\mu_H \approx m_e$, the error involved in the case of hydrogen in taking the revolving mass to be m_e (the actual mass of the electron) instead of μ_H will only be a marginal one. But, since μ_ρ is only half the actual mass of the electron, *i.e.,* m_e, the error involved in the case of positronium, in taking the revolving mass to be m_e instead of μ_ρ will obviously be quite considerable.

Incidentally, the fact of the reduced mass of positronium being half that of hydrogen is amply borne out by examining their respective spectra. For, the frequency (v) of a spectral line in a hydrogen spectrum is, *in accordance with Bohr's theory,* given by the relation

$$V = \frac{2\pi^2 e^2 \mu}{h^3}\left(\frac{1}{n^2} - \frac{1}{m^2}\right),$$

where h is the wen known *Planck's constant* (= 6.625 × 10^{-27} *erg-sec*), μ the reduced mass of hydrogen, (*i.e.,* equal to μ_H) and n and m, simple integers, such that m>n.

So that, $v \;\alpha\; \mu_H \;\alpha\; m_{e,}$ [because $\mu = \mu_H \approx m_e$]

Now, positronium too, being hydrogen-like in structure, gives a pattern of spectral lines similar to that of hydrogen, with the frequencies of its spectral lines also given by the above relation, where μ is now equal μ_ρ the reduced mass of positronium. We, therefore,

have $v \;\alpha\; \mu_\rho \;\alpha\; m_{e/2}$ [because, as we know $\mu = \mu_\rho \approx me/2$]
indicating clearly that *the frequencies of the spectral lines of positronium are half those of hydrogen.*

And, since wavelength is inversely proportional to frequency ($\because$ c = vλ and therefore, λ = c/v), it follows that *the wavelengths of the spectral lines given by positronium are twice those of the spectral lines given by hydrogen.*

This being an experimental fact, it stands fully confirmed that $\mu p = 1/2\, \mu_{H,}$ *i.e.,* the reduced mass of positronium is half the reduced mass of hydrogen.

ELECTROSTATIC SELF ENERGY OF A CHARGED BODY

Like the gravitational self energy of a material body, the electrostatic self energy of a charged body too is defined as its electrostatic potential energy or the work done in bringing infinitesimal charges, initially lying infinite distance apart, on to it until the charge on it builds up to its given value Q. Again, therefore, if n be the number of such infinitesimal charges, we have, as explained under above, *electrostatic self energy of the charged body in air, i.e.,*

$$U_3 = \frac{1}{2}\sum_{i=1}^{n}\sum_{j=1}^{n}\frac{Q_iQ_j}{r_{ij}},$$

the + ve sign of the expression indicating that here this much energy has to be supplied in thus assembling the charge (Q) on the body.

ELECTROSTATIC SELF ENERGY OF A CHARGED SPHERE

Here, obviously, the sphere may be of (i) a *conducting material* (say, a metal) or (ii) a *non-conducting material.* We shall consider both the cases separately.

(i) *When the Sphere is of a Conducting Material* : In this case, as we know, the charge on the sphere resides only on its outer surface and is not, therefore, distributed uniformly throughout its volume. It thus functions as a charged spherical shell.

To obtain an expression for its self energy (or P.E.), consider infinitesimal charges, dQ each, to be brought to it from their initial positions infinite distance apart. Then, work done in increasing the charge on the sphere from Q to (Q + dQ) is clearly equal to the *potential of the sphere* × dQ = (Q/C) × dQ, where C is the capacitance of the conducting sphere.

Therefore work done in raising the charge on the sphere from 0 to Q,

i.e., self energy of the sphere. $U_s = \int_0^Q \frac{Q}{C}dQ = \frac{Q^2}{2C}$.

Since for a spherical conductor, C = R, we have

self energy of the conducting sphere $= \frac{1}{2}\frac{Q^2}{R}$.

(ii) *When the sphere is of a non-conducting material:* Here, the charge on the sphere is distributed uniformly throughout its volume. So that, proceeding exactly as in the case of gravitational self energy of a uniform solid sphere where the mass of the sphere is distributed uniformly throughout it volume, and taking ρ as the volume density of the charge on the sphere (so that $Q = \frac{4}{3}\pi R^3\rho$), we have *electrostatic self energy of the non-conducting sphere* $= \frac{3}{5}\frac{Q^2}{R}$

CLASSICAL RADIUS OF THE ELECTRON

We have just seen under above how the self energy of a charged sphere works out to be $\frac{1}{2}\left(\frac{Q^2}{R}\right)$ if it be a *conducting one* and $\frac{3}{5}\left(\frac{Q^2}{R}\right)$ it be a *non-conducting one.*

In the absence of any accurate knowledge about the structure of an electron, we cannot possibly decide on any one of these two expressions as representing the self energy of the electron. We, therefore, tentatively take it to be $\approx \frac{Q^2}{R}$, a factor common to both the expressions.

Since, the *charge on an electron* is e and *its classical radius*, r_0 we have Q = e and R = r_0. So that, *self energy of the electron,* $\approx \frac{Q^2}{R}$

Now, in accordance with *Einstein's mass-energy relation*, the energy E, associated with a mass m is given by $E = mc^2$, where c is the velocity of light in free space, equal to 3×10^{10} cm/sec.

We, therefore, have $e^2/r_0 = mc^2$, whence, $r_0 = e^2/mc^2$.

Since e (the charge on an electron) = 4.8×10^{10} *esu* and m (mass of an electron) = 9.1×10^{-28} gm, we have *classical radius of the electron.*

$$r_0 = \frac{\left(4.8 \times 10^{-10}\right)^2}{9.1 \times 10^{-28} \times \left(3 \times 10^{10}\right)^2} \approx 2.80 \times 10^{-13} \text{ cm.}$$

This length r_0 is regarded to be a *fundamental length.*

FUNDAMENTAL LENGTHS AND NUMBERS

In physics we come across quite a few constants, dimensional as well as non-dimensional, various combinations of which give what are regarded as *fundamental lengths and numbers.*

Among familiar examples of *dimensional constants* are G, the *gravitational constant*, having dimensions $M^{-1} L^3T^{-2}$, c, the *velocity of light*, having dimensions LT^{-1} or M^0LT^{-1} and h, *Planck's constant*, having dimensions of energy × time or ML^2T^{-1} etc.

And, although the electronic charge e has no dimensions in *mass, length* and *time,* e^2 has the dimensions of *energy* × *length* or $ML^2T^2 \times L = ML^3T^{-2}$, because, as we have seen in above, e^2/r_o is taken to be the self energy of an electron and has thus the dimensions of energy. Obviously, therefore, e^2 has dimensions of *energy* × *length*.

Among the *non-dimensional constants* may be mentioned π and *Reynold's number* k. These non-dimensional constants as well as expressions like $T\sqrt{\frac{g}{l}}$ in the case of a simple pendulum (which too have no dimensions and are pure numbers) are usually referred to as *numerics*.

Now, let us see how various combinations of these enable us to obtain other fundamental lengths and numbers.

The *fine structure constant* α. The name *fine structure* for this constant is derived from its having been used first in connection with seemingly single spectral lines observed to consist of (or split into) thinner lines, thus revealing their *fine structure*. Its value can be obtained from e, $\hbar$ and c, where $\hbar = h/2\pi$ (and read as h bar). Thus,

$$\alpha = \frac{e^2}{\hbar c} = \frac{e^2}{\left(\frac{h}{2\pi}\right)c} = \frac{2\pi e^2}{hc} = \frac{1}{137.04}.$$

So that, dimensions of α are those of $\frac{\text{energy} \times \text{length}}{(\text{energy} \times \text{time})(\text{length} / \text{time})}$, *i.e.*, it has no dimensions in M,L and T.

It is thus a *non-dimensional constant* (or *numeric*) and its great importance lies in the fact that quite a number of important fundamental lengths can be obtained by dividing the classical radius of an electron, r_0 by various powers of α.

Thus, for example, one such fundamental length we often meet with in Quantum physics is the *Crompton wavelength of an electron*, λe (read as λ bar and equal to λ/2π). It is obtained by dividing r_e by the first power of α Thus,

$$\textit{Crompton wavelength of an electron}, \lambda e = \frac{r_0}{\alpha} = \frac{e^2}{mc^2} \bigg/ \frac{2\pi e^2}{hc},$$

[∵ $r_0 = e^2/mc^2$, where m is the mass of the electron.]

$$\text{Or, } \lambda e = \left(\frac{e^2}{mc^2}\right)\left(\frac{hc}{2\pi e^2}\right) = \frac{h}{2\pi mc} = 3.86 \times 10^{-11} \text{cm}.$$

Another important fundamental length thus obtained is the *Bohr's radius of the ground state of hydrogen* which is in effect, the *radius of the hydrogen atom*, a_0. It is given by r_0/α^2. So that,

$$\alpha_0 = \frac{r_0}{\alpha^2} = \frac{e^2/mc^2}{\left(2\pi e^2/hc\right)^2} = \frac{e^2}{mc^2} \times \frac{h^2c^2}{4\pi^2 e^4} = \frac{h^2}{4\pi^2 me^2} = 0.529 \times 10^{-8}\ \text{cm}.$$

Yet another important length we may mention here is what is called the *gravitational length of a body,* (R_0), such that the gravitational self energy of the body (of mass M) is given by M^2G/R_0.

To obtain the value of R_0 for the given body, we equate its self energy M^2G/R_0 against Mc^2.

So that, $M^2G/R_0 = Mc^2$, whence, $R_0 = MG/c^2$.

Sometimes, Crompton wavelength of an electron is taken to be $2\pi\lambda c$.

By way of an example, let us calculate the *gravitational length of the sun.* Taking the mass of the sun to be $\approx 2 \times 10^{33}$gm, we have

$$R_0 = \frac{2 \times 10^{33} \times 6.7 \times 10^{-8}}{\left(3 \times 10^{10}\right)^2} \approx 10^5\ \text{cm},$$

which, it may be noted, is very much smaller than the radius of the sun $\approx 7 \times 10^{10}$ cm.

Similarly, let us calculate the *gravitational length of our known universe,* whose mass can be obtained from the total number of nucleons (*i.e., protons and neutrons*) in it, which is estimated to be 10^{80}. Since each nucleon weighs nearly 10^{24}gm, the *total mass of the universe* (M) comes to $10^{80} \times 10^{-24} = 10^{56}$gm.

$$\therefore \quad R_0 = \frac{10^{56} \times 6.7 \times 10^{-8}}{\left(3 \times 10^{10}\right)^2} \approx 10^{23}\ \text{cm},$$

which is near about the same as its estimated radius.

It is interesting to observe that in accordance with the general theory of relativity, *no photons can escape from the surface of a body whose radius* (R) *is smaller than, or equal to, the gravitational length* (R_0) for it and it is, therefore, non-luminous and hence invisible.

Thus, the reason why the sun is intensely visible is that its radius $R \approx 7 \times 10^{10}$ cm being very much larger than its gravitational length

$R_0 \approx 10^5$ cm, there is an enormously large number of photons escaping from it, which make it so dazzlingly luminous.

CASE OF GRAVITATIONAL FORCE (IN A TWO-BODY PROBLEM)

We have seen under above how a two-body problem, in general, where the two bodies, distant r from each other, interact through a central force, may be reduced to a one-body problem by means of the relation $F = \mu(d^2r/dt^2)$. Let us consider it now in the particular case of *gravitational force as the central force.*

As we know, the gravitational force of attraction between two particles or bodies of masses m_1 and m_2 is given by

$F = -G\dfrac{m_1m_2}{r_2}\hat{r}$ where $\hat{r}$ is the unit vector along r, the distance of particle m_1 from particle m_2. We thus have

$$F = Fr = \mu\left(d^2r/dt^2\right) = -G\frac{m_1m_2}{r^2}r,$$

Here also, the motion of particle m_1 with respect to particle m_2 would be the same as that of a particle of *reduced mass* μ [of m_1 and m_2, *i.e.*, $m_1m_2/(m_1 + m_2)$] under the force of attraction of a fixed mass or a fixed centre m_2 at a vector distance r from it, and may thus be obtained by *replacing* m_1 by μ.

We cannot, however, replace m_1 by μ in the expression for force above, *i.e.*, we cannot take the central force between the two particles to be $-G\dfrac{\mu m_2}{r^2}\hat{r}$ for the simple reason that the force between the two particles is not merely a function of the distance r between them but also involves the masses of both the particles. The actual masses (m_1 and m_2) of the two particles have, therefore, to be used in the expression for force. So that,

$$\mu\frac{d^2r}{dt^2} = -G\frac{m_1m_2}{r^2}\hat{r}.\text{Or,}\frac{d^2r}{dt^2} = \frac{1}{\mu}\frac{G\,m_1m_2}{r^2}\hat{r}.$$

Or, substituting the value of μ, we have

$$\frac{d^2r}{dt^2} = -\left(\frac{m_1+m_2}{m_1m_2}\right)\frac{G\,m_1m_2}{r^2}\hat{r} = -G\frac{m_1+m_2}{r^2}r$$

$$= -F'\hat{r}, \text{say, where, } F' = G\left(\frac{m_1 + m_2}{r^2}\right),$$

This is clearly the equation of motion of a particle of unit mass, or negligible mass, compared with $(m_1 + m_2)$, at a vector distance r (equal to the distance between the two particles) from a fixed mass $(m_1 + m_2)$ exerting a force of attraction on it.

This is the reason why, when calculating the angular velocity of a satellite, of mass m, going round the earth, of mass M, we must take the total mass (M + m) as the value of the attracting mass and not merely the mass of the earth M. Of course, if the mass of the satellite (m) be negligible compared with that of the earth, we may take M as the attracting mass without causing any appreciable error in the result. In cases, however, where the mass of the satellite is not negligible, as for example the earth and the other planets going round the sun, or the moon going round the earth, this will cause considerable error and is not therefore, permissible.

EQUIVALENT ONE-BODY PROBLEM

We have just seen in the article above how, in the case of two bodies of masses m_1 and m_2, interacting through gravitational force as the central force along the line joining them, we may determine the motion of m_1 relative to m_2 by taking m_2 to lie at the fixed origin of an inertial frame and replacing m_1 by their reduced mass μ, *i.e.*, we can take

$$\mu\frac{d^2r}{dt^2} = -G\frac{m_1 m_2}{r^2}\hat{r},$$

where r is the unit vector along r, thus reducing the two-body problem effectively to a one-body problem.

Now, as we know, *for a particle moving about a fixed centre of force, the angular momentum remains conserved.* So that, *the angular momentum in our effective or equivalent one-body problem, viz.,*

$$J = r \times \mu\frac{dr}{dt} \qquad \text{...(i)}$$

must also remain conserved in both magnitude and direction.

Of course, the value of the angular momentum is its usual value for the system of the two bodies whose position vectors with respect to their centre of mass ($R_{c.m.}$) are r_1 and r_2, *i.e.*,

$$J = r_1 \times m_1\, dr_1/dt + r_2 \times M_2\, dr_2/dt. \qquad \text{...(ii)}$$

Since the reduced mass $\mu = m_1m_2/(m_1 + m_2)$, we may write relation (i) in the form

$$J = \frac{m_1m_2}{m_1+m_2}(r_1 - r_2\left(\frac{dr_1}{dt} - \frac{dr_2}{dt}\right)$$

In a centre-of-mass frame, the centre of mass being at rest, $dR_{c.m}/dt = 0$ and, as we have seen before,

$$\frac{dr_1}{dt} = -\frac{m_2}{m_1}\frac{dr_2}{dt} \text{ and } \frac{dr_2}{dt} = -\frac{m_1}{m_2}\frac{dr_1}{dt}.$$

So that, in that case,

$$J = \frac{m_1m_2}{m_1+m_2}(r_1 - r_2)\left(\frac{m_1}{m_2}\frac{dr_1}{dt} - \frac{m_2}{m_1}\frac{dr_2}{dt}\right)$$

Now, angular momentum is said to be conserved when both its magnitude and direction remain constant. Thus, in order that J in expression (i) above may remain constant (or conserved), $r \times dr/dt$ must necessarily remain constant in direction. In other words, the motion must be in a plane, as indicated in Fig. 2.29.

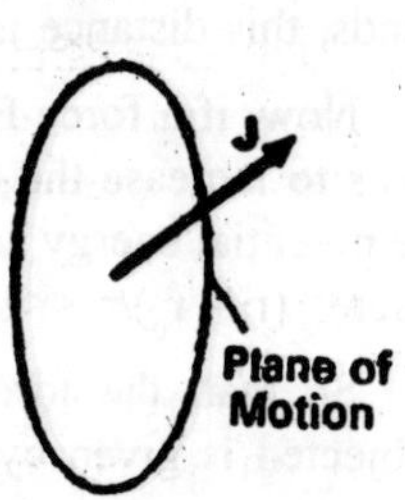

Fig. 2.29

It is found to be easier and more useful to describe the motion of a particle in a plane in terms of *the polar coordinates* r and θ, where r is the unit vector from the fixed origin directed towards the particle (p) and θ, the unit vector in the plane, perpendicular to r, as shown in Fig. 2.30.

The velocity of the particle P will thus have a component along r as also a component along $\hat{\theta}$. So that,

$$\frac{dr}{dt} = \frac{dr}{dt}\hat{r} + r\,\omega\,\hat{\theta}.$$

We may, therefore, write expression (i) for J above as

$$J = r \times \mu\left[\frac{dr}{dt}\hat{r} + r\omega\hat{\theta}\right]$$

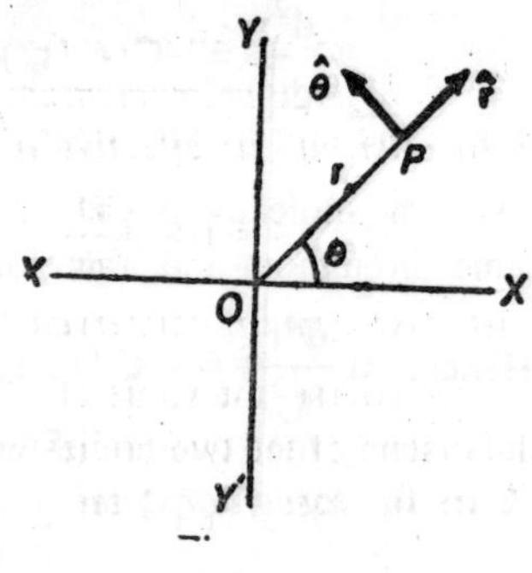

Fig. 2.30

Or, since $r \times \hat{r} = 0$ and $\hat{r} \times \hat{\theta} = \hat{z}$, the unit vector normal to the plane of motion of the particle, we have

$J = \mu r^2 \omega \hat{z}$. Or, $J = \mu r^2 \omega$.

We shall make use of this equivalent one-body problem later in deducing Kepler's Laws.

VIBRATION OF A DIATOMIC MOLECULE

Suppose we have a non-rotating molecule, consisting of two atoms, a distance r_o apart, in their stable equilibrium position, this distance being really the distance between their nuclei, or their *inter-nuclear distance,* since the mass of an atom is almost wholly concentrated in its nucleus. And, since the atoms are linked together by what are called valency bonds, this distance is also referred to as *bond length* of the molecule.

Now, if a force F be applied along the line Joining the two atoms, so as to increase the distance between the two atoms a little to, say, r, the potential energy acquired by the molecule will be $U = 1/2\ (Cr - r_o)^2$, where $(r - r_o)/r << 1$ and C, a constant.

So that, the force to which the two atoms of the molecule are subjected is given by $F = -dU/d(r - r_o) = -C(r - r_o)$.

But we know that the force between the two atoms is gravitational and, therefore, a *central force,* and that in the event of free vibration, although the atoms are in motion, their centre of mass remains constant. Reducing the two-body problem to one-body problem, therefore, we have $F = \mu\ (d^2r/dt^2)$, where μ. is the *reduced mass* of the two atoms (or the molecule) and r, the vector distance between them.

Substituting this value of F in the expression above, we thus have

$$\mu \frac{d^2 r}{dt^2} = -C(r - r_o)r,$$

where r is the unit vector along the direction of r,

Since the molecule is a non-rotating one, the direction of r remains the same throughout and, therefore, $d^2r/dt^2 = (d^2r/dt^2)r$.

Hence, $\mu \dfrac{d^2 r}{dt^2}\hat{r} = -C(r - r_o)\,r$. Or, $\mu \dfrac{d^2 r}{dt^2} = -C(r - r_o)$,

which is clearly the equation of motion of a simple harmonic oscillator, with C as the *force constant.*

The diatomic molecule thus vibrates as a simple harmonic oscillator of *angular frequency* $\omega_0 = \sqrt{\frac{C}{\mu}}$.

ORBITS

Let us first consider the general case of the orbit of a particle of mass m moving under the action of a *central force* F exerted by another *massive* particle of mass M fixed in position at the point O (Fig. 2.31), with r as the position vector of m with respect to O.

Since the force is directed towards O, it has no moment about 0. Hence, by the law of conservation of angular momentum, its angular momentum J about O remains conserved in both magnitude and direction, *i.e., it must move in a plane.*

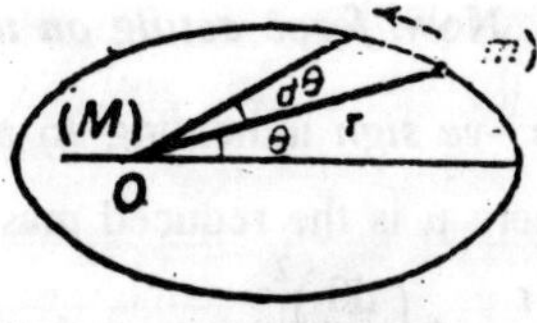

Fig. 2.31

Thus, $J = mvr = m\omega r^2 = m\,(d\theta/dt)r^2 =$ *constant.*

This, it will be noted, is the same result as obtained under (on equivalent one-body problem).

If we denote *angular momentum per unit mass, i.e.,* J/m, by h, we have

$$h = r^2\left(\frac{d\theta}{dt}\right) = \textit{constant}, \quad \text{..(i)}$$

where r and θ are the *polar coordinates* of the particle.

Now, if the position vector (or the radius vector) turns through an infinitesimal angle dθ in time dt, it sweeps out an area $ds = 1/2\ r^2 d\theta$ and, therefore, its areal velocity

$$\frac{ds}{dt} = \frac{1}{2}r^2\frac{d\theta}{dt} \quad \text{...(ii)}$$

Since $r^2(d\theta/dt) = h$, it follows that the *areal velocity of the particle* = h/2.

We thus arrive at the *important result* that, *in motion under a central force, the areal velocity is constant.*

Let us now try to determine the *orbit of the particle.*

We see from relation (i) above that $d\theta/dt$ (or ω) = h/r^2. If, therefore, we put $1/r = u$, we have $d\theta/dt = hu^2$.

So that, $$\frac{dr}{dt} = \frac{d}{dt}\left(\frac{1}{u}\right) = -\frac{1}{u^2}\frac{du}{dt} = -\frac{du}{d\theta}\frac{d\theta}{dt} = \frac{1}{u^2}\frac{du}{d\theta}hu^2$$

or, $$\frac{dr}{dt} = -h\frac{du}{d\theta} \qquad \text{...(iii)}$$

And, therefore, $$\frac{d^2r}{dt^2} = -h\frac{d^2u}{d\theta}\frac{d\theta}{dt} = -h^2u^2\frac{d^2u}{d\theta^2} \qquad \text{...(iv)}$$

Now, *force acting on the particle,* $F = -\mu\left[\frac{d^2r}{dt^2} - r\left(\frac{d\theta}{dt}\right)^2\right]$,

the *–ve sign* indicating an attractive force,

where μ is the reduced mass of the system of particles M and m and $\frac{d^2r}{dt^2} - r\left(\frac{d\theta}{dt}\right)^2$, the acceleration along the radius vector, or the *radial acceleration.*

Since particle M is much more massive than m, we have $\mu \approx m$

So that, $$F = -m\left[\frac{d^2r}{dt^2} - r\left(\frac{d\theta}{dt}\right)^2\right] = -m\left(-h^2u^2\frac{d^2u}{d\theta^2} - rh^2u^4\right)$$

Or, $$F = m\left(h^2u^2\frac{d^2u}{d\theta} + \frac{1}{u}h^2u^4\right) = m\left(h^2u^2\frac{d^2u}{d\theta^2} + h^2u^3\right)$$

$$= mh^2u^2\left(\frac{d^2u}{d\theta^2} + u\right),$$

whence, $$\frac{d^2u}{d\theta} + u = \frac{F}{m\,h^2\,u^2}.$$

Or, if we put F/m, *i.e., force per unit mass = p,* we have

$$\frac{d^2u}{d\theta} + u = \frac{P}{h^2u^2}. \qquad \text{...(v)}$$

This is the *differential equation of the orbit of a particle moving under an attractive central force P per unit mass.*

If we solve this equation, obtaining *u* as a function of θ, we have the equation of the orbit in polar coordinates.

We shall now express relation (v) above in terms of energy. For, if we consider P to be a function of r alone, the work done in passing from one position to another will be quite independent of the path taken and the system will thus be *conservative,* with the principle of energy easily applicable. So that, if T, U and E be the *kinetic, potential* and *total energies* respectively, *per unit mass,* we have

$$T + U = E$$

or $$T = E - U \qquad \text{...(vi)}$$

Now, $T = \frac{1}{2}\left[\left(\frac{dr}{dt}\right)^2 + r^2\left(\frac{d\theta}{dt}\right)^2\right] = \frac{1}{2}h^2\left[\left(\frac{du}{d\theta}\right)^2 + u^2\right]$

And since P = –grad U, the magnitude of its inward component P is given by P = dU/dr, and therefore, $U = \int_{r_0}^{r} P dr$, where r_0 is some constant.

Equation (vi) thus becomes $\frac{1}{2}h^2\left[\left(\frac{du}{d\theta}\right)^2 + u^2\right] = E - u.$

Or, $$\left(\frac{du}{d\theta}\right)^2 + u^2 = \frac{2(E-U)}{h^2}. \qquad \text{...(vii)}$$

This equation is really the same as equation (v) above, as may be easily seen on differentiation, if only we remember that U, being a function of r, is also a function of u.

Particular Case of Inverse Square Law Forces

If the force acting on the moving particle towards the fixed particle or the origin O be an inverse square law force, we have *force per unit mass towards the centre offerce, i.e.,* $P = K/r^2$, where K is a constant.

Our differential equation (v) for the orbit now becomes $(d^2u/d\theta^2) + u = K/h^2$, the general solution of which is

$$u = \frac{K}{h^2} + A\cos(\theta - \theta_0), \qquad \text{...(viii)}$$

where A and θ_0 are constants of integration.

This, in polar coordinates, is the equation of the most general orbit described under a central inverse square law force.

Now, the *potential energy per unit mass,*

$$U = \int P dr = \int\left(\frac{K}{r^2}\right)dr = -\frac{K}{r} = -Ku,$$

the constant of integration having been so chosen that U = 0 at ∞.

From relations (vii) and (viii), therefore, we have

$$A^2 + \frac{K^2}{h^4} + 2A\frac{K}{h^2}\cos(\theta - \theta_0) = \frac{2}{h^2}\left[E + \frac{K^2}{h^2} + KA\cos(\theta - \theta_o)\right]$$

Or $$A^2 + \frac{K^2}{h^4} + \frac{2AK\cos(\theta - \theta_0)}{h^2} = \frac{2E}{h^2} + \frac{2K^2}{h^4} + \frac{2KA\cos(\theta - \theta_0)}{h^2}.$$

Or $$A^2 = \frac{K^2}{h^4} + \frac{2E}{h^2},$$

thus obtaining the value of constant A in terms of the total energy and angular momentum per unit mass.

If we rotate the base line $\theta = 0$ to make $\theta_0 = 0$ and A' > 0, relation (viii) above will assume the form

$$u = \frac{K}{h^2}\left(1 + \sqrt{1 + \frac{2Eh^2}{K^2}}\cos\theta\right) \qquad \text{...(ix)}$$

From the focus-directrix property of a conic, we know that its equation in polar coordinates may be written as

$$u = \frac{1}{l}(1 + e\cos\theta). \text{ Or, } \frac{l}{r} = (1 + e\cos\theta), \quad [\because\ u = 1/r.] \qquad \text{...(x)}$$

where l is the *semi latus rectum* (*i.e.*, half the focal chord parallel to the directrix) and e, the *eccentricity*, with θ measured from the perpendicular dropped on the directrix from the focus.

The conic may be of any one of the following types:

(i) *ellipse,* if e < 1, (ii) *parabola*, if e = 1, (iii) *hyperbola*, if e > I, (in this case, relation (x) gives only the branch adjacent to the focus) and (iv) circle, if e = 0.

It may be noted that relations (ix) and (x) become identical if we choose

$$l = h^2/K \text{ and } e = \sqrt{1 + \frac{2Eh^2}{2}}. \qquad \text{...(xi)}$$

Thus, we arrive at the result that *the orbit described by a particle, attracted by a central inverse square law force towards a fixed point or origin, is a conic having the fixed point or centre of force as focus,* (its semi latus rectum and eccentricity being given by relation (xi) in terms of E and h),

The orbit may take the following forms:

(i) If E < 0 or –ve, it is an *ellipse*, (ii) if E = 0, it is a *parabola* and (iii) if E > 0, it is a hyperbola.

Actually, ***most of the orbits of the planets are ellipses***. A body describing a parabolic or a hyperbolic orbit would pass out of the solar system, never to return. We shall, therefore, consider here only the constants of the elliptical orbit.

Constants of the elliptical orbit: It can be seen at once from relation (x) above that *the shape and size of the orbit are given by the constants l and e but not its orientation in space*. These two constants are related to the constants E and h in the manner shown by expression (xi). So that, we have five constants in all appearing in our various relations deduced above, viz., K, l, e, E and h. Of these, K (the intensity of the force-centre) has obviously a fixed or constant value and is to be known once for all, but the other four constants take on different values for different orbits. Of these four also, only two are really independent, as can be seen from relation (xi). We may, however, use any two of them, as an independent pair to suit our convenience in any given case.

In fact, it is much better to use a and e as our *fundamental constants* instead of *l* and e, where a is the *semi major axis* of the ellipse. Its relationship with *l* is given by

$$l = \frac{b^2}{a} = a\left(1 - e^2\right),$$

where b is the *semi minor axis of the ellipse.*

The constants a and e are usually referred to as *geometrical constants* and constants E and h, as *dynamical constants,* their inter-relationship being as follows:

$$a = -K/2E;\ e = \sqrt{1 + \left(\frac{2Eh^2}{K^2}\right)};\ E = -\frac{K}{2a}$$

and
$$h = \sqrt{Ka\left(1 - e^2\right)}. \qquad \text{...(xii)}$$

Speed of the particle : We have *energy per unit mass* given by

$$E = \frac{1}{2}v^2 - \frac{k}{r}.$$

Or, substituting for E from relation (xii) above,

$$-K/2a = \frac{1}{2}v^2 - K/r, \text{ whence, } v^2 = K\left(\frac{2}{r} - \frac{1}{a}\right) \qquad \text{...(xiii)}$$

Periodic time of the particle : We have seen how the *areal velocity*

$$ds/dt = \frac{1}{2} r^2\theta = h/2.$$

Therefore, time taken by the particle to move right round the orbit, or to sweep the entire area ($s = \pi ab$) of the ellipse

$$= \pi\, ab/\frac{h}{2} = 2\,\pi\, ab/h = 2s/h.$$

And this is obviously, the *time-period of the particle*, τ, say.

So that, $\tau = 2\,\pi\, ab/h = 2\, s/h.$

Or, to express it in terms of the geometrical or the dynamical constants, we note that $b = a\sqrt{1-e^2}$ and from relation (xii) above,

$$h^2 = Ka\,(1-e^2).\ \text{so that,}\quad (1-e^2) = h^2/Ka.$$

$$\therefore \tau^2 = \frac{4\pi^2 a^2 b^2}{h^2} = \frac{4\pi^2 a^2\left[a^2\left(1-e^2\right)\right]}{h^2} = \frac{4\pi^2 a^4}{h^2}\left(\frac{h^2}{Ka}\right) = \frac{4\pi^2 a^3}{K},$$

whence, $$\tau = 2\pi\sqrt{\frac{a^3}{K}}.$$

Or, *in terms of energy*, $$\tau = \frac{2\pi K}{\sqrt{(-2E)^3}}, \qquad \text{...(xiv)}$$

remembering that $E < 0$ for an elliptical orbit.

Thus, the expression for time-period of the particle contains either one geometrical constant (a) or one dynamical constant (E). It will easily be seen from the two expressions for τ that *for orbits, having the same semi major axis, the peviodic time is the same and so also for orbits with the same total energy.*

Circular orbits and their stability: For a circular orbit, clearly, r and, therefore, u is constant. Hence, the possible radii of circular orbits are determined by the relation $u = P(u)/h^2u^2$

Thus, *with force per unit mass, P, directed towards the centre of force, we can obtain a circular orbit of any radius by projecting the particle at right angles to the radius vector (or position vector} with a velocity such that* $h^2 = P(u)/u^3$ ...(xv)

The question, however, arises whether these circular orbits will all be stable, *i.e.*, whether, due to any slight temporary disturbance, which will almost always be there, the orbit would continue to be close to the original circular orbit or will deviate far from it.

Let us put $u = u_0$ and $h = h_0$ *for a circular orbit*, so that relation (xv) becomes

$$h_0^2 = P(u_0)/u_0^3.$$

To study the effect of a disturbance, let us put $u = u_0 + \xi$ where ξ and its derivatives are supposed to be small and so is $(h - h_0)$.

Substituting these values in relation (v) above, we have

$$\frac{d^2\xi}{d\theta^2} + u_0 + \xi = \frac{P(u_0+\xi)}{h^2(u_0+\xi)^2}, \qquad \text{...(xvi)}$$

which, on expansion in powers of ξ, gives

$$\frac{P(u_0+\xi)}{h^2(u_0+\xi)^2} = \frac{1}{h^2 u_0^2}\left(1+\frac{\xi}{u_0}\right)^{-2}(P_0 + \xi P_0' + \ldots)$$

$$= \frac{P_0}{h^2 u_0^2}\left[1+\xi\left(\frac{P'_0}{P_0} - \frac{2}{u_0}\right) + \ldots\right]$$

where $P' = dP/du$ and the subscript 0 indicates evaluation for $u=u_0$.

Then, neglecting higher than the first order of small quantities, relation (XVI) takes the form

$$d^2\xi/d\theta^2 + a\xi = B, \qquad \text{...(.xvii)}$$

where constant $A = 1 - \frac{P_0}{h_0^2 u_0^2}\left(\frac{P_0'}{p_0} - \frac{2}{u_0}\right) = 3 - \frac{u_0 p_0'}{p_0}$, ...(xviii)

and we need not bother about the value of constant B in which we are not at present interested.

Now, expression (xvii) has three solutions (for ξ) corresponding to $A > 0$, $A < 0$ and $A = 0$. Of these, the only solution which remains permanently small corresponds to $A > 0$, the other two increasing indefinitely with θ.

Clearly, therefore, a circular orbit (of radius l/u_0) will be stable *only* if

$$A > 0. \text{Or}, \left(3 - \frac{u_0 P_0'}{P_0}\right) > 0. \text{Or}, \frac{u_0 P_0'}{P_0} < 3.$$

Now, suppose the force obeys the *inverse* n^{th} *power law, i.e.,*

$$P = K/r^n = Ku^n. \qquad \text{(...xix)}$$

Then, clearly, $uP'/P = n$. So that, circular orbits under the force directed towards the centre of force, *i.e.,* under the attractive force given by relation (xix) above) will be stable *only if* $n < 3$.

It follows, therefore, that *circular orbits will be stable for a force varying inversely as the distance or the square of the distance (i.e., inverse square law force) and will be unstable for the inverse cube (or a higher power) law.*

DEDUCTION OF NEWTON'S LAW OF GRAVITATION FROM KEPLER'S LAWS

We have deduced *Kepler's laws* from Newton's laws of motion and gravitation. Let us now take up the reverse problem of deducing Newton's law of gravitation from *Kepler's laws,* for that is how, historically, the law of gravitation was deduced by *Newton.*

Kepler's *second law* tells us that $h = J/m$, the *angular momentum per unit mass* $= r^2\,(d\theta/dt) =$ *constant*. Hence the force on the planet must be directed towards the sun.

And from the *first law*, as we have seen, the equation of the orbit of the planet around the sun may be written as

$$u = \frac{1}{l}\,(1 + e\cos\theta).$$

So that, in accordance with relation (v) *force per unit mass of the planet* is given by $P = h^2u^2\,(d^2u/d\theta^2 + u) = h^2u^2/l$.

Now, *force per unit mass means acceleration.* Therefore, *radial acceleration* of every planet towards the sun is, say,

$$a_r = -\frac{h^2u^2}{l} = -\frac{h^2/l}{r^2} = -\frac{K}{r^2}, \text{ where } K = h^2/l, \text{ a constant.}$$

Thus, *the acceleration, and hence the force acting on a planet,, is inversely proportional to the square of its distance from the sun,* the negative sign merely indicating that it is directed inwards, towards the sun, or that it is a force of attraction.

And, the time-period of a planet, in accordance with Kepler's *third law,* is given by $\tau = \pi ab\,\frac{h}{2}$. Or, $\tau^2 = 4\pi^2a^2b^2/h^2$, where a and b are the semi-major and minor axes of the elliptical orbit respectively.

Since $b^2/a = l$, the *semi-latus rectum* of the elliptical orbit, we have

$$b^2 = al.$$

And, therefore, $\tau^2 = (4\pi^2 l/h^2)\ a^3 = (4\pi^2/K)a^3$,

indicating that $\tau \propto a^3$ *for every planet*, or that $4\pi^2/K$, and hence K, is constant for every planet and thus quite *independent of its nature.*

Hence, if m and M be the masses of a planet and the sun respectively, the force of attraction by the sun on the planet is, say, $F = -Km/r^2$ and its reaction, *i.e.*, the force of attraction by the planet on the sun, say, $F' = -KM/r^2$, where K and k are *constants.*

Since by Newton's third law of motion, $F = F'$, we have

$$Km/r^2 = kM/r^2.$$

Or, $Km = kM$. Or, $K/M = k/m$, a constant, say, G.

So that, $$K = MG.$$

Substituting this value of K in the expression for F we, therefore, have $$F = -\frac{Mm}{r^2}G,$$

showing that the force of attraction between the planet and the sun is *directly proportional to the product of their masses and inversely proportional to the square of the distance between them, which is Newton's law of gravitation*

KEPLER'S LAWS

Kepler, on the basis of the data on the motion of planet *Mars* collected by his chief, *Tycho Brake*, at the Royal observatory at *Prague*, succeeded after 22 years of ceaseless work in evolving the famous three laws, known after him—the first two in the year 1609 and the third, ten years later, in 1619. These laws have an historical importance in that they provided the original experimental evidence on the validity of Newton's laws of mechanics as also his theory of gravitational attraction. The three laws may be stated thus:

First Law

The path of a planet is an elliptical orbit around the sun, with the sun at one of its foci. This is aptly known as the *law of elliptical orbits* and obviously gives the shape of the orbit of a planet around the sun.

Second Law

The radius vector, drawn from the sun to a planet, sweeps out equal areas in equal time, i.e., its areal velocity (or the area swept out by it per unit time) is constant. This is referred to as the *law of areas* and gives the relationship between the orbital speed of the planet and its distance from the sun.

Third Law

The square of a planet's year, i.e., its time-period (or its time of one complete round of the sun) is proportional to the cube of the semi-major axis of its orbit. This is known as the *harmonic law* and gives the relationship between the size of the orbit of a planet and its time of revolution.

Let us discuss these laws on the basis of the theory of orbits we have dealt with in.

We know that in the case of an inverse square law attractive force, the *force per unit mass of the planet towards the centre of the force, (in this case, the sun) is given by* $P = K/r^2$, *where r is now the distance of the planet form the sun* and K, a constant equal to MG, M being *the mass of the sun* and G, the *gravitational constant.*

($\because F = MmG/r^2$ and $\therefore P = K/r^2 = F/m\ MG/r^2$, whence, $K = MG$). where the sun may be imagined at O.

Our differential equation (v) for the orbit thus becomes

$$\frac{d^2u}{d\theta^2} + u = \frac{MG}{h^2},$$

the general solution for which is $u = MG/h^2 + A\cos(\theta - \theta_0)$, where A and θ_0 are constants of integration.

Now θ_0 can be made equal to 0 by rotating the base line $\theta = 0$, with $A > 0$. Hence, the above solution becomes $u = MG/h^2 + A\cos\theta$.

Or, $$1/r = MG/h^2 + A\cos\theta.$$

This may be put as, $$\frac{h^2/MG}{r} = 1 + \frac{h^2A}{MG}\cos\theta.$$

This equation is of the same form as equation (x), of a conic in polar coordinates, *viz.*, $u = 1/l\,(1 + e\cos\theta)$. Or, $l/r = 1 + e\cos\theta$, with l, the *semi latus rectum, equal to* h^2/MG and e, the *eccentricity, equal to* h^2A/MG.

So that, *with e, i.e.,* h^2A/MG *less than* 1, *the orbit of the planet around the sun is an ellipse.* This then, establishes *Kepler's First Law.*

Or, if we obtain the *value of e in terms of the total energy of the system,* we arrive at the same result.

$$l = h^2/K \text{ and } e = \sqrt{\frac{(1+2\,Eh^2)}{K^2}}.$$

So that, here, $l = h^2/MG$ and $e = \sqrt{\dfrac{(1+2\,Eh^2)}{M^2G^2}}$, where E *represents the total energy per unit mass.*

In case E_m be taken to be the total energy of the planet (or the system), we have $E = E_m/m$ in the expression above. Also, remembering that h = J/m and K = MG, we have

$$l = h^2K = J^2m^2MG \text{ and } e = \sqrt{\frac{(1+2\,E_m h^2)}{mM^2G^2}}$$

$$= \sqrt{\frac{(1+2\,E_m J^2)}{mM^2G^2}}, \text{ whence, } E_m = -\frac{m^3M^2G^2}{2J^2}\left(1-e^2\right).$$

So that, if $e < 1$, the total energy E_m is negative, *i.e., the planet is bound in an orbit of the form of a closed ellipse.*

For a circular orbit, since e = 0, we have $E_m = -\dfrac{m^3M^2G^2}{2J^2}$.

Alternatively, we could proceed *on the basis of the equivalent one-body problem as follows:*

If μ be the *reduced mass* of the system of the planet and the sun, we have

$$\mu\frac{d^2r}{dt^2} = -\frac{MmG}{r^2} + \frac{\mu v^2}{r} = \frac{\mu\omega^2 r^2}{r} - \frac{MmG}{r^2}$$

Or, $$\mu\frac{d^2r}{dt^2} = \mu\omega^2 r - \frac{MmG}{r^2} \quad \text{...(i)}$$

But. as we know, $J = \mu r^2\omega$ ...(ii)

So that, $\omega^2 = J^2/\mu^2r^4$. ...(iii)

$\therefore$ $$\mu\frac{d^2r}{dt^2} = \frac{\mu J^2 r}{\mu^2r^4} - \frac{MmG}{r^2} = \frac{J^2}{\mu r^3} - \frac{MmG}{r^2}. \quad \text{...(iv)}$$

Solving relation (iv) for r as a function of r (θ) of the angle θ, we have

$$\frac{dr}{dt} = \frac{dr}{d\theta}.\frac{d\theta}{dt} = \frac{dr}{d\theta}\omega = \frac{dr}{d\theta}.\frac{J}{\mu r^2}. \qquad ...(v)$$

Differentiating again, we have

$$\frac{d^2r}{dt^2} = \frac{d^2r}{d\theta^2}\left(\frac{J}{\mu r^2}\right)^2 + \frac{dr}{d\theta}.\frac{J}{\mu}\frac{d}{dt}\left(\frac{1}{r^2}\right).$$

Or, $$\frac{d^2r}{dt^2} = \frac{d^2r}{d\theta^2}\left(\frac{J}{\mu r^2}\right)^2 - \frac{2}{r^3}.\frac{J}{\mu}.\left(\frac{dr}{d\theta}\right)^2.\frac{J}{\mu r^2}. \qquad ...(vi)$$

[Using $J = \mu r^2\omega$ and relation (v).

Now, let us put the function $u(\theta) = \frac{1}{r}(\theta)$. Then, on differentiating, we have

$$\frac{du}{d\theta} = -\frac{1}{r^2}\frac{dr}{d\theta}; \text{ and } \frac{d^2u}{d\theta^2} = -\frac{1}{r^2}\frac{d^2r}{d\theta^2} + \frac{2}{r^3}\left(\frac{dr}{d\theta}\right)^2 \qquad ...(vii)$$

Comparing relations (vi) and (vii) we see that

$$\frac{d^2r}{dt^2} = -\frac{1}{r^2}\left(\frac{J}{\mu}\right)^2\frac{d^2u}{d\theta^2}. \qquad ...(viii)$$

We may, therefore, write relation (iv) as

$$-\mu\left[\frac{1}{r^2}\left(\frac{J}{\mu}\right)^2\frac{d^2u}{d\theta^2}\right] = \frac{1}{r^3}\frac{J^2}{\mu} - \frac{1}{r^2}MmG.$$

Remembering that $1/r = u$, we have

$$-u^2\frac{J^2}{\mu}\frac{d^2u}{d\theta^2} = u^3\frac{J^2}{\mu} - u^2 MmG.$$

Or $$-\frac{d^2u}{d\theta^2} = u - \frac{\mu MmG}{J^2}.$$

Or, $$\frac{d^2u}{d\theta^2} + u = \frac{\mu MmG}{J^2}, \qquad ...(ix)$$

the solution of which is

$$u = \frac{\mu MmG}{J^2} + A\cos\theta. \text{ Or, } \frac{1}{r} = \frac{\mu MmG}{J^2} + A\cos\theta.$$

Or, $$\frac{J^2/\mu MmG}{r} = 1 + \frac{J^2 A}{\mu MmG}\cos\theta,$$

which is an equation similar to that of a general conic section

$$\frac{l}{r} = 1 + e\cos\theta, \text{ where, } l = \frac{J^2}{\mu MmG} \text{ and } e = \frac{J^2 A}{\mu MmG}$$

Thus for $e < 1$, *i.e.*, $(J^2A/\mu MmG) < 1$, *the planet will describe an elliptical orbit round the sun.*

Again, let us obtain the value of e in terms of the total energy E_m of the system.

We have $E_m = \frac{1}{2}mv^2 - \frac{mM}{r}G$, where m is the *mass of the planet.*

Now, at *minimum and maximum* distances from the sun, *i.e., at* positions respectively called *perihelion* and *aphelion*, the velocity vector is perpendicular to the position vector r and $\theta = 0$ and π respectively. So that, at these points, $v = rd\theta/dt = h/r = J/\mu r$. Hence

$$E_m = \frac{J^2}{2\mu}\left(\frac{1}{r_{min}}\right)^2 - \frac{mMG}{r_{min}} = \frac{J^2}{2\mu}\left(\frac{1}{r_{max}}\right)^2 - \frac{mMG}{r_{max}}. \quad ...(i)$$

Therefore, from the relation, $l/r = 1 + e\cos\theta$, we have

$$\frac{l}{r_{min}} = 1 + e \text{ and } \frac{l}{r_{max}} = 1 - e, \quad ...(ii)$$

which gives $e = \frac{r_{max} - r_{min}}{r_{max} + r_{min}}$, a convenient relation to remember.

Substituting relation (ii) in (i) and solving out, we obtain, as before,

$$e = \sqrt{1 + \frac{2E_m J^2}{\mu M^2 m^2 G^2}}$$

and $$E_m = -\frac{\mu M^2 m^2 G^2}{2J^2}\left(1 - e^2\right).$$

So that, again, for $e < 1$, the total energy E_m is constant and negative and the planet describes a closed ellipse around the sun.

For $e = 0$, the orbit would be a circle and the total energy of the system would be $E_m = -\mu M^2 m^2 G^2/2J^2$.

It will be seen that if m be $<< M$, $\mu \approx m$ and the expressions for e and E_m respectively become

$$\sqrt{\frac{(1+2E_mJ^2)}{m^3M^2G^2}} \text{ and } -\left(\frac{m^3M^2G^2}{2J^2}\right)$$

$(1 - e^2)$, as obtained above.

Now, at the very outset we have shown how, in the case of a body moving under a central force, its areal velocity is constant. Therefore, since a planet moves around the sun under the central gravitational force directed towards the sun, its areal velocity must remain constant. This, then, takes care of *Kepler's second law.* In fact, we have proceeded on this very assumption in deducing Kepler's first law above.

Finally, as shown (under constants of the elliptical orbit), the time-period of a body describing an ellipse around a fixed body, and, therefore, here, of a planet describing an elliptical orbit around the sun, is given by

$$\tau = \frac{\text{area of ellipse}}{\text{areal velocity}} = \frac{\pi ab}{h/2}. \text{ Or, } \tau^2 = \frac{4\pi^2 a^2 b^2}{h^2}.$$

Again, $l = h^2/K = h^2/MG$. Also, $l = b^2/a$.

So that, $b^2/a = h^2/MG$.

Or, $b^2 = h^2a/MG$.

$\therefore$ substituting this value of b^2 in the expression for τ^2 above, we have,

$$\tau^2 = \frac{4\pi^2 a^2}{h^2}\left[\frac{h^2 a}{MG}\right] = \frac{4\pi^2 a^3}{MG}. \qquad \text{...(iii)}$$

Or $\tau^2 \propto a^3$,

since $4\pi^2/MG$ is clearly a constant for all planets.

And, this is *Kepler's third law,* a consequence of which is that *all planets, with the same major axis of their elliptical orbits have the same periodic time.*

This relation (iii) actually enables us to *weigh* the sun, as it were, if we know the time-period and the major axis of the orbit of any planet around it.

The end of the chapter, it has been done from the time-period of the earth around the sun, its orbit being assumed circular. Relation (iii), however, tells us that we can do so even for an elliptical orbit if we use the *semi-major axis* of the orbit in place of the *radius* of the circular orbit.

SOLVED EXAMPLES

Example 1:

(a) State and explain Newton's law of universal gravitation.

(b) Define universal gravitational constant and obtain its dimensions. Give its numerical value.

(c) Discuss the evidence that the gravitational constant G is a universal constant.

(d) Show how the law of gravitation is verified from the moon's period of revolution round the earth.

Solution:

(a) *Newton's Law of Gravitation :* It is stated as follows : *Every particle of matter in the universe attracts every other particle with a force which is directly proportional to the product of their masses and inversely proportional to the square of the distance between them, and acts along the line joining the two particles.*

Thus, If m_1 and m_2 are the masses of two particles placed at a distance r from each other, the magnitude of the force of attraction $\vec{F}$ exerted by each particle on the other is given by

$$F \propto \frac{m_1 m_2}{r^2}$$

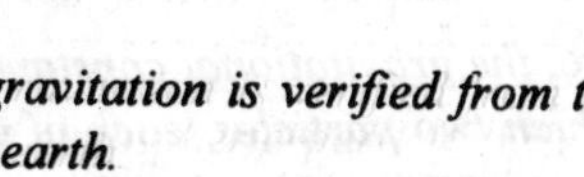

or $$F = -G\frac{m_1 m_2}{r^2}, \quad ...(i)$$

Fig. 2.32

and acts along the line joining the particles. The negative sign indicates that the force is of 'attraction'. G is a universal constant known as gravitational constant.

The characteristics of gravitational forces between two particles are:

(i) they are always attractive,

(ii) they are central that is, act along the line joining the particles,

(iii) they form an action-reaction pair and

(iv) they are completely independent of the presence of other particles or the intervening medium.

On the other hand, the other kinds of forces between material particles, such as electrical, may be attractive as well as repulsive and do depend on the presence of other particles and on the intervening medium.

(b) *Gravitational Constant* : The Newton's law of gravitation is

$$F = -G\frac{m_1 m_2}{r^2}.$$

If we put $m_1 = m_2 = 1$

and $r = 1$, we have

$$G = F.$$

Therefore, *the gravitational constant G is numerically equal to the force with which two particles, each of unit mass. and placed at a unit distance apart, attract each other.* Its accepted value is 6.673×10^{-11} N-m²/kg².

Dimensions of G : From the above eq., we have

$$G = F\frac{r^2}{m_1 m_2}.$$

Writing the dimensions of various quantities, we have

$$[G] = \frac{[MLT^{-2}][L^2]}{[M][M]} = [M^{-1}L^3T^{-2}].$$

The universal constant G, which is a scalar, is different from the acceleration due to gravity $\vec{g}$ which is a vector, $\vec{g}$ is neither universal nor constant.

(c) *Universality of G* : In the law of gravitation, the constant of proportionality G is supposed to be the same for all pairs of bodies in all places at all times at all distances of separation, and also independent of the presence of other bodies or the properties of the intervening space. The correctness of this assumption depends on the correctness of the deductions using it for the motion of the planets in the solar system and of bodies falling near the surface of the earth. There is so far no evidence against this assumption. Hence G is regarded as a "universal" constant.

(d) *Verification from Moon's Period* : Newton verified his inverse-square law by calculating the period of revolution of the moon about the earth in the following manner:

The moon revolves round the earth in a nearly circular orbit of radius r, which is nearly 60 times the radius of the earth. Its centripetal acceleration a directed toward the centre of the earth is given by

$$\alpha = r\omega^2,$$

where ω is the angular velocity. Now, the gravitational acceleration on the earth's surface is 9.8 m/s². The moon is 60 times as far from the centre of the earth as a body on the earth's surface. Therefore, if inverse-square law is correct, the gravitational acceleration on moon must be $\frac{9.8\,\text{m/s}^2}{(60)^2}$. Thus

$$\frac{9.8\,\text{m/s}^2}{(60)^2} = r\omega^2$$

But r is known to be 3.8×10^8 m.

$$\therefore \quad \frac{9.8\,\text{m/s}^2}{3600} = (3.8 \times 10^8\ \text{m})\,\omega^2$$

$$\text{or} \qquad \omega = 2.65 \times 10^{-6}\ \text{rad/s}.$$

Therefore, the time of one revolution

$$T = \frac{2\pi}{\omega} = \frac{2\pi}{2.65\times10^{-6}}\ \text{second}$$

$$= \frac{2\pi}{2.65\times10^{-6}\times3600\times24} = 27.4\ \text{days}.$$

The nearness of this result to the observed value of 27.3 days is a very strong evidence for the truth of the Newton's inverse-square law.

Example 2:

(a) Explain the terms 'gravitational field', 'gravitational attraction' and 'gravitational potential'.

(b) Establish relation between gravitational attraction and gravitational potential.

Or

Prove that the gravitational attraction is equal to the negative gradient of gravitational potential.

Solution:

(a) *Gravitational Field* : Every particle of matter exerts a force of attraction on every other particle. This is called the gravitational force or attraction and is given by Newton's law of gravitation. *The space surrounding the attracting particle within which its gravitational force of attraction can be experienced is called its gravitational field.*

Gravitational Attraction : The intensity of gravitational field, or the gravitational attraction, at a point in the field is the force experienced by a unit mass placed at that point, provided the unit mass itself does not produce any change in the field. Thus, if there be a gravitational field due to a particle of mass M, the attraction F at a point distant r from the particle is

$$F = -G\frac{M \times I}{r^2} = -\frac{GM}{r^2}.$$

Attraction is a vector quantity.

Gravitational Potential : When a mass moves in a gravitational field, work is done. If it moves against the gravitational field, work is done by some outside agent against the gravitational attraction, but if it moves in the direction of the field, the work is done by the field itself.

The gravitational potential at a point in a gravitational field is defined as the work done when a unit mass moves from infinity to that point.

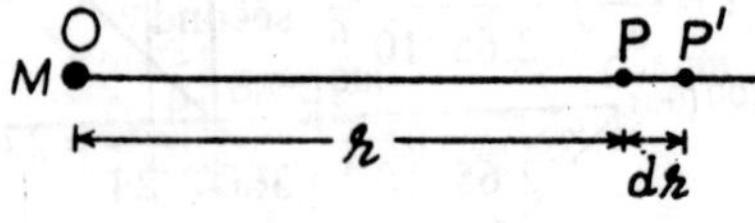

Fig. 2.33

Let M be an attracting point-mass situated at O. Let a unit mass be placed at a point P distant r from O. The gravitational attraction exerted on the unit mass is $F = -\frac{GM}{r^2}$, which acts toward O. Now, suppose the unit mass moves against this attraction from P to P′ through a small distance dr. The work done against the gravitational attraction

$$= -(F)(dr) = \frac{GM}{r^2}dr.$$

Thus, the work done as the unit mass moves from P to infinity will be

$$\int_r^{\infty} \frac{GM}{r^2} dr = GM\left[-\frac{1}{r}\right]_r^{\infty} = \frac{GM}{r}$$

Hence the work done as the unit mass moves *from infinity* to P will be – GM/r. This will be the gravitational potential V at a distance r from the attracting mass M. That is,

$$V = -\frac{GM}{r}.$$

Potential is a scalar quantity. The negative sign here denotes that the potential increases in the direction of r increasing *i.e., away* from the attracting mass. The potential at infinity is considered to be zero and becomes increasingly negative in a direction toward the attracting mass.

(b) *Relation between Attraction and Potential :* Let F be the gravitational attraction at a point. The work done as a unit mass moves from this point to another point in the direction of the field through an infinitesimal distance dr

= attraction × distance moved

= F (– dr).

By definition, this is equal to the difference of potential dV between the two points.

$$\therefore \qquad dV = F(-dr)$$

$$\text{or} \qquad F = -\frac{dV}{dr}.$$

dV/dr, the rate of change of potential with distance, is called the "potential gradient". Thus, *the attraction at any point in a gravitational field is equal to the negative potential gradient at that point.*

Example 3:

Prove that the gravitational potential produced by a ring of mass M and radius R on an axial point distant r from the centre is

$$V = -\frac{GM}{\left(R^2 + r^2\right)^{1/2}}$$

Solution:

Gravitational Potential due to a Ring : Let O be the centre of a ring of mass M and radius R. Let P be the axial point, distant r from O, at which the gravitational potential is required.

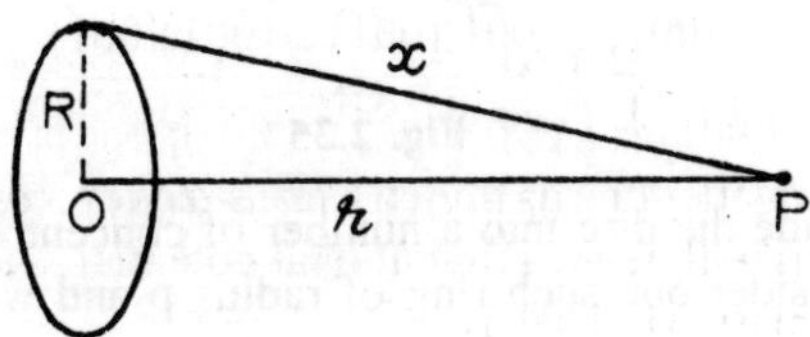

Fig. 2.34

Every element of the ring is at the same distance x (say) from the point P. Therefore, the potential at P due to the entire ring is given by

$$V = -\frac{GM}{x}.$$

But $x = \sqrt{R^2 + r^2}.$

$$\therefore \quad V = -\frac{GM}{\sqrt{R^2 + r^2}}$$

The gravitational attractions

$$F = -\frac{dV}{dr} = -\frac{d}{dr}\left[-GM\,(R^2 + r^2)^{-1/2}\right]$$

$$= GM\frac{d}{dr}\left[(R^2 + r^2)^{-1/2}\right]$$

$$= GM\left(-\frac{1}{2}\right)(R^2 + r^2)^{-3/2}\,(2r)$$

$$= -GM\frac{r}{\left(R^2 + r^2\right)^{3/2}}.$$

Example 4:

Deduce expressions for the gravitational potential and attraction due to a thin circular disc at a point along its axis.

Solution:

Potential and Attraction due to a Disc : Let O be the centre of a thin circular disc of radius R and mass per unit area σ. Let P be the axial point, distant r from O, at which the gravitational potential and attraction are required.

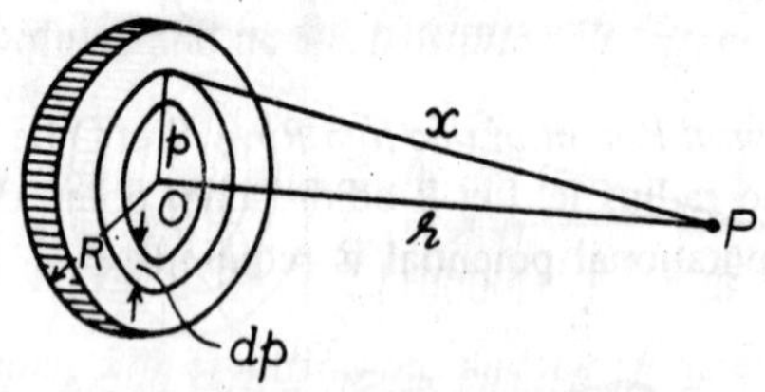

Fig. 2.35

Let us divide the disc into a number of concentric rings with O as centre and consider one such ring of radius p and width dp. The area of this ring

$$= \text{circumference} \times \text{width} = 2\pi p\,(dp).$$

$\therefore$ its mass = 3π p (dp) σ.

Every element of this ring is at the *same distance* x (say) from P. Therefore, the potential dV at P due to this ring is given by

$$dV = -G\frac{2\pi P(dp)\sigma}{x} = -G\frac{2\pi p(dp)\sigma}{\left(r^2+p^2\right)^{1/2}}.$$

The potential V at P due to the whole disc is obtained by integrating this expression between p = 0 and p = R.

$$\therefore \quad V = \int_0^R -G\frac{2\pi p(dp)\sigma}{\left(r^2+p^2\right)^{1/2}}$$

$$= -2\pi\, G\, \sigma \int_0^R \frac{p}{\left(r^2+p^2\right)^{1/2}} dp$$

$$= -2\pi\, G\, \sigma \left[\left(r^2+p^2\right)^{1/2}\right]_0^R$$

$$= -\,2\pi\, G\, \sigma\, [(r^2+R^2)^{1/2} - r].$$

The attraction F at P is equal to the negative potential gradient at P. Therefore

$$F = -\frac{dV}{dr} = -\frac{d}{dr}\left[-2\pi\, G\, \sigma \left\{\left(r^2+R^2\right)^{1/2} - r\right\}\right]$$

$$= 2\pi\, G\, \sigma \left[\frac{1}{2}\left(r^2+R^2\right)^{-1/2}(2r) - 1\right]$$

$$= 2\pi\, G\, \sigma \left[\frac{r}{\left(r^2+R^2\right)^{1/2}} - 1\right]$$

$$= -2\pi\, G\, \sigma \left[1 - \frac{r}{\left(r^2+R^2\right)^{1/2}}\right].$$

Example 5:

Deduce expressions for the gravitational potential and attraction due to a thin uniform spherical shell at a point (a) outside, (b) inside the shell. Give a graphical representation also.

Solution:

(a) *Potential of Shell at External Point :* Let P be the centre of a thin spherical shell of radius R. Let P be the point, distant r from O, at which the gravitational potential is required. Let σ be the mass per unit area of the shell.

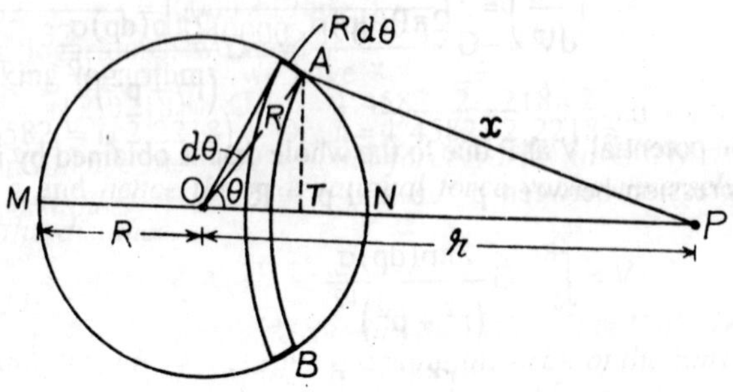

Fig. 2.36

Let us divide the shell into a number of circular coaxial rings and consider one such ring AB of radius AT = R sin θ and thickness R dθ. The area of this ring = circumference × thickness = 2π R sin θ (R dθ). Therefore, its mass = $2\pi R \sin\theta\, (R\, d\theta)\, \sigma = 2\pi R^2 \sigma \sin\theta\, d\theta$.

Every element of this ring is, to a first approximation, at the same distance AP = x, from P. Therefore, the potential dV at P due to this ring is given by

$$dV = -G\frac{\text{mass of the ring}}{x} = -G\frac{2\pi R^2 \sigma \sin\theta\, d\theta}{x}. \qquad \text{...(i)}$$

In ΔAOP, from the cosine law, we have

$$\cos\theta = \frac{R^2 + r^2 - x^2}{2Rr}.$$

Differentiating, we get

$$-\sin\theta\, d\theta = \frac{0 + 0 - 2x\, dx}{2Rr}$$

or $$\sin\theta\, d\theta = \frac{x\, dx}{Rr}.$$

Substituting for sin θ dθ in eq. (i), we get

$$dV = -G\frac{2\pi R^2 \sigma}{x}\left(\frac{x\, dx}{Rr}\right) = -G\frac{2\pi R\sigma}{r}dx.$$

The potential V due to the whole shell is obtained by integrating this expression between N (where x = r – R) and M (where x = r + R).

$$\therefore \quad V = \int_{r-R}^{r+R} -G\frac{2\pi R\sigma}{r}dx$$

$$= -G\frac{2\pi R\sigma}{r}[x]_{r-R}^{r+R}$$

$$= -G\frac{2\pi R\sigma}{r}2R = -G\frac{4\pi R^2\sigma}{r}.$$

If M be the mass of the shell, then $M = 4\pi R^2\sigma$.

$$\therefore \quad V = -G\frac{M}{r}.$$

Attraction : As the shell is symmetrical with respect to the point P, the attraction F at P is along PO and equal to the potential-gradient at P, *i.e.*,

$$F = -\frac{dV}{dr}$$

$$= -\frac{d}{dr}\left(-G\frac{M}{r}\right) = -G\frac{M}{r^2}.$$

Thus, the *shell exerts attraction at an external point as if its entire mass were concentrated at its centre.*

(b) *Potential at Internal Point* : Now, suppose that the point P lies inside the shell at a distance r from O. As before, the potential at P due to the ring is

$$dV = -G\frac{2\pi R\sigma}{r}dx$$

The potential V due to the whole shell is obtained by integrating this expression between N (where x = R − r) and M (where x = R+r).

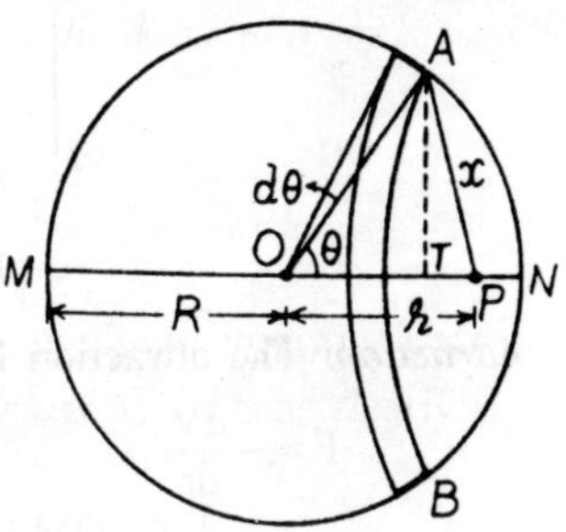

Fig. 2.37

$$\therefore \quad V = \int_{R-r}^{R+r} -G\frac{2\pi R\sigma}{r}dx.$$

$$= -G\frac{2\pi R\sigma}{r}[x]_{R-r}^{R+r}$$

$$= -G\frac{2\pi R\sigma}{r}2r = -G\,4\pi R\sigma$$

$$= -G\frac{M}{R},$$

where $M = 4\pi R^2\sigma$. This is independent of r and is the same as at a point on the surface of the shell. Thus, *the potential inside the shell is constant, equal to its value on the surface.*

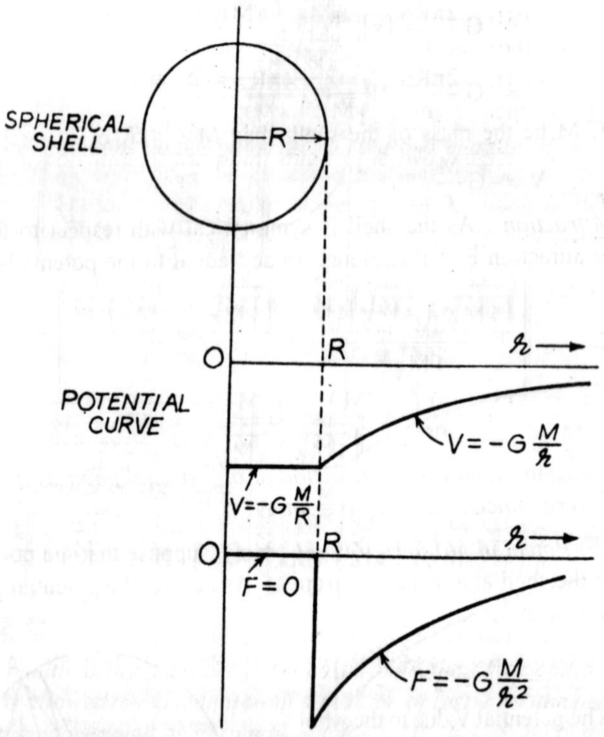

Fig. 2.38

Attraction: The attraction F at P is given by

$$F = -\frac{dV}{dr}$$

$$= -\frac{d}{dr}\left(-\frac{GM}{R}\right) = 0.$$

Thus, *the attraction at an internal point of the shell is zero everywhere.*

Example 6:

One uniform shell of mass m_2 lies inside, and concentric with, a larger uniform shell of mass m_1. Obtain the gravitational field due to the system (i) at a point out-side the two shells, (ii) at a point in the space between the two shells and (iii) at a point inside the smaller shell.

Solution:

Let represent the two shells with their common centre O, and let points P_1, P_2 and P_3 lie outside the two shells, in between the two shells and inside the inner shell at distances r_1, r_2 and r_3, respectively from O.

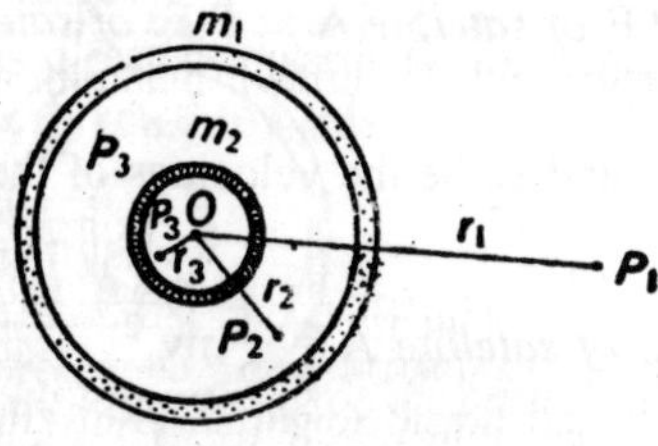

Fig. 2.39

(i) Since point P_1 lies outside both the shells, at distance r_1 from their common centre 0, we have *intensity of the gravitional field at P due to the system*

= *intensity due to one shell + intensity due to the other i.e.,*

$$E = -\frac{m_1}{r_1^2}G - \frac{m_2}{r_1^2}G = -\left(\frac{m_1 + m_2}{r_1^2}\right)G.$$

(ii) Point P_2 lies inside the outer, but outside, the inner shell. The field at p_2 due to the former is, therefore, *zero* and due to the latter, $-(m_2/r_2^2)G$.

Therefore, *intensity at P_2 due to the system, as a whole, i.e.,*

$$E = O - \left(\frac{m_2}{r_2^2}\right)G = -\frac{m_2}{r_2^2}G.$$

(iii) Point P_3 lies inside both the shells, so that the field there due to either shell is *zero*. Hence the *intensity of the gravitational field at P_3 due to the system as a whole = 0.*

Example 7:

Two Satellites A and B of the same mass are orbiting the earth at altitudes R and 3R respectively, where R is the radius of the earth. Taking their orbits to be circular, obtain the ratios of their kinetic and potential energies.

Solution:

Clearly, *distance of satellite A from the centre of the earth* = R + R = 2R and that of satellite B = R + 3R = 4R.

If, therefore, m be the *mass a f each satellite* and M, the *mass of the earth, we have*

PE. of *satellite* A = mMG/2R and P.E. *of satellite* B = mMG/4R

∴ P.E *of satellite* A : : P.E. *of satellite* B : : mMG/2R : mMG/4R or as 2:1.

If v_1 and v_2 be the velocities of the two satellites respectively, we have

K.E. *of satellite* A $= \frac{1}{2} m v_1^2$

and K.E. *of satellite* B $= \frac{1}{2} m v_2^2$

Or, $v_1 = \sqrt{\frac{MG}{2R}}$, and $v_2 = \sqrt{\frac{MG}{4R}}$, we have

KE. *of satellite* $A = \frac{1}{2}\frac{mMG}{2R} = \frac{mMG}{4R}$

and K.E *of satellite* $B = \frac{1}{2}\frac{mMG}{4R} \neq \frac{mMG}{8R}$.

∴ K.E. *satellite* A: K.E. *of satellite* B $:: \frac{mMG}{4R} : \frac{mMG}{8R}$, *i.e.*, as 2:1.

Example 8:

From a larger metallic sphere of radius R and mass M, a smaller sphere is scooped out such that the spherical hollow thus formed just touches the surface of the larger sphere on one side and its centre on the other, as shown in Fig.. Obtain an expression for the gravitational field due to the hollowed sphere at a point P lying on a line passing through the centres of the larger sphere and the spherical hollow at a distance r from the centre of the former.

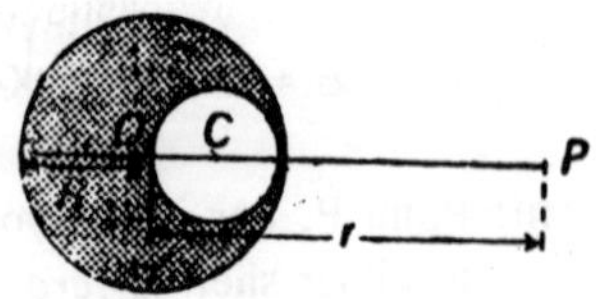

Fig. 2.40

Solution:

Here, clearly, *the diameter of the spherical hollow formed* = R and, therefore, *its radius, i.e., the radius of the smaller sphere removed* = R/2.

Hence *its volume* $= \frac{4}{3}\pi\left(\frac{R}{2}\right)^3 = \frac{4}{3}\pi R^3/8 = \frac{1}{8}$ of the volume of the larger sphere $\left(\frac{4}{3}\pi R^3\right)$ and, therefore, *its mass* = M/8.

If 0 be the centre of the larger sphere and C, that of the spherical hollow, we have distance OP = r (given) and distance CP = r – R/2 = (2r – R)/2.

Clearly, intensity of the gravitational field at P due to the hollowed spheres = intensity at P due to the complete larger sphere of radius R– intensity of the gravitational field at P due to the smaller sphere of radius R/2, i.e.,

$$E = -\frac{MG}{r^2} - \left(-\frac{MG}{8}.\frac{1}{CP^2}\right) = -MG\left(\frac{1}{r^2} - \frac{1}{8}.\frac{4}{(2r-R)^2}\right)$$

$$= -MG\left(\frac{1}{r^2} - \frac{1}{2(2r-R)^2}\right).$$

Example 9:

A satellite of mass M_s is orbiting the earth in a circular orbit of radius R. It starts losing energy slowly at a constant rate C due to friction. If M_e and R_e denote the mass and radius of the earth respectively, show that the satellite falls on the earth in a time

$$t = \frac{GM_eM_s}{2C}\left(\frac{1}{R_e} - \frac{1}{R}\right).$$

Solution:

Let the velocity of the satellite in its orbit of radius R be v_1. Then, clearly,

$$\frac{M_s v_1^2}{R} = \frac{M_s M_e}{R^2}G, \text{ whence, } v_1^2 = M_eG/R.$$

(Because centripetal force on the satellite = force of attraction due to the earth),

Similarly, if v_2 be the velocity of the satellite when its orbit touches the earth, *i.e.*, when the radius of its orbit is R_e, we have

$$\frac{M_s v_2^2}{R_e} = \frac{M_s M_e}{R_e^2}, \text{ whence, } v_2^2 = M_eG/R_e.$$

∴ K.E. *of the satellite in orbit of radius*

$$R = \frac{1}{2}M_s v_1^2 = \frac{1}{2}M_sM_e\,G/R$$

and K.E. *of the satellite in orbit of radius* R_e (when it touches the earth)

$$= \frac{1}{2}M_s v_2^2 = \frac{1}{2}M_sM_eG/R_e.$$

Now, P.E. *of the satellite when at distance R from the centre of the earth*

$$= -M_sM_eG/R$$

and *its P.E. when at a distance R_e from the centre of the earth (i.e., when it touches the earth)*$= -M_sM_eG/R_e$

$\therefore$ *total energy of the satellite when in orbit of radius*

R = *its* P.E + *its* K.E.

$$-\frac{M_sM_eG}{R}+\frac{1}{2}\frac{M_sM_eG}{R}=\frac{1}{2}-\frac{M_sM_eG}{R}.$$

and total energy of the satellite when on the surface of the earth.

$$=-\frac{M_sM_eG}{R_e}+\frac{1}{2}\frac{M_sM_eG}{R_e}=-\frac{1}{2}\frac{M_sM_e}{R_e}G.$$

Hence, *loss of energy in falling to the earth*

$$=-\frac{1}{2}\frac{M_sM_eG}{R}-\left(-\frac{1}{2}\frac{M_sM_eG}{R_e}\right)$$

$$=\frac{1}{2}M_sM_eG\left(\frac{1}{R_e}-\frac{1}{R}\right)$$

If the satellite takes time t to fall to the earth, clearly, *energy* lost by it = Ct.

$$\therefore\ Ct=\frac{1}{2}M_sM_eG\left(\frac{1}{R_e}-\frac{1}{R}\right),\text{whence, } t=\frac{M_sM_eG}{2C}\left(\frac{1}{R_e}-\frac{1}{R}\right).$$

Example 10:

(a) Derive expressions for the gravitational potential and attraction due to a solid sphere of uniform density at an (i) external point, (ii) internal point.

(b) Show that the gravitational potential at the surface of a solid sphere is two-thirds of the potential at the centre.

Solution:

(a) (i) *Potential due to a Solid Sphere at External Point* : Let O be the centre of a sphere of radius R and density ρ. Let P be a point outside the sphere at which the potential is required. Let OP = r.

Let us divide the sphere into a large number of thin concentric shells of masses m_1, m_2, m_3 ...., etc. The potential at the point P due to the shell of mass m_1 is

$$= -G\frac{m_1}{r},$$

where r is the distance of P from the centre O of the shell.

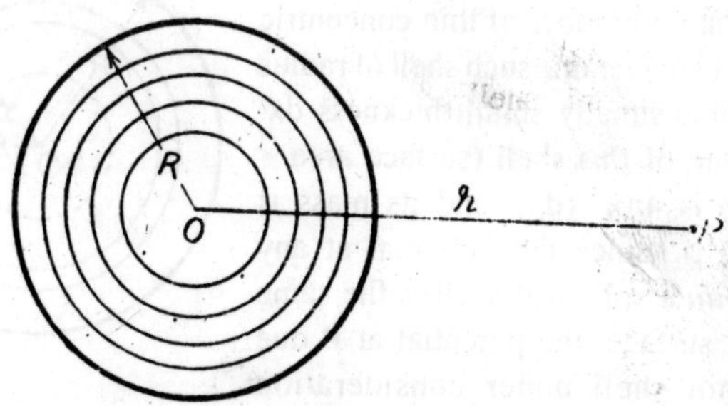

Fig. 2.41

Now, potential is a *scalar* quantity. Therefore, the potential V due to the whole sphere is equal to the sum of the potentials due to all the shells.

$$\therefore \quad V = -\frac{Gm_1}{r} - \frac{Gm_2}{r} - \frac{Gm_3}{r} \ldots$$

$$= -\frac{G}{r}(m_1 + m_2 + m_3 + \ldots).$$

But $(m_1 + m_2 + m_3 + \ldots) = M$, the mass of the sphere.

$$\therefore \quad V = -G\frac{M}{r}$$

Attraction : The gravitational field (attraction) F at a point is equal to negative potential-gradient at that point, that is,

$$F = -\frac{dV}{dr}$$

$$= -\frac{d}{dr}\left(-G\frac{M}{r}\right) = -G\frac{M}{r^2}.$$

Thus, *the solid sphere exerts attraction at an external point as if its entire moss were concentrated at its centre.*

(ii) *Potential at Internal Point* : Now, suppose the poin* P lies inside the sphere at a distance r from the centre O. If we draw a concentric sphere through P, the point P will be external for the inner solid sphere of radius r, and internal for the outer spherical shell of internal radius r and external radius R.

The mass of the inner solid sphere is (4/3) $\pi r^3 \rho$. Therefore, the potential V_1 at P due to this sphere is, from eq. (i), given by

$$V_1 = -G\frac{(4/3)\,\pi r^3 \rho}{r} = -\frac{4}{3}\pi G \rho r^2. \qquad ...(ii)$$

Let us now find the potential at P due to the outer spherical shell. Let us divide the shell into a number of thin concentric shells and consider one such shell of radius x and infinitesimally small thickness dx. The volume of this shell (surface area × thickness) is $4\pi x^2$ (dx), and its mass is $4\pi x^2$ (dx) ρ. Since the potential at any point *within* a spherical shell is the same as on the surface, the potential at P due to the thin shell under consideration will be

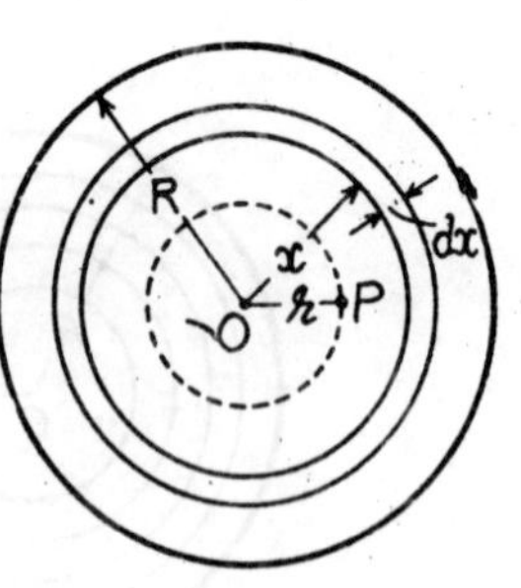

Fig. 2.42

$$= -G\frac{4\pi x^2 (dx)\rho}{x} = -G\,4\pi x\,(dx)\rho.$$

The potential V_2 at P due to the whole shell of internal radius r and external radius R is obtained by integrating this expression between the limits x = r and x = R. Thus

$$V_2 = \int_r^R -G\,4\pi x (dx)\rho = -4\pi\,G\rho\left[\frac{x^2}{2}\right]_r^R$$

$$= -4\pi\,G\rho\left(\frac{R^2}{2} - \frac{r^2}{2}\right)$$

$$= -\frac{4}{3}\pi\,G\rho\left(\frac{3R^2}{2} - \frac{3r^2}{2}\right). \qquad ...(iii)$$

Since the potential is a scalar quantity, the total potential V at P is obtained by adding the potentials given by eq. (ii) and (iii). Thus,

$$V = V_1 + V_2$$

$$= -\frac{4}{3}\pi\,G\rho r^2 - \frac{4}{3}\pi\,G\rho\left(\frac{3R^2}{2} - \frac{3r^2}{2}\right)$$

$$= -\frac{4}{3}\pi\,G\rho\left(r^2 + \frac{3R^2}{2} - \frac{3r^2}{2}\right)$$

$$= -\frac{4}{3}\pi\,G\rho\left(\frac{3R^2}{2} - \frac{r^2}{2}\right)$$

$$= -\frac{4}{3}\pi\, G\rho R^3 \left(\frac{3R^2 - r^2}{2R^3}\right).$$

If M be the mass of the sphere, then $M = \left(\frac{4}{3}\right)\pi\, R^3\rho$.

$$\therefore \qquad V = -GM\left(\frac{3R^2 - r^2}{2R^3}\right)$$

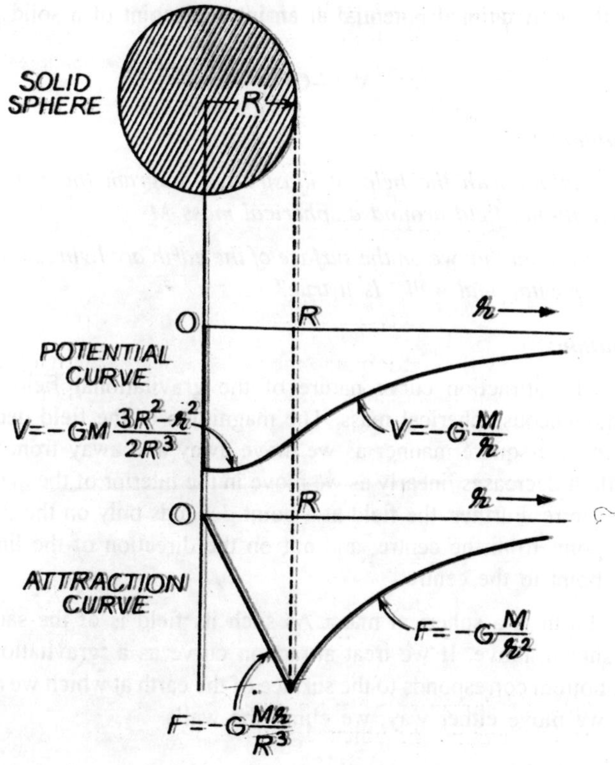

Fig. 2.43

Attraction : The attraction F is given by

$$F = -\frac{dV}{dr} = -\frac{d}{dr}\left(-GM\frac{3R^2 - r^2}{2R^3}\right)$$

$$= \frac{GM}{2R^3}(0 - 2r)$$

$$= -G\frac{Mr}{R^3}.$$

Thus, *the gravitational attraction at a point inside a solid sphere is proportional to its distance from the centre.*

(b) *Potential at the Centre and on the Surface:* The general expression for the gravitational potential at an internal point of a solid sphere is

$$V = -GM\left(\frac{3R^2 - r^2}{2R^3}\right).$$

Example 11:

Explain with the help of illustrative diagram the nature of the gravitational field around a spherical mass M.

It is said that we on the surface of the earth are living at the bottom of a gravitational well'. Is it true?

Solution:

The attraction curve nature of the gravitational field around a homogeneous spherical mass. The magnitude of the field decreases in an inverse-square manner as we move away and away from the mass, while it decreases linearly as we move in the interior of the mass toward the centre. Further, the field at a point depends only on the distance of the point from the centre, and not on the direction of the line joining the point to the centre.

Earth is a spherical mass. As such its field is of the same nature as shown above. If we treat attraction curve as a 'gravitational' well, the bottom corresponds to the surface of the earth at which we are living. As we move either way, we climb the walls.

Example 12:

Find the gravitational potential and attraction due to a spherical shell bounded by spheres of radii R_1 and R_2 at a point (i) outside the shell, (ii) inside the shell and (iii) between the two surfaces.

Solution:

(i) *Potential due to a Thick Spherical Shell at a point outside the Shell :* Let us consider a thick spherical shell whose centre is O (Fig. 2.11), internal radius is R_1 and external radius is R_2. Let ρ be the density of its material. Let ρ be the point at which the potential is required. Let OP = r.

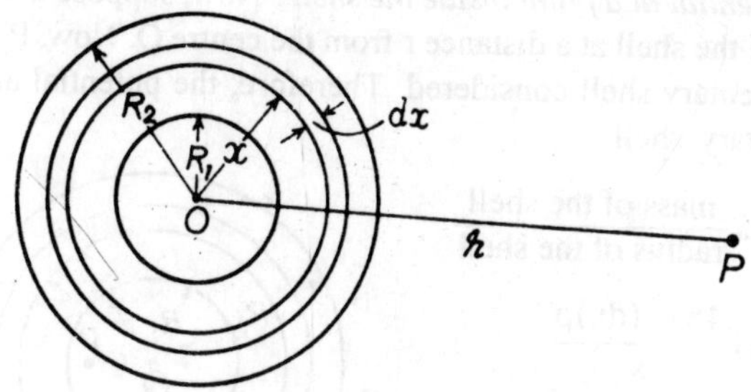

Fig. 2.44

Let us divide the thick shell into a large number of thin concentric shells and consider one such shell of radius x and infinitesimally small thickness dx. The volume of the shell is (surface area × thickness) $4\pi x^2$ (dx) and its mass is $4\pi x^2$ (dx) ρ. Hence the potential at P due to this shell is

$$dV = -G\frac{4\pi x^2(dx)\rho}{r},$$

where r is the distance at P from the centre O of this shell.

The potential V due to the whole thick shell is found by integrating the above expression between $x = R_1$ and $x = R_2$.

$$\therefore \quad V = \int_{R_1}^{R_2} -G\frac{4\pi x^2(dx)\rho}{r}$$

$$= -G\frac{4\pi\rho}{r}\left[\frac{x^3}{3}\right]_{R_1}^{R_2}$$

$$= -G\frac{4}{3}\frac{\pi\rho}{r}\left(R_2^3 - R_1^3\right). \qquad \text{...(i)}$$

But $\frac{4}{3}\pi\left(R_2^3 - R_1^3\right)\rho = M$, the mass of the shell.

$$\therefore \quad V = -G\frac{M}{r}.$$

Attraction : The attraction F is given by

$$F = -\frac{dV}{dr}$$

$$= -\frac{d}{dr}\left(-G\frac{M}{r}\right) = -G\frac{M}{r^2}.$$

(ii) *Potential at a point inside the shell* : Now, suppose that the point P lies inside the shell at a distance r from the centre O. Now, P is internal to the elementary shell considered. Therefore, the potential at P due to the elementary shell

$$= -G \frac{\text{mass of the shell}}{\text{radius of the shell}}$$

$$= -G \frac{4\pi x^2 (dx) \rho}{x}$$

$$= -G\, 4\pi x\, (dx)\, \rho.$$

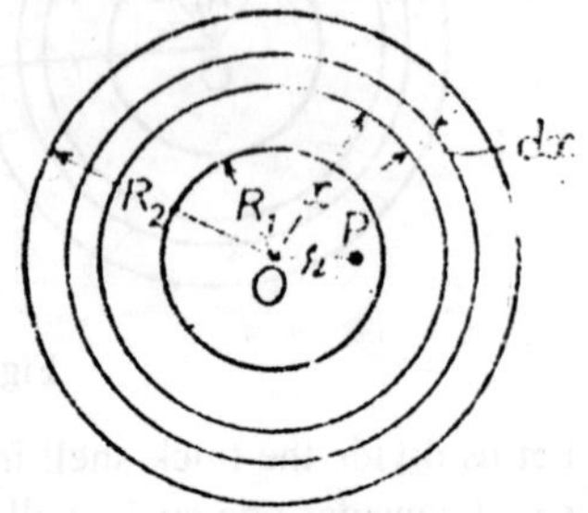

The potential V due to the whole thick shell is obtained by integrating the expression between the limits $x = R_1$ and $x = R_2$.

Fig. 2.45.

$$\therefore \quad V = \int_{R_1}^{R_2} -G\, 4\pi x\, (dx)\, \rho$$

$$= -G\, 4\pi\rho \left[\frac{x^2}{2} \right]_{R_1}^{R_2}$$

$$= -G\, 4\pi\rho \frac{1}{2} \left(R_2^2 - R_1^2 \right). \qquad \text{...(ii)}$$

Since the mass of the shell $M = \frac{4}{3}\pi \left(R_2^3 - R_1^3 \right)\rho$, we have

$$4\pi\rho = \frac{3M}{\left(R_2^3 - R_1^3 \right)}.$$

Putting this result in eq. (ii), we have

$$V = -G \frac{3M}{\left(R_2^3 - R_1^3 \right)} \frac{1}{2} \left(R_2^2 - R_1^2 \right)$$

$$= -\frac{3}{2} GM \frac{(R_2 - R_1)(R_2 + R_1)}{(R_2 - R_1)\left(R_2^2 + R_1 R_2 + R_1^2 \right)}$$

$$= -\frac{3}{2} GM \left(\frac{R_2 + R_1}{R_2^2 + R_1 R_2 + R_1^2} \right),$$

which is constant.

Attraction : $F = -\dfrac{dV}{dr}$

$$= -\frac{d}{dr}\left(-\frac{3}{2}GM\frac{R_2 + R_1}{R_2^2 + R_1R_2 + R_1^2}\right)$$

$$= 0.$$

Thus, *the attraction inside the shell is zero.*

(iii) *Potential at a point between the two surfaces* : Now, suppose the point P lies between the two surfaces, at a distance r from the centre O.

Let us draw an imaginary spherical surface of radius r with O as centre. It is clear from the figure that the point P is external to the thick shell of internal radius R_1 and external radius r. Therefore, the potential at P due to this shell is, from eq. (i),

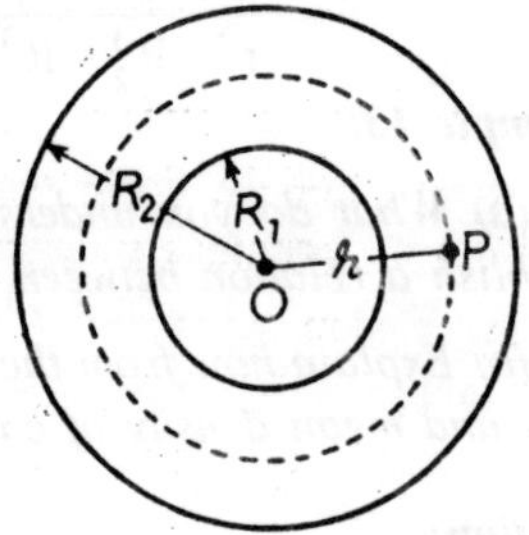

Fig. 2.46

$$V_1 = -G\frac{4}{3}\frac{\pi\rho}{r}\left(r^3 - R_1^3\right).$$

Also, the point P is internal to the thick shell of internal radius r and external radius R_2. Therefore, the potential at P due to this shell is, from eq. (ii),

$$V_2 = -G\,4\pi\rho\frac{1}{2}\left(R_2^2 - r^2\right).$$

As potential is a scalar quantity, the total potential at P is the sum of V_1 and V_2. Thus

$$V = V_1 + V_2$$

$$= -G\frac{4}{3}\frac{\pi\rho}{r}\left(r^3 - R_1^3\right) - G\,4\pi\rho\frac{1}{2}\left(R_2^2 - r^2\right)$$

$$= -G\frac{2}{3}\frac{\pi\rho}{r}\left[2\left(r^3 - R_1^3\right) + 3r\left(R_2^2 - r^2\right)\right]$$

$$= -G\frac{2}{3}\frac{\pi\rho}{r}\left(3rR_2^2 - r^3 - 2R_1^3\right). \qquad \text{...(iii)}$$

Now, the mass of the shell, $M = \frac{4}{3}\pi\left(R_2^3 - R_1^3\right)\rho.$

$$\therefore \qquad \rho = \frac{3M}{4\pi\left(R_2^3 - R_1^3\right)}.$$

Putting the value of ρ in eq. (iii), we obtain

Attraction : $F = -\frac{dV}{dr}.$

$$= -\frac{d}{dr}\left[-\frac{GM}{2r}\left(\frac{3rR_2^2 - r^3 - 2R_1^3}{R_2^3 - R_1^3}\right)\right]$$

$$= -\frac{d}{dr}\left[-\frac{GM}{2}\left(\frac{3R_2^2 - r^2 - 2R_1^3/r}{R_2^3 - R_1^3}\right)\right]$$

$$= -\frac{GM}{2\left(R_2^3 - R_1^3\right)}\left(0 - 2R + \frac{2R_1^3}{r^2}\right)$$

$$= \frac{GM}{r^2}\frac{r^3 - R_1^3}{R_2^3 - R_1^3}.$$

Example 13:

(a) What do you understand by acceleration due to gravity g? Establish a relation between g and G.

(b) Explain how from the knowledge of gravitational constant, the mass and mean density of earth may be calculated.

Solution:

(a) *Acceleration due to Gravity* : Earth attracts all bodies towards its centre. This property of the earth is called 'gravity' and the force with which it attracts a body is called the 'force of gravity' acting on that body. Thus, 'when a body falls freely toward the earth's surface, the force of gravity $\vec{F}$ produces an acceleration $\vec{g}$ in it, given by

$$\vec{g} = \frac{\vec{F}}{m},$$

where m is the mass of the body. This acceleration is called "Acceleration due to gravity". Its magnitude g is independent of the mass, size, shape and composition of the body: It is directed radially in toward the centre of the earth.

Relation between g and G : The earth's attraction for the bodies is a special case of universal gravitational attraction. If earth be assumed to be a homogeneous solid sphere, its attraction on external bodies is same as if its mass were concentrated at its centre. Therefore, it M be the mass and R the radius of the earth, its attraction on a body of mass m placed on its surface is

$$G\frac{Mm}{R^2}$$

This force of attraction is the 'weight' mg of the body. Therefore,

$$mg = G\frac{Mm}{R^2}$$

or $$g = \frac{GM}{R^2}.$$

This is the relation between g and G.

(b) *Mass and Mean Density of Earth* : From the last expression, we can write

$$M = \frac{gR^2}{G}.$$

If ρ be the mean density of earth, then

$$M = \frac{4}{3}\pi R^3 \rho = \frac{gR^2}{G}$$

or $$\rho = \frac{3g}{4\pi RG}.$$

Thus, M and ρ for the earth can be evaluated, if G, g and R are known.

The known values are $G = 6.67 \times 10^{-11}$ N-m²/kg², $R = 6.37 \times 10^6$ m and $g = 9.80$ m/s². Thus

$$M = \frac{(9.80)\left(6.37\ 10^6\right)^2}{6.67\times10^{-11}} = 5.96 \times 10^{24} \text{ kg}$$

and $$\rho = \frac{3\times9.80}{4\times3.14\times\left(6.37\times10^6\right)\times\left(6.37\times10^{-11}\right)}$$

$$= 5.51 \times 10^3 \text{ kg/m}^3.$$

Example 14:

(a) Define gravitational potential energy.

Show that the energy of a mass m placed at a distance r from a gravitating mass M is $-\frac{GMm}{r}$.

(b) Find the gravitational potential energy and the gravitational potential of a particle of mass m near the earth's surface.

Solution:

(a) Gravitational Potential Energy : The gravitational potential energy of a particle placed in a gravitational field is measured by the

amount of work done in displacing the particle from a reference position to its present position. The reference position is chosen at an infinite distance from the attracting mass where the potential energy of the particle is taken as zero.

Thus, if a particle of mass m be at a distance r from a gravitating (attracting) mass M, its potential energy will be given by

$$U(r) = -W_{\infty r} + U(\infty)$$

$$= -W_{\infty r}, \qquad [\because U(\infty) = 0]$$

where $W_{\infty r}$ is the work done by the gravitational force on the particle as it moves from infinity to a distance r. If F(r) be the gravitational force acting on the particle, when it is at a distance r from the gravitating mass, then

$$U(r) = -W_{\infty r} = -\int_{\infty}^{r} F(r)\,dr.$$

But $F(r) = -\dfrac{GMm}{r^2}$ (minus sign, because the force is attractive).

$$\therefore \quad U(r) = -\int_{\infty}^{r} -\frac{GMm}{r^2}\,dr$$

$$= \left[-\frac{GMm}{r}\right]_{\infty}^{r} = -\frac{GMm}{r}.$$

This is the gravitational potential energy of the particle at the point. The minus sign indicates that the potential energy is negative at any finite distance *i.e.,* it decreases (from zero) as the particle is brought (from infinity) toward the attracting mass.

(b) *Gravitational Potential Energy at Earth's Surface* : The gravitational potential energy of a mass m in the field of a mass M at a distance r from it is

$$U(r) = -\frac{GMm}{r}.$$

The earth behaves for all external points as if its mass M (say) were concentrated at its centre. Therefore, a mass in near its surface may be considered at a distance equal to earth's radius R from M. Hence the potential energy of m due to earth is

$$U(R) = -\frac{GMm}{R}.$$

The earth's surface potential is

$$V = \frac{U(R)}{m} = -\frac{GM}{R}.$$

Example 15:

(a) What do you understand by the term 'gravitational self-energy' of a body or a system of particles? Show that the gravitational self-energy of a system of n particles, each of mass m at an average distance r from each other is given by $-\frac{1}{2}Gn(n-1)\frac{m^2}{r}$

(b) Deduce an expression for the gravitational self-energy of a star of mass M and radius R if the density is taken as uniform.

Solution:

(a) Gravitational Self-Energy : The gravitational self-energy of a body (or a system of particles) is defined as the work done in assembling the body (or system of particles) from infinitesimal elements (or particles) that are initially an infinite distance apart.

The gravitational potential energy of two particles of masses m_1 and m_2 separated by distance r_{12} is $-G\frac{m_1 m_2}{r_{12}}$ (negative sign, because the gravitational force is attractive). The potential energy of n particles due to their mutual gravitational attraction is equal to the sum of the potential energies of all pairs of particles, *i.e.*,

$$U_s = -G \sum_{\substack{\text{all pairs} \\ j \neq i}} \frac{m_i m_j}{r_{ij}}$$

This expression can be written as

$$U_S = -\frac{1}{2}G \sum_{i=1}^{i=n} \sum_{\substack{j=1 \\ j\neq i}}^{j=n} \frac{m_i\, m_j}{r_{ij}}$$

Here we have counted each pair twice (once for example as 3, 4 and once as 4, 3), hence the factor 1/2 is introduced for correcting the over-counting. Further, we have to exclude the case j = i, which is not a pair at all.

Let us now consider a system of n particles, each of same mass m and separated from each other by the same average distance r. The self-energy is then

$$U_S = -\frac{1}{2}G \sum_{i=1}^{n} \sum_{\substack{j=1 \\ j\neq i}}^{n} \left(\frac{m^2}{r}\right)_{ij}.$$

Infact, on the right-hand side i comes n times while j comes (n – 1) times. Thus

$$U_S = -\frac{1}{2} G\, n(n-1) \frac{m^2}{r}.$$

(b) *Gravitational Self-energy of a Star (Uniform Sphere)* : Let us treat the star as a uniform sphere of mass M and radius R. For this, let us first consider the energy of interactions between a solid spherical core of radius rand a surrounding spherical shell of thickness dr. The mass of the core is (4/3) $\pi r^3 \rho$ (where ρ is density) and the mass of the shell is $4\pi r^2$ (dr) ρ. The gravitational potential energy of the shell due to the presence of the core is

$$U_{shell} = -G \frac{\left(\frac{4}{3}\pi r^3 \rho\right)\left(4\pi r^2\, dr\, \rho\right)}{r}$$

$$= -\frac{1}{3} G\, (4\pi\rho)^2\, r^4\, dr,$$

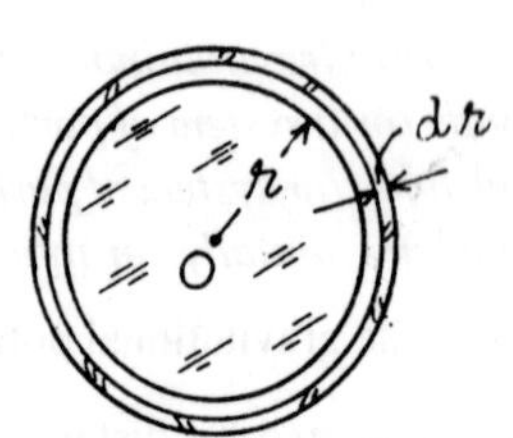

Fig. 2.47

the mass of the core being treated as concentrated at its centre. The solid sphere (star) can be built up by adding successive shells to the core until the core has radius R. Initially the core has zero radius. Therefore, the self-energy of the star, U_{star}, is given by the integral of the above expression between r = 0 and r = R.

$$U_{star} = -\frac{1}{3} G (4\pi\rho)^2 \int_0^R r^4 dr$$

$$= -\frac{1}{3} G (4\pi\rho)^2 \left[\frac{r^5}{5}\right]_0^R = -\frac{3}{5} G \left(\frac{4\pi}{3} R^3 \rho\right)^2 \frac{1}{R}.$$

$\frac{4\pi}{3} R^3 \rho$ is the mass M of the star.

$$\therefore \quad U_{star} = -\frac{3}{5} \frac{GM^2}{R}.$$

This is the gravitational energy of the star. It is negative because the gravitational force is attractive. We shall have to do positive work against gravity to separate the atoms of the star, taking each atom to infinity.

Example 16:

Deduce an expression for the speed with which a body should be projected vertically upward from earth's surface so as to reach a height h from the surface of the earth.

Solution:

Launching Speed of a Projectile : The gravitational potential energy of a particle of mass m on earth's surface is

$$U(R) = -\frac{GMm}{R},$$

where M is the mass of earth (supposed to be concentrated at its centre) and R is the radius of earth (distance of the particle from the centre of the earth). The gravitational potential energy of the same particle at a height h *from earth's surface* would be

$$U(R+h) = -\frac{GMm}{(R+h)}.$$

$\therefore$ change in potential energy

$$U(R + h) - U(R) = -GMm\left(\frac{1}{R+h}-\frac{1}{R}\right)$$

$$= -GMm\left[\frac{-h}{(R+h)R}\right] = \frac{GM}{R^2}\frac{mh}{\left(1+\frac{h}{R}\right)}.$$

But $\frac{GM}{R^2} = g$ (acceleration due to gravity on earth's surface).

$$\therefore \quad U(R+h) - U(R) = \frac{mgh}{\left(1+\frac{h}{R}\right)}.$$

This difference must come from the initial kinetic energy given to the particle in launching it. Suppose the particle is launched with speed v, so that its initial kinetic energy is $\frac{1}{2}mv^2$. Then

$$\frac{1}{2}mv^2 = \frac{mgh}{1+\frac{h}{R}}$$

or

$$v = \sqrt{\frac{2gh}{1+\left(\frac{h}{R}\right)}}.$$

Example 17:

(a) What is escape velocity. Calculate escape velocity for earth.

Or

Calculate the velocity with which a body is to be projected from the surface of the earth so that it will never return.

(b) Calculate escape velocity in terms of earth's density.

Solution:

(a) *Escape Velocity* : The minimum initial velocity needed for a particle projected upward, so as to escape from the earth for ever is called the 'escape velocity' for earth.

We know that the gravitational potential energy of a particle of mass m on the surface of the earth, is given by

$$U(R) = -\frac{GMm}{R}.$$

where M is the mass of the earth (supposed to be concentrated at its centre) and R is the radius of the earth (distance of the particle from the centre of the earth). Therefore, the amount of work required to move the particle from the surface of the earth to infinity would be GMm/R, Clearly, a particle of kinetic energy greater than this, if projected upward, would escape from the earth never to return, provided the resistance of earth's atmosphere is neglected. Thus, the minimum velocity v_e above which escape is possible is given by

$$\frac{1}{2}mv_e^2 = \frac{GMm}{R}$$

or $$v_e = \sqrt{\frac{2GM}{R}}. \qquad ...(i)$$

This is the expression for the escape velocity for earth in terms of mass and radius of earth.

Now, the acceleration due to gravity on earth's surface is given by

$$g = \frac{GM}{R^2}$$

or $$GM = gR^2.$$

Making this substitution in eq. (i) we get

$$v_e = \sqrt{2gR}. \qquad ...(ii)$$

This is the alternative expression for escape velocity. It shows that *the escape velocity from the surface of a planet is directly proportional to the square, root radius of that planet.*

Substituting the known values of g are R for earth in eq. (ii), we get

$$v_e = \sqrt{2\times(9.8\,m/s^2)\times(6.4\times10^6\,m)}$$

$$= 11.2 \times 10^3 \text{ m/s} = 11.2 \text{ km/s}.$$

Should a projectile be given this initial speed, it would escape from the earth. For initial speeds less than this, the projectile will return.

The escape velocity for the moon is much smaller. Therefore, the molecules in the moon attain speed by thermal agitation which is greater than the escape velocity for the moon. On the other hand, the escape velocity for the sun is much larger and even the lightest molecules (hydrogen) cannot escape from the sun's atmosphere.

(b) *Escape velocity in terms of Earth's Density* : Eq. (i) is

$$v_e = \sqrt{\frac{2GM}{R}}$$

If ρ be the mean density of earth, then $M = \frac{4}{3}\pi R^3 \rho$

$$\therefore \qquad v_e = R\sqrt{\left(\frac{8}{3}\right)\pi G\rho}.$$

Example 18:

The masses and radii of the earth and the moon are M_1, R_1 and M_2, R_2 respectively. Their centres are d apart Show that the minimum speed with which a particle of mass m should be projected from a point midway between the two centres so as to escape to infinity is $2\sqrt{\frac{G(M_1 + M_2)}{d}}$.

Solution:

The particle m is at a distance d/2 from the centre of the earth (mass M_1) as well as from the centre of the moon (mass M_2). Therefore, the total gravitational potential energy of the particle due to the earth and the moon is

$$-\frac{GM_1m}{d/2} - \frac{GM_2m}{d/2} = -\frac{2G(M_1 + M_2)m}{d}.$$

Clearly, the work required to move the particle from the point midway between the earth and the moon to infinity would be 2 G (M_1 + M_2) m/d. Thus, the minimum velocity of projection for the particle m to escape is given by

$$\frac{1}{2}mv_e^2 = \frac{2G(M_1 + M_2)m}{d}$$

or
$$v_e = 2\sqrt{\frac{G(M_1 + M_2)}{d}}.$$

Example 19:

A projectile is fired vertically upward from earth's surface with a velocity of magnitude k v_e, where v_e is escape velocity and k < 1. Show that in the absence of air resistance the height to which it will reach, measured from earth's centre, is $R/(1 - k^2)$, where R is earth's radius.

Solution:

Suppose the projectile of mass m and projected with velocity k v_e, reaches a height h, measured from earth's centre. Its kinetic energy is

$$K = \frac{1}{2} m (k v_e)^2.$$

But v_e (escape velocity) = $\sqrt{2gR}$, where R is earth's radius.

$$\therefore \quad K = \frac{1}{2} mk^2 \times 2gR = mk^2 gR. \qquad \text{...(i)}$$

This is equal to the work required to carry the projectile to the height h.

Let r be the instantaneous height of the ascending projectile from earth's centre. The gravitational force on the projectile is

$$F = G\frac{Mm}{r^2},$$

where M is earth's mass. The work required to raise the projectile from r to r + dr is

$$dW = F\,dr = G\frac{Mm}{r^2}dr.$$

Therefore, the work required to raise the projectile from earth's surface (r = R) to the height h measured from earth's centre (r = h) is

$$W = \int_R^h \frac{GMm}{r^2}dr$$

$$= GMm\left[-\frac{1}{r}\right]_R^h = GMm\left[-\frac{1}{h} + \frac{1}{R}\right]$$

$$= GMm\frac{h - R}{hR} = \frac{GM}{R^2}\frac{m(h - R)R}{h}.$$

But $\dfrac{GM}{R^2} = g.$

$$\therefore \quad W = g\frac{m(h - R)R}{h}.$$

Equating it to the kinetic energy K, given by eq. (i), we have

$$mk^2gR = g\frac{m(h-R)R}{h}$$

or $$k^2 = \frac{h-R}{h} = 1 - \frac{R}{h}$$

or $$\frac{R}{h} = 1 - k^2$$

or $$h = \frac{R}{(1-k^2)}$$

This is the required expressio

Example 20:

Assuming that the interior of the earth can be treated as a homogeneous spherical mass in hydrostatic equilibrium, express the pressure within the earth as a function of distance r from the centre. Taking radius of the earth, $R = 6.3 \times 10^8$ cm and its uniform density, $\rho = 5.5$ gm/cm^3, calculate the pressure at the centre of the earth.

Solution:

Let R be the *radius of the earth* and r, the distance of a point P from its centre O where the pressure is to be determined.

Consider a shell of radius x and thickness dx, where $r < x < R$. Clearly, *volume of the shell* $= 4\pi x^2 dx$ and hence *its mass* $= 4\pi x^2 dx\rho$.

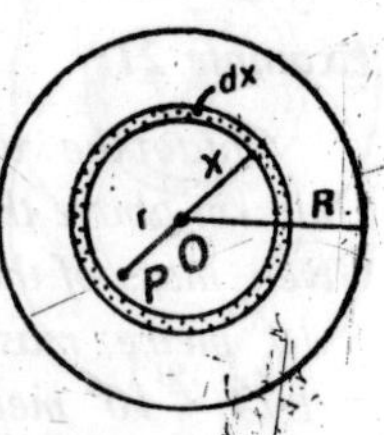

Fig. 2.48

And, *mass of the solid sphere of radius x enclosed by the shell* $= = \frac{4}{3}\pi x^3 \rho$.

Obviously, *gravitational pull exerted by the sphere on the shell inwards*

$$= \frac{\text{mass of the sphere} \times \text{mass of the shell}}{x^2} G$$

$$= -\frac{4\pi x^3 \rho \times 4\pi x^2 dx\rho}{3 \times x^2} G = -\frac{16}{3}\pi^2\rho^2 G x^3 dx.$$

Since the whole spherical mass (of the earth) is in hydrostatic equilibrium, the shell does not move inwards due to this force. This clearly means that an equal and opposite force $(16/3\pi^2\rho^2 G x^3 dx)$ must be acting radially outwards on its surface.

Since the area of the inner surface of the shell $= 4\pi x^2$, we have *outward pressure on the shell at distance x from the centre, say,*

$$dp = \frac{16}{3}\pi^2\rho^2 Gx^3\, dx/4\pi x^2$$

i.e., $$dp = \frac{4}{3}\pi\rho^2 Gxdx$$

Now, the solid sphere of radius x also attracts all other shells outside the shell of radius x, *i.e.*, all other shells of radii greater than x and less then R.

These shells too do not move inwards under its force of attraction, showing that there must be an equal and opposite force acting on their inner sides, pushing them outwards.

Hence, to determine the *total pressure p at the point P*, distant r from O, we must integrate the expression for dp between the limits x = r and x = R. Thus,

total pressure at P, *i.e.*, $p = \frac{4}{3}\pi\rho^2 G\int_r^R xdx = \frac{4}{3}\pi\rho^2 G\left[\frac{x^2}{2}\right]_r^R$

$$= \frac{2\pi}{3}\rho^2 G(R^2 - r^2).$$

Since at O, r = 0, we have pressure at the centre of the earth

$$= \frac{2\pi}{3}\rho^2 GR^2 = \frac{2\pi}{3}(5.5)^2(6.67\times10^{-8})(6.3\times10^8)^2 = 1.7\times10^{12}\text{ dyne/cm}^2.$$

Example 21:

Obtain the values of the escape velocity for an atmospheric particle 1000 km above the surface of (i) the earth, (ii) the moon, (iii) the sun. Given, mass of the earth = 5.98 × 10^{24} kg, radius of the earth = 6.37 × 10^6 metre; mass of the moon = 7.34 × 10^{22} kg, radius of the moon = 1.74 × 10^6 metre; mass of the sun = 1.99 × 10^{30} kg, radius of the sun = 6.96 × 10^8 metre; and gravitational constant G = 6.67 × 10^{-11} N – m^2/kg^2.

Solution:

We know that

$$\text{Escape velocity } V_6 = \sqrt{\frac{2MG}{R}}.$$

(i) *In the case of the earth.* M = 5.98 × 10^{24} kg and R = *radius of the earth* + 1000 km = 6.37 × 10^3 km + 1000 km = 7.37 × 10^3 km = 7.37 × 10^6 + 73.7 × 10^5 *metre.*

∴ *escape velocity of the atmospheric particle, i.e.,*

$$V_e = \sqrt{\frac{2 \times 5.98 \times 10^{24} \times 6.67 \times 10^{-11}}{73.7 \times 10^5}}.$$

$$= \sqrt{\frac{2 \times 5.98 \times 6.67}{73.7 \times 10^8}} = 1.04 \times 10^4 \text{ m/sec.}$$

(ii) *In the case of the moon,* M = 7.34 × 10^{22} kg and R = *radius of the moon* + 1000 km = 1.74 × 10^3 km + 1000 km = 27.4 × 10^5 *metre.*

∴ *escape velocity of the atmospheric particle, i.e.,*

$$V_e = \sqrt{\frac{2 \times 7.34 \times 10^{22} \times 6.67 \times 10^{-11}}{27.4 \times 10^5}} = \sqrt{\frac{2 \times 7.34 \times 6.67}{27.4 \times 10^6}}$$

$$= 1.89 \times 10^3 \text{ m/sec.}$$

(iii) *In the case of the sun,* M = 1.99 × 10^{30} kg and R = *radius of the sun* + 1000 km = 6.96 × 10^5 km + 1000 km = 69.7 × 10^7 *metre.*

∴ *escape velocity of the atmospheric particle, i.e.,*

$$V_e = \sqrt{\frac{2 \times 1.99 \times 10^{30} \times 6.67 \times 10^{-11}}{69.7 \times 10^7}} = \sqrt{\frac{2 \times 1.99 \times 6.67}{69.7 \times 10^{12}}}$$

$$= 6.172 \times 10^5 \text{ m/sec.}$$

Example 22:

Three electrical charges A, B, C, of values + 40, – 70, –12 esu respectively are placed such that distances AB, BC, CA are 10, 30, 30 cm respectively. Calculate the work required to separate the three charges to infinity.

Solution :

Let the three charges A, B, C be placed

The work required to separate them to infinity will obviously be equal to the potential energy of the system, with its sign reversed.

Now, *potential energy of the system,*

$$U = U_{AB} + U_{BC} + U_{CA} = \frac{AB}{r_{AB}} + \frac{BC}{r_{BC}} + \frac{AC}{r_{AC}}.$$

Or $$U = \frac{(+40)(-70)}{10} + \frac{(-70)(-12)}{30} + \frac{(+40)(-12)}{30}$$

$$= -280 + 28 - 16 = -268 \text{ ergs}$$

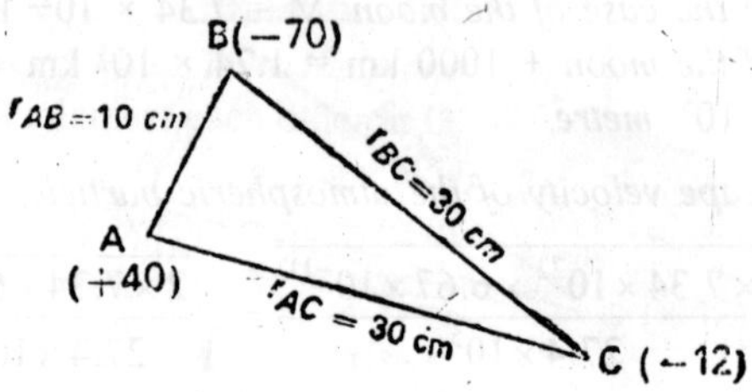

Fig. 2.49

Therefore, *work required to separate the three charges to infinity = 268 ergs.*

Example 23:

Determine the potential energy of a galaxy consisting of n stars, where $n = 1.6 \times 10^{11}$, each equal to the mass of the sun and with an average distance $r = 10^{21}$ metres between each pair of stars. Ignore the self energy of each star. (Take mass of the sun $\approx 2 \times 10^{30}$ kg and $G = 7 \times 10^{-11}$ N – M/kg^2).

Solution:

As we have *gravitational potential energy or self energy of n stars* is given by $$U_s = -\frac{1}{2} G \sum_{i=1}^{n} \sum_{\substack{j=1 \\ \neq i}}^{n} \frac{m_i m_j}{r_{ij}}.$$

Here, $m_i = m_j = m$ and there are n equal terms in $\sum_{i=1}^{n}$ and $(n - 1)$ terms in $\sum_{\substack{j=1 \\ \neq i}}^{n}$

So that, $\sum_{i=1}^{n}$ and $\sum_{\substack{j=1 \\ \neq i}}^{n} = (n - 1)$

$$\therefore \qquad U_s = -\frac{1}{2}Gn(n-1)\frac{m^2}{r}.$$

Substituting the given values, therefore, we have *gravitational potential energy of the galaxy*

$$= -\frac{1}{2}\frac{7\times10^{-11}\times1.6\times10^{11}\times\left(1.6\times10^{11}-1\right)\times\left(2\times10^{30}\right)^2}{10^{21}}$$

$$\approx -4\times10^{51} \text{ joules.}$$

Example 24:

Calculate the self energy of the sun, taking its mass to be equal to 2 × 10^{30} kg and its radius to be very nearly 7 × 10^{8} metre. (G = 7 × 10^{-11} N–m/kg^2).

If its radius contract by 1 km per year, without affecting its mass, calculate the rate at which it radiates out energy.

Solution:

We know that the self energy of a solid sphere is given by

$$U_s = -\frac{3}{5}\frac{M^2}{R}G.$$

$$\therefore \textit{ self energy of the sun} = -\frac{3}{5}\frac{\left(2\times10^{30}\right)^2}{7\times10^{8}}\times7\times10^{11} = -\frac{12}{5}\times10^{41}$$

$$= -2.4\times10^{41} \text{ joules.}$$

Now, *rate of change of energy of the sun, i.e.,* $\frac{dU_3}{dt} = \frac{dU_3}{dR}.\frac{dR}{dt}$

$$= \frac{3}{5}\frac{M^2G}{R^2}.\frac{dr}{dt}.$$

Here, $\frac{dR}{dt} = \frac{1000}{365\times24\times60\times60}$ metre/sec (∵ 1 km = 1000 metres)

∴ *rate of energy radiated out by the sun*

$$= \frac{3}{5}\frac{M^2G}{R^2}\frac{dR}{dt} = \frac{2.4\times10^{41}}{7\times10^{8}}\times\frac{1000}{365\times24\times3600}$$

$$= 1.087\times10^{28} \text{ joules/sec.}$$

Example 25:

Charges of + 100 esu are placed at the corners of a square of 10 cm ride and a charge of $-50\sqrt{2}$ *esu at the point of intersection of its diagonals. Calculate (i) the resultant force on each charge, (ii) the potential energy of the system, (iii) work done in separating the charge to infinity.*

If the charges We initially infinite distance apart, what would be the work done in assembling them into their present configuration?

Solution:

Let ABCD be the square of side 10 cm, at the four corners of which charges of 100 *esu* each are placed and let a charge of $-50\sqrt{2}$ *esu* be placed at the point of intersection O of its diagonals.

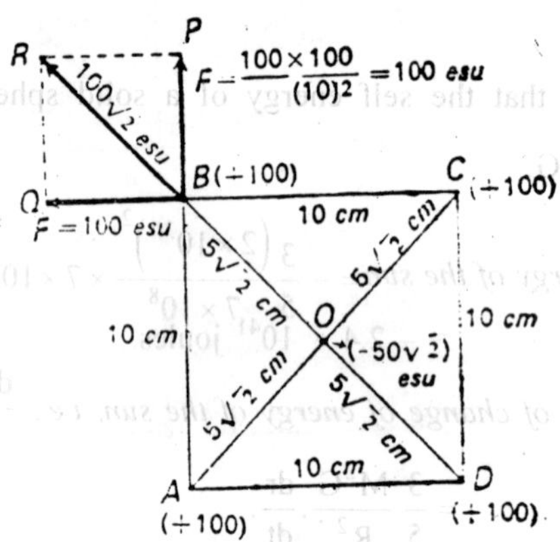

Fig. 2.50

Then, clearly, diagonal AC = BD $= \sqrt{10^2 + 10^2} = 10\sqrt{2}$ cm.

And, therefore, AO = OC = OB = OD $10\sqrt{\frac{2}{2}} = 5\sqrt{2}$ cm.

(i) Hence, *force on the charge at O due to the charge at A*

$$= (100)\frac{(-50\sqrt{2})}{(5\sqrt{2})^2} = -\frac{5000\sqrt{2}}{50} = -100\sqrt{2} \text{ dynes along OA.}$$

And, *force on the charge of O due to the charge at C*

$$= (100)\frac{\left(-50\sqrt{2}\right)}{\left(5\sqrt{2}\right)^2} = -\frac{5000\sqrt{2}}{50} = -100\sqrt{2} \textit{ dynes along } \text{OC},$$

The two, being *equal and opposite,* cancel out.

Similarly, forces on the charge at O due to the charges at B and D are $-100\sqrt{2}$ *dynes* and $-100\sqrt{2}$ *dynes* along OB and OD respectively and they too, being *equal and opposite,* cancel out.

So, t*he resultant force on the charge at O = 0.*

Now, force on the charge at B due to the charge at A = 100 × 100·/$(10)^2$ *dynes along* AB.

Let it be represented in magnitude and direction by BP.

Similarly, force on the charge at B due to the charge at C = 100 × 100/$(10)^2$ = 100 dynes along CB. Let it be represented by BQ.

Then, the *resultant* force *on the charge at B due to the charges at A and C* is represented by $BR = \sqrt{100^2 + 100^2} = 100\sqrt{2}$ *dynes, along* OB.

And, force on the charge at B due to the charge at

$$O = \frac{100\left(-50\sqrt{2}\right)}{\left(5\sqrt{2}\right)^2} = -100\sqrt{2} \textit{ dynes } \text{along BO}.$$

These too, again being equal and opposite, the resultant force on the charge at B is zero.

Similarly, it can be shown that the resultant force on the charges at A, C and D is also zero.

In other words, *the resultant force on each charge of the system is zero.*

(ii) The *potential energy of the system of charges* is

$$= \frac{(100)(100)}{10} + \frac{(100)(100)}{10} + \frac{(100)(100)}{10} + \frac{(100)(100)}{10}$$

$$-\frac{(100)\left(50\sqrt{2}\right)}{5\sqrt{2}} - \frac{(100)\left(5\sqrt{2}\right)}{5\sqrt{2}} - \frac{(100)\left(50\sqrt{2}\right)}{5\sqrt{2}} - \frac{(100)\left(50\sqrt{2}\right)}{5\sqrt{2}}$$

$$= \frac{4(100)(100)}{10} - \frac{4(100)(50\sqrt{2})}{5\sqrt{2}} = 4000 - 4000 = 0.$$

(iii) Since the work done in separating the charges to infinity is equal to the potential energy of the system, with its sign reversed, we have

work done in separating the charges to infinity = 0.

Finally, *work done in assembling the charges into their present configuration,* being equal to the work done in separating them to infinity from their present configuration, is also equal to *zero.*

Example 26:

Calculate the electrostatic self energy of a (i) conducting, (ii) non-conducting sphere of radius 10 cm carrying a charge of 100 esu.

Solution:

(i) In the case of a *conducting sphere,* the charge resides only on its surface and it, therefore, *behaves as a charged spherical shell.* Its electrostatic potential energy or self energy is thus given by $U_s = - Q^2/2R$.

Substituting the given values of Q and R, therefore, we have

electrostatic self energy of the conducting sphere

$$= (100)^2/2\ (10) = 10000/20 = 500\ ergs,$$

(ii) In the case of a *non-conducting sphere,* the charge is spread over its entire volume and its *electrostatic self energy* is given by $U_s = 3/5\ (Q^2/R)$.

So that, *electrostatic self energy of the non- conducting sphere*

$$= \frac{3}{5} \times (100)^2/10 = \frac{3}{5}\ (10000/10) = 600\ ergs.$$

Example 27:

A meteor of mass 500 kg falls to the surface of the earth. How does the potential energy of the meteor-earth system change? If the meteor be supposed to start from rest, with what velocity does it strike the earth? (Radius of the earth = 6.37 × 10^6 metre; g = 9.80 m/sec^2)

Solution:

Clearly, potential energy (or self energy) of the meteor-earth system will decrease by MmG/R when the meteor falls to the surface of the earth, where M is the *mass of the earth,* m, the *mass of the meteor* and R, the *radius of the earth,* or the distance of the meteor (now lying on the surface of the earth) from the centre of the earth.

Now, as we know, $MG/R^2 = g$ and, therefore, $MG/R = (MG/R^2)R = gR$.

$\therefore$ *loss in P.E. of meteor-earth system* $= mgR = 500 \times 9.8 \times 6.37 \times 10^6 = 3.121 \times 10^{10}$ J.

Obviously, *loss in P.E. of the system = gain in K.E. of the meteor.*

Since the meteor is supposed to start from rest, the gain in its K E. $= 1/2\, mv^2$, where v is the *velocity with which it strikes the earth.* We, therefore, have

$1/2\, mv^2 = mgR$. Or, $v^2 = \sqrt{2gR}$ = *velocity of escape from the surface of the earth*

$$= \sqrt{2 \times 9.8 \times 6.37 \times 10^6} = 11.17 \times 10^3 \text{ m/sec} \approx 11.20 \text{ km/sec.}$$

The meteor thus strikes the earth with a velocity 11.20 km/sec.

Example 28:

The radius of the Moon's orbit r is 240,000 miles and the period of revolution is 27 days; the diameter of the Earth is 8000 miles and the value of gravity on its surface is 32 ft/sec². Verify the statement that the gravitational force varies inversely as the square of the distance.

Solution:

Clearly, velocity of the Moon,

$$v = \frac{\text{distance covered}}{\text{time taken}} = \frac{2\pi \times 240000 \times 1760 \times 3}{27 \times 24 \times 60 \times 60} \text{ ft/sec.}$$

$\therefore$ centripetal acceleration of the Moon towards the centre of the earth, *i.e.*,

$$gM = \frac{v^2}{r} = \left(\frac{2\pi \times 24 \times 176 \times 3 \times 10^5}{27 \times 24 \times 36 \times 10^2}\right)^2 \times \left(\frac{1}{24 \times 176 \times 3 \times 10^5}\right)$$

$$= \frac{4\pi^2 \times 176 \times 3 \times 10}{(27 \times 36)^2 \times 24} = 0.009189 \text{ ft/sec}^2.$$

Let the gravitational force be inversely proportional to the nth power of distance. Then, if the acceleration due to gravity on the surface of the Earth be g_E (equal to 32 ft / sec²), we have

$$g_M/g_E = \left(\frac{R}{r}\right)^n, \text{ where R is the } \textit{radius of the Earth,}$$

$$\text{Or,} \quad \frac{0.009189}{32} = \left(\frac{4000 \times 1760 \times 3}{240000 \times 1760 \times 3}\right)^n = \left(\frac{1}{60}\right)^n$$

Or, taking logarithms, we have

$$\bar{4}.4582 = n(\bar{2}.2218). \quad \text{Or, } n = \bar{4}.4582 / \bar{2}.2218 \approx 2.$$

Thus, g and *hence the gravitational force varies inversely as the square of the distance.*

Example 29:

Estimate the mass of the sun, assuming the orbit of the Earth round the sun to be a circle. The distance between the sun and the Earth is 1.49×10^{13} cm and G = 6.66 × 10^{-8}c.g.s units.

Solution:

Here, clearly, the *force of gravitational attraction* between the Sun and the Earth = MmG/r^2, where M is the *mass of the Sun. m, the mass of the Earth* and r, the distance between them or the *radius of the Earth's orbit round the Sun.*

This, then, is the *centripetal force* mv^2/r acting on the Earth as it goes round the Sun with a uniform speed v. So that, we have

$$MmG/r^2 = mv^2/r, \qquad \text{whence,} \quad M = v^2r^2/rG = v^2r/G.$$

Obviously, $v = 2\pi r/T$, where T the time taken by the Earth to make one complete round of the Sun, *i.e.*, 365 *days* or 365 × 24 × 3600 *sec.*

$$\text{We, therefore, have} \quad M = \frac{(2\pi r / T)^2 r}{G} = \frac{4\pi^2 r^3}{T^2 G}$$

Or, substituting the values of r, T, and G, we have

mass of the Sun.

$$M = \frac{4\pi^2 (1.49 \times 10^{13})^2}{(365 \times 24 \times 3600)^2 \times 6.66 \times 10^{-8}} = 19.72 \times 10^{24} \text{ gm.}$$

Example 30:

Show that the intensity and potential at any point on the surface of the earth are g and gR respectively, assuming the earth to be a uniform sphere.

Solution:

Let the earth be a uniform sphere of radius R and mass M. Then, clearly, *intensity of the gravitational field at any point on its surface,*

$$E = \frac{M}{R^2}G.$$

Now, weight of a body of mass m on the surface of the earth or the gravitational force on it due to attraction by the earth = mg = $-mMG/R^2$, where g is the acceleration due to gravity at the place.

$\therefore$ *force on unit mass on the surface of the earth or intensity of the gravitational field on the surface of the earth, i.e.,* $E = -\frac{M}{R^2}G = g.$

This is the reason why g is also quite often referred to as *intensity of gravity.*

And, since *potential at any point on the surface of the earth,*

i.e., $$V = -\frac{M}{R}G,$$

we have $$V = -\left(\frac{M}{R^2}G\right)R = gR.$$

Example 31:

Two bodies of masses M_1 and M_2 are placed distance d apart. Show that at the position where the gravitational field due to them is zero, the potential is given by

$$V = -\frac{G}{d}\left(M_1 + M_2 + 2\sqrt{M_1M_2}\right).$$

Solution:

Let the gravitational field E be *zero* at a point *distant* x *from mass* M_1, and, therefore, distant (d–x) *from mass* M_2. Then, clearly,

$$-\frac{M_1}{x^2}G = -\frac{M_2}{(d-x)^2}G.$$

Or $$\frac{x}{d-x} = \sqrt{\frac{M_1}{M_2}}.$$

Or $$\frac{x}{d} = \frac{\sqrt{M_1}}{\sqrt{M_1}+\sqrt{M_2}},$$

whence, $$x = \frac{d\sqrt{M_1}}{\sqrt{M_1}+\sqrt{M_2}}$$ and, therefore, $(d-x) = d - \dfrac{d\sqrt{M_1}}{\sqrt{M_1}+\sqrt{M_2}}$

Or, $$d - x = \frac{d\sqrt{M_1}}{\sqrt{M_1}+\sqrt{M_2}}.$$

Hence, *potential at the point due to the two masses*

$$= -\left(\frac{M_1}{x}G + \frac{M_2}{(d-x)}G\right)$$

$$= -G\left[\frac{M_1\left(\sqrt{M_1}+\sqrt{M_2}\right)}{d\sqrt{M_1}} + \frac{M_2\left(\sqrt{M_1}+\sqrt{M_2}\right)}{d\sqrt{M_2}}\right]$$

$$= -\frac{\sqrt{M_1}+\sqrt{M_2}}{d}G\left(\sqrt{M_1}+\sqrt{M_2}\right).$$

Or, *potential at the point*

$$= -\frac{G}{d}\left(\sqrt{M_1}+\sqrt{M_2}\right)^2 = -\frac{G}{d}\left(M_1 + M_2 + 2\sqrt{M_1M_2}\right).$$

Example 32:

A smooth straight tunnel is bored through the earth and a small particle is allowed to move in it from a position of rest. Find the periodic time of one vibration. Given that G = 6.67 × 10^{-8} C.G.S. units and the mean density of earth = 5.6 gm per c.c.

Solution:

Let a tunnel AB be bored through the earth and let a particle of mass m be placed in it at P, such that its distance from the centre O of the earth is r and that from the mid-point C of the tunnel, x. And, let the angle OPC be θ.

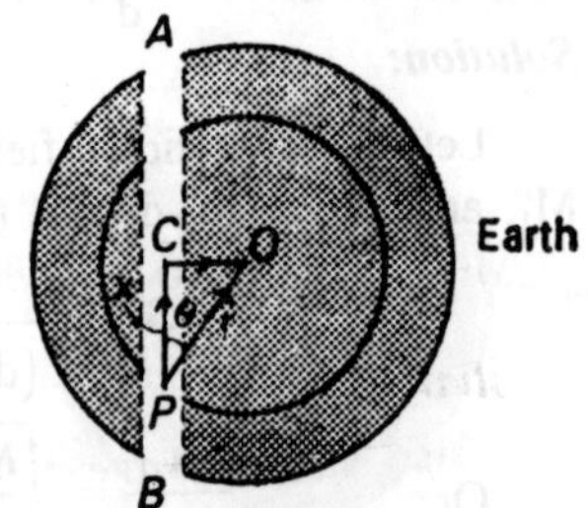

Fig. 2.51

As we know, the only force of attraction on the particle will be due to a sphere of radius OP = r. (shown dotted in the figure), on the surface of which the particle lies, there being no effect on it due it to the outer shells around this sphere. So that, *gravitational force acting on the particle, i.e.,*

$$F = -\frac{m(\text{mass of sphere of radius r})}{r^2}G,$$

the –ve sign indicating that it is directed towards O, the centre of the sphere.

Or, $$F = -\frac{m\left(\frac{4}{3}\pi r^3\rho\right)}{r^2}G = -\frac{4}{3}\pi r\rho mG.$$

And, clearly, *component of this force acting on the particle along PC*

$$= F\cos\theta = -\frac{4}{3}\pi r\rho mG\cos\theta = -\frac{4}{3}\pi r\rho mG\times\frac{x}{r} = -\frac{4}{3}\pi\rho mGx,$$

directed towards C, the mid-point of the tunnel.

∴ *acceleration of the particle towards*

$$C = \frac{\text{force}}{\text{mass}} = -\frac{\frac{4}{3}\pi\rho mGx}{m} = \frac{4}{3}\pi\rho Gx$$

$$= -\mu x, \text{ say, where } \frac{4}{3}\pi\rho G = \mu = \text{constant.}$$

Thus, the acceleration of the particle is proportional to its displacement (x) from C and is directed towards it. It, therefore, executes a *simple harmonic motion* about C and its *time-period* is given by

$$T = 2\pi\sqrt{\frac{1}{\mu}} = 2\pi\sqrt{\frac{3}{4\pi\rho G}} = \sqrt{\frac{3\pi}{\rho G}} = \sqrt{\frac{3\pi}{5.6\times6.67\times10^{-8}}} = 5023\text{ sec}$$

$$= 83.72 \text{ min} \approx 84 \text{ min},$$

Example 33:

Estimate the temperature of the interior of the sun, taking its mass to be nearly equal to 2 × 10^{33} gm, its radius to be nearly equal to 7 × 10^{10}cm., G = 7 × 10^{-8} C.C.S. units and the average mass of an atom in the sun to be equal to 3 × 10^{-24} gm.

Solution:

We know that the potential energy (or self energy) of the sun, *i.e.*

$$U_s = -\frac{3}{5}\frac{M^2}{R}G,$$ where M is its *mass* and R its *radius*.

In accordance with the *virial theorem*, we have *average K.E. of the atoms in the sun* = –1/2 (*average P.E. of the sun*)

Now, *K.E. of the atom* = 3/2 kT, where k is the familiar *Boltzmann constant* = 1.38×10^{-16} erg/^{0}K and T, the average temperature of the sun, measured is degrees kelvin (or K).

If, therefore, there be N atoms in all in the sun, their *average* K.E. = 3/2 NkT.

We, thus, have $$\frac{3}{2}\text{NkT} = -\frac{1}{2}\left(-\frac{3}{5}\frac{M^2}{R}G\right) = \frac{3M^2G}{10R},$$

Whence, $$T = \frac{3M^2G}{10R} \times \frac{2}{3Nk} = \frac{M^2G}{5RNk} = \frac{MMG}{5RNk}.$$

• Clearly, M/N is the *average mass of an atom in the sun* = 3×10^{-24}gm.

So that, $$T = \frac{MG}{5Rk}\left(3\times10^{24}\right) = \frac{2\times10^{33}\times7\times10^{-8}\times\left(3\times10^{-24}\right)}{5\times7\times10^{10}\times1.38\times10^{-16}}.$$

$$= \frac{6\times10^7}{6.9} \approx \left(10^7\right)^0\text{K}.$$

Thus, *the average temperature of the interior of the sun* = $(10^7)^0$ K.

Example 34:

Calculate the gravitational potential energy of a system of 8 masses of 10 kg each placed at the corners of cube of each edge equal to 0.25 metre, (G = 6.67×10^{11} N – m/kg^2).

Solution:

Figure 2.37 shows a cube of each edge, 0.25 metre in length, with a mass of 10 kg placed at each corner. The gravitational potential energy or self energy of the system is thus obviously given by

$$U_s = -G\sum_{i>j}^{8}\sum_{j=1}^{8}\frac{mi\,m}{rij}, \text{ where } mi = mj = 10\text{kg}.$$

Now, each of the *twelve* distances, $r_{21}, r_{41}, r_{51}, r_{32}, r_{62}, r_{42}, r_{73}, r_{84}, r_{65}, r_{85}, r_{76}$ and r_{87} (*all edges of the cube*) is equal to r, each of the *twelve* distances, $r_{31}, r_{61}, r_{81}, r_{42}, r_{52}, r_{72}, r_{63}, r_{63}, r_{54}, r_{74}, r_{75}$ and r_{36} (*all face-diagonals of the cube*) is equal to $\sqrt{r^2+r^2} = \sqrt{2}r,$ and each of the *four distances*, r_{71}, r_{82}, r_{53} and r_{64} (all long diagonals, or distances between opposite corners, of the cube) is equal to $\sqrt{2r^2+r^2} = \sqrt{3r^2} = \sqrt{3}r.$

$$\therefore U_3 = -Gm^2\left(\frac{12}{r}+\frac{12}{\sqrt{2}r}+\frac{4}{\sqrt{3}r}\right)$$

$$= -\frac{4Gm^2}{r}\left(3+\frac{3}{\sqrt{2}}+\frac{1}{\sqrt{3}}\right) = -\frac{4Gm^2}{r}(5.7).$$

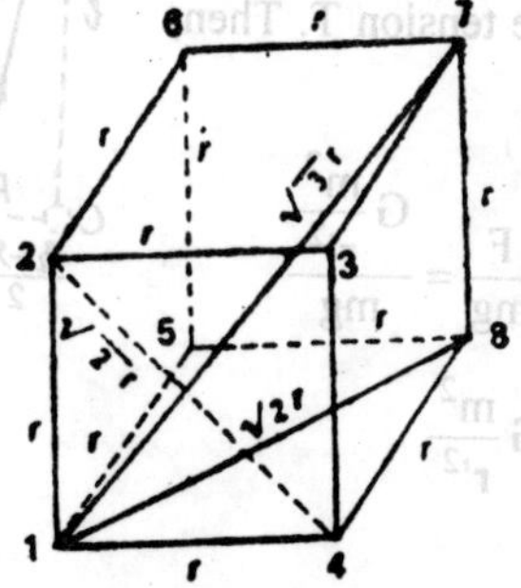

Fig. 2.52

Substituting the given values of G, m and r, therefore, we have

gravitational potential energy of the system,

$$U_s = -\frac{4\times 6.67\times 10^{-11}\times(10)^2}{0.25} = -1.067\times 10^{-7} \text{ joule.}$$

Example 35:

Two balls of mass m each are hung side by side by two long threads of same length l. If the distance between the upper ends is r, show that the distance r' between the centres of the balls is given by

$$gr'^2\ (r - r') = 2lGm.$$

Solution:

The problem is shown in Fig. Each ball is in equilibrium under the following three forces ;

(i) weight of the ball, mg

(ii) force of gravitational attraction

$$F = G\frac{mm}{r'^2}$$

(iii) tension in the thread, T.

These three forces can be represented by the sides of a triangle, say ACP, such that AC represents the weight mg, CP represents the force F and PA represents the tension T. Then, clearly

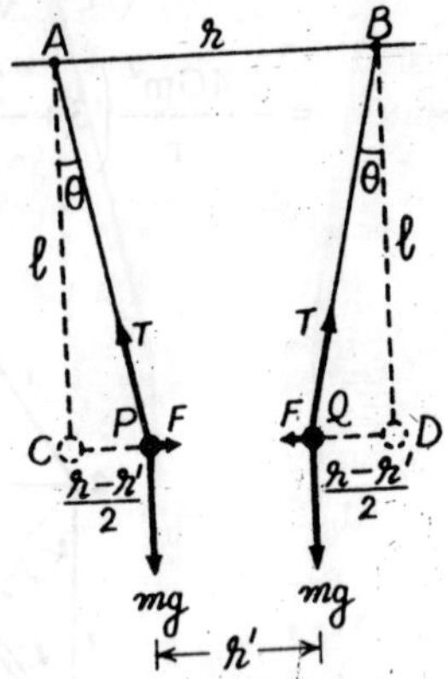

$$\tan\theta = \frac{\frac{r-r'}{2}}{l} = \frac{F}{mg} = \frac{G\frac{m^2}{r'^2}}{mg}$$

or $$\frac{r-r'}{2}mg = lG\frac{m^2}{r'^2}$$

Fig. 2.53

Example 36:

Two particles of masses m_1 and m_2, initially at rest at infinite distance from each other, move under the action of mutual gravitational pull. Show that at any instant their relative velocity of approach is $\sqrt{\frac{2G(m_1+m_2)}{R}}$, where R is their separation at that instant.

Solution:

The gravitational force of attraction on m_1 due to m_2 at a separation r is

$$F_1 = \frac{Gm_1m_2}{r^2}.$$

Therefore, the acceleration of m_1 is

$$a_1 = \frac{F_1}{m_1} = \frac{Gm_2}{r^2}.$$

Similarly, the acceleration of m_2 due to m_1 is

$$a_2 = -\frac{Gm_1}{r^2},$$

the negative sign being put as a_2 is directed opposite to a_1. The relative acceleration of approach is

$$a = a_1 - a_2 = \frac{G(m_1+m_2)}{r^2} \quad \text{...(i)}$$

If v is the relative velocity, then

$$a = \frac{dv}{dt} = \frac{dv}{dr}\frac{dr}{dt}.$$

But $-\frac{dr}{dt} = v$ (negative sign shows that r decreases with increasing t).

$$\therefore \qquad a = -\frac{dv}{dr}v. \qquad \text{...(ii)}$$

From eq. (i) and (ii), we have

$$v\,dv = -\frac{G(m_1 + m_2)}{r^2}dr.$$

Integrating, we get

$$\frac{v^2}{2} = \frac{G(m_1 + m_2)}{r} + C.$$

At $r = \infty$, $v = 0$ (given), and so $C = 0$.

Example 37:

Three identical bodies of mass M each are located at the vertices of an equilateral triangle with side L. At what speed must they move if they all revolve under the influence of one another's gravity in a circular orbit circumscribing the triangle while still preserving the equilateral triangle?

Solution:

Let A, B, C be the three masses and O the centre of the circumscribing circle. The radius of this circle is

$$R = \frac{L}{2}\sec 30° = \frac{L}{2} \times \frac{2}{\sqrt{3}} = \frac{L}{\sqrt{3}}.$$

Let v be the speed of each mass M along the circle. Let us consider the motion of the mass at A. The force of gravitational attraction on it due to the masses at B and C are

$$\frac{GM^2}{L^2} \text{ along AB}$$

and $$\frac{GM^2}{L^2} \text{ along AC.}$$

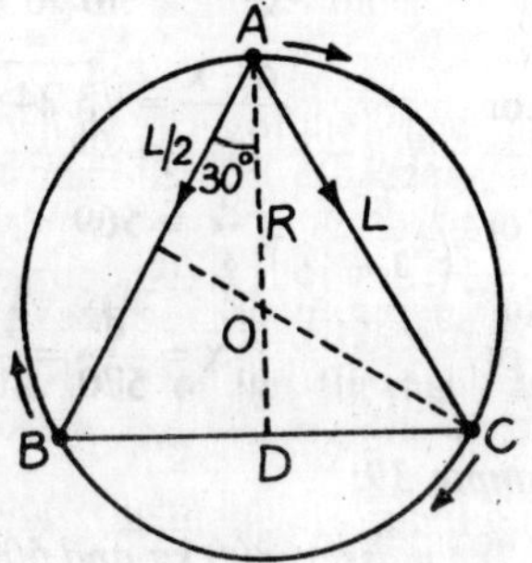

Fig. 2.54

The resultant force is therefore

$$2\frac{GM^2}{L^2}\cos 30° = \frac{\sqrt{3}GM^2}{L^2} \text{ along AD.}$$

This, for preserving the triangle, must be equal to the necessary centripetal force. That is,

$$\frac{\sqrt{3}\,GM^2}{L^2} = \frac{Mv^2}{R} = \frac{\sqrt{3}\,Mv^2}{L} \qquad \left[\because R = \frac{L}{\sqrt{3}}\right]$$

or
$$v = \sqrt{\frac{GM}{L}}.$$

Example 38:

How far from the earth must a body be along a line toward the sun so that the sun's gravitational pull balances the earth's? The mass of the sun is 3.24×10^5 times the mass of earth. The distance between sun and earth is 1.5×10^{11} m.

Solution:

Let d be the distance between earth and sun. Suppose the body is placed at a distance x from earth where the sun's gravitational pull balances the earth's. Then, we have

$$\frac{GM_e m}{x^2} = \frac{GM_s m}{(d-x)^2},$$

where M_e and M_s are the masses of earth and sun respectively. This gives

$$M_e\,(d - x)^2 = M_s x^2$$

or
$$\frac{(d-x)^2}{x^2} = \frac{M_s}{M_e} = 3.24 \times 10^5 \text{ (given)}$$

or
$$\frac{d-x}{x} = \sqrt{3.24 \times 10^5} = 569$$

or
$$\frac{d}{x} = 569 + 1 = 570$$

or
$$x = \frac{d}{570} = \frac{1.5 \times 10^{11}\,m}{570} = 2.63 \times 10^8 \text{ m.}$$

Example 39:

The masses, 800 kg and 600 kg, are at a distance 0.25 meter apart. Compute the gravitational potential and field at a point distant 0.20 meter from the 800-kg mass and 0.15 meter from the 600-kg mass. ($G = 6.67 \times 10^{11}$ N-m²/kg²).

Solution:

We have to compute the potential and field at the point P in Fig.

Potential at P due to 800 kg is

$$V_A = -G\frac{800\text{kg}}{0.20\text{m}} = -(4000\text{kg/m})G$$

and that due to 600 kg is

$$V_B = -G\frac{600\text{kg}}{0.15\text{m}} = -(4000\text{kg/m})G.$$

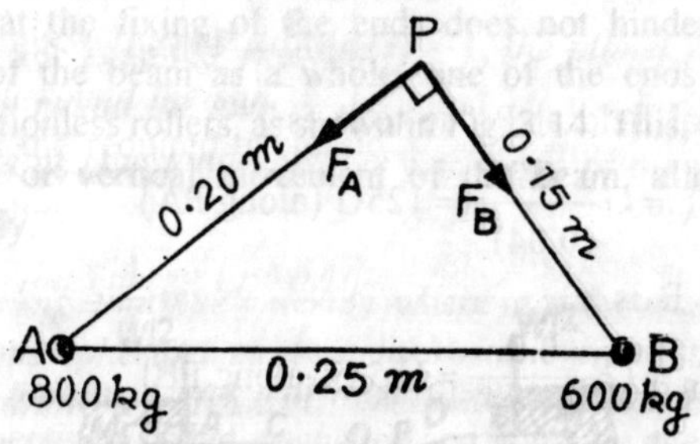

Fig. 2.55

Since the potential is scalar, the potential at P is

$$\begin{aligned} V &= V_A + V_B \\ &= -(4000 \text{ kg/m})\, G - (4000 \text{ kg/m})\, G \\ &= -(8000 \text{ kg/m})\, G = -(8000 \text{ kg/m}) \times (6.67 \times 10^{-11} \text{ N-m}^2/\text{kg}^2) \\ &= -5.34 \times 10^{-7} \text{ N m/kg} = -5.34 \times 10^{-7} \text{ J/kg.} \end{aligned}$$

Now, the field (attraction) at P due to 800 kg is

$$F_A = G\frac{800}{(0.20)^2} = 20000\text{ G, along PA}$$

and that due to 600 kg is

$$F_B = G\frac{600}{(0.15)^2} = \frac{80000}{3}G\text{, along PB.}$$

The distances AB, AP and BP show that $\angle APB$ is 90°.

Therefore, the resultant field is

$$\begin{aligned} F &= \sqrt{F_A^2 + F_B^2} = G\sqrt{(20000)^2 + \left(\frac{80000}{3}\right)^2} \\ &= 6.67 \times 10^{-11} \times \frac{100000}{3} \\ &= 2.22 \times 10^{-6} \text{ N/kg.} \end{aligned}$$

Example 40:

Masses of 0.2 kg and 0.8 kg are placed 0.12 meter apart, find the gravitational field and potential at a point distant 0.04 meter from the smaller mass and collinear with them. Compute the work required to shift a unit mass from this point to another point distant 0.04 meter from the larger mass along the same line. ($G = 6.67 \times 10^{-11}$ $N\text{-}m^2/kg^2$)

Solution:

Gravitational field at P due to A is

$$F_A = G\frac{0.2}{(0.04)^2} = 125G \text{ (along PA)}$$

and that due to B is

$$F_B = G\frac{0.8}{(0.08)^2} = 125\,G \text{ (along PB)}.$$

Since F_A and F_B are equal and opposite, the resultant field at P is zero.

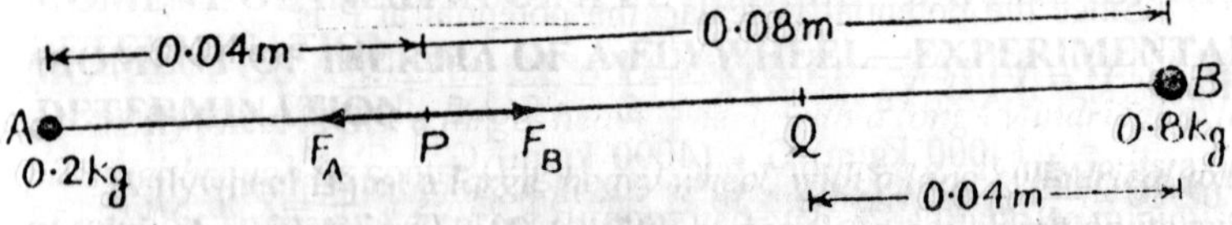

Fig. 2.56

Now, potential at P due to A is

$$V_A = -G\frac{0.2}{0.04} = -5.0G$$

and that due to B is

$$V_B = -G\frac{0.8}{0.08} = -10G.$$

Since potential is a scalar quantity, the resultant potential at P is

$$V_p = V_A + V_B$$

$$= -\,5.0\,G - 10\,G = -\,15\,G$$

$$= -\,15 \times 6.67 \times 10^{-11} = -\,1.0 \times 10^{-9}\ \text{J/kg/}$$

If we take a point Q such that QB = 0.04 m, then the resultant potential at Q would be

$$V_Q = -G\left[\frac{0.2}{0.08} + \frac{0.8}{0.04}\right] = -22.5G.$$

The work required to shift a unit mass from P to Q will be given by

$$W = V_Q - V_P = -22.5\ G - (-15\ G) = -7.5\ G$$
$$= -7.5 \times 6.67 \times 10^{-11} = -5.0 \times 10^{-10}\ \text{J/kg.}$$

Example 41:

Two concentric spherical shells of uniform density of masses M_1 and M_2 are situated as shown in figure. Find the force on a particle of mass m when the particle is located at (a) r = a, (b) r = b and (c) r = c. The distance r is measured from the centre of the shells.

Solution:

(a) The point a is external to both the shells M_1 and M_2, which behave as if their mass is concentrated at the centre. The magnitude of the gravitational fields at a due to M_1 and M_2 are $\frac{GM_1}{a^2}$ and $\frac{GM_2}{a^2}$, and both are directed along the line joining the (common) centre and the point. Hence the magnitude of the total field is

$$\frac{G(M_1 + M_2)}{a^2}$$

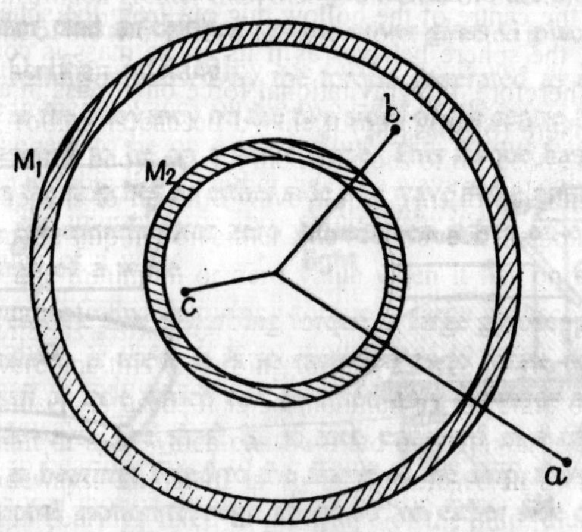

Fig. 2.57

The force on a particle of mass m placed at a is therefore

$$\frac{G(M_1 + M_2)m}{a^2}.$$

(b) The point b is external to the shell M_2 and internal to M_1. There in no field at an internal point due to a shell. Hence the field at b is due only to M_2 (which behaves as if its entire mass is at its centre) and is

$$\frac{GM_2}{b^2}.$$

The force on the particle m is therefore

$$\frac{GM_2m}{b^2}.$$

(c) The point c is internal to both M_1 and M_2. Hence, the field, and also the force on m placed at c, would be *zero*.

Example 42:

A solid sphere of lead has mass M and radius R. A spherical hollow is dug out from it, as shown in Fig, its boundary passing through the centre O and also touching the boundary of the solid sphere. Deduce the gravitational force on a mass m placed at P, which is distant r from O along the line of centres.

Solution:

Let O′ be the centre of the hollow dug out from the sphere. For an external point, the sphere behaves as if its entire mass is concentrated at its centre. Therefore, the gravitational force on a mass m at P due to the original sphere (of mass M) is

$$F = G\frac{Mm}{r^2}, \text{ along PQ.}$$

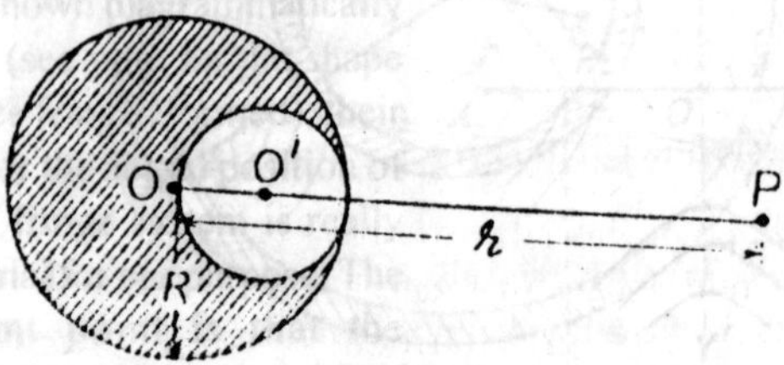

Fig. 2.58

The diameter of the smaller sphere (which would be cut off) is R, so that its radius OO' is R/2. The force on m at P due to this sphere of mass M' (say) would be

$$F' = G\frac{M'm}{\left(r - \frac{R}{2}\right)^2}, \text{ along PO'}. \qquad \left[\because \text{distance PO'} = r - \frac{R}{2}\right]$$

As the radius of this sphere is half of that of the original sphere, we have

$$M' = \frac{M}{8}.$$

$$\therefore \qquad F' = G\frac{Mm}{8\left(r - \frac{R}{2}\right)^2}, \text{ along PO'}.$$

As both F and F' point along the same direction, the force due to the hollowed sphere is

$$F - F' = \frac{GMm}{r^2} - \frac{GMm}{8r^2\left(1 - \frac{R}{2r}\right)^2}$$

$$= \frac{GMm}{r^2}\left\{1 - \frac{1}{8\left(1 - \frac{R}{2r}\right)^2}\right\}.$$

Example 43:

Derive an expression for the gravitational field inside a sphere of radius R, when the mass density at a point is $\rho = a + br^2$, where r is the distance of the point from the centre of the sphere.

Solution:

Let O the centre of the sphere of radius R, and P the *internal* point at which the gravitational field is required. Let OP = r.

Let us draw a concentric sphere through P. The point P is now external for the outer solid sphere of radius r, which alone exerts gravitational attraction at P. The outer spherical shell of internal radius r and external radius R, for which P is internal, exerts no attraction at P.

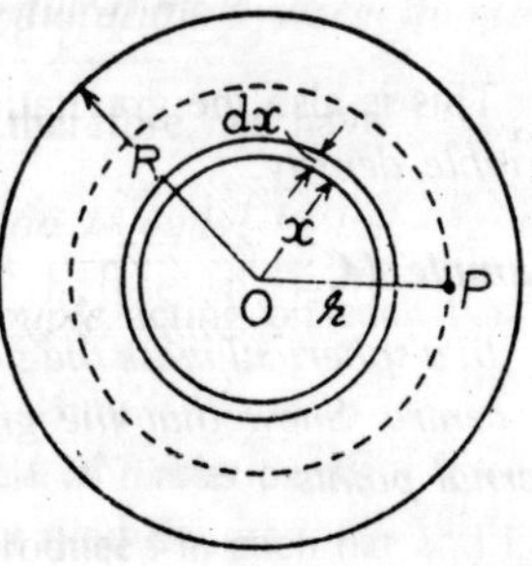

Fig. 2.59

Let us evaluate the mass of the inner solid, sphere. For this, we divide it into a large number of thin concentric shells and consider one such shell of radius x and infinitesimally small thickness dx. The volume of this elementary shell is $4\pi x^2\, dx$. The density at a distance x from O is $a + bx^2$ (as given) and can be considered uniform over the thickness dx. Thus, the mass of the elementary shell is

$$\text{volume} \times \text{density} = 4\pi x^2\, dx \times (a + bx^2).$$

Therefore, the mass of the inner solid sphere of radius r is

$$M' = 4\pi \int_0^r (a + bx^2)x^2 dx$$

$$= 4\pi \left[\frac{ax^3}{3} + \frac{bx^5}{5}\right]_0^r$$

$$= 4\pi \left(\frac{ar^3}{3} + \frac{br^5}{5}\right)$$

$$= 4\pi r^3 \left(\frac{a}{3} + \frac{br^2}{5}\right).$$

The solid sphere of radius r behaves for the external point P as if its mass were concentrated at its centre. Therefore, the gravitational field at P due to it is

$$F = -\frac{GM'}{r^2}$$

$$= -\frac{G}{r^2}\left\{4\pi r^3 \left(\frac{a}{3} + \frac{br^2}{5}\right)\right\}$$

$$= -4\pi Gr \left(\frac{a}{3} + \frac{br^2}{5}\right).$$

This is also the gravitational field at P due to the entire sphere of *variable density*.

Example 44:

In a spherical mass the density varies inversely as the distance from the centre. Show that the gravitational field is the same at any two internal points.

Solution:

Do your self

Example 45:

(a) Imagine a smooth straight tunnel to exist between two diametrically-opposite points on the surface of the earth. Show that a body dropped into it will perform simple harmonic motion. Find an expression for the time-period.

(b) If a letter were delivered through this tunnel, what would be the time-interval between posting at one end and delivery at the other? Take the density of earth as 5.51×10^3 *kg/m*^{3} *and* $G = 6.67 \times 10^{-11}$ *N-m*^{2}*/kg*^{2}*.*

Solution:

(a) Let O be the centre of the earth and P the position of the particle at any instant. Let OP = r. The gravitational force on the particle is that due to a sphere of radius r, since the outer shell exerts no force on the particle.

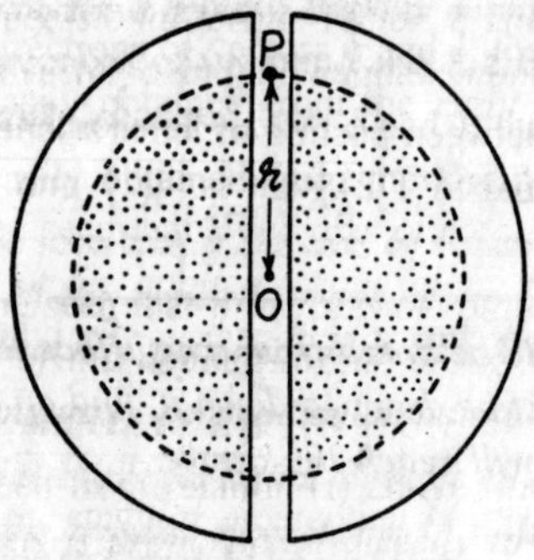

Fig. 2.60

Let us treat the earth as a uniform solid sphere of mass M and radius R. The gravitational field at an internal point distant r from the centre is $-G\frac{Mr}{R^3}$, so that the force on the particle of mass m at P is

$$F = -G\frac{Mm}{R^3}r.$$

The minus sign indicates that the force is attractive and directed toward the center of the earth, O. The acceleration at P is

$$a = \frac{F}{m} = -\frac{GM}{R^3}r = -\omega^2 r,$$

where $\omega^2 = GM/R^3$ (constant). The acceleration is thus proportional to the displacement from O and directed toward it. Hence the motion of the particle is simple harmonic. The period of this S.H.M. is

$$T = \frac{2\pi}{\omega} = 2\pi\sqrt{\frac{R^3}{GM}}. \qquad ...(i)$$

If ρ be the uniform density of earth, then $M = \left(\frac{4}{3}\right) \pi R^3\rho$. Thus

$$T = 2\pi\sqrt{\frac{3}{4\pi G\rho}} = \sqrt{\frac{3\pi}{G\rho}} \qquad ...(ii)$$

Eq. (i) and (ii) are alternative expressions for the time-period.

(b) Putting $\rho = 5.51 \times 10^3$ kg/m^3

and $G = 6.67 \times 10^{-11}$ N-m^2/kg^2 in eq. (ii), we get

$$T = \sqrt{\frac{3 \times 3.14}{(6.67 \times 10^{-11})(5.51 \times 10^3)}}$$

$$= 5050 \text{ s} = 84.2 \text{ min.}$$

The time-interval between posting of letter at one end and delivery at the other will be equal to T/2 , that is, 42.1 min.

Example 46:

If a frictionless hole be bored from the surface to the center of the earth and a small object be dropped down it, find an expression for the velocity with which it will reach the centre.

Solution:

As proved in the last problem, the particle will perform S.H.M. of period

$$T = \frac{2\pi}{\omega} = 2\pi\sqrt{\frac{R^3}{GM}} \qquad ...(i)$$

Now, we know that the gravitational constant G is related to the acceleration due to gravity g by

$$G = \frac{GM}{R^2}$$

or $\qquad GM = gR^2$

Substituting this value of GM in eq. (i), we get

$$T = \frac{2\pi}{\omega} = 2\pi\sqrt{\frac{R}{g}}.$$

The velocity of the particle at a distance x from the mean position (earth's centre) is given by

$$v = \omega\sqrt{R^2 - x^2}.$$

The velocity while passing through the centre of earth (x = 0) is

$$v = \omega R = \sqrt{\frac{g}{R}} \times R = \sqrt{gR}.$$

Example 47:

Two points on the surface of the earth are joined by a straight smooth tunnel, not passing through the center of the earth. A particle is dropped inside the tunnel. Show that it will execute a simple harmonic motion. Find the time required for crossing the tunnel and show that it is independent of the length and direction of the tunnel.

Solution:

Let AB a tunnel through the earth. Let a particle of mass m be placed in it at P, such that its distance from the center O of the earth is r and from the mid-point C of the tunnel is x. Let the angle OPC be θ.

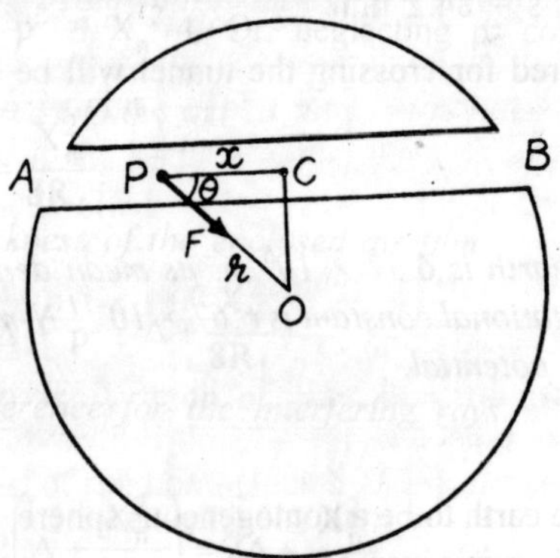

Fig. 2.61

Let us treat the earth as a uniform solid sphere. We know that the gravitational force F on a particle of mass m placed internally at a distance r from the centre of a solid sphere of mass M and radius R is given by

$$F = -G\frac{Mm}{R^3}r,$$

and is directed toward the centre of the sphere. The component of this force along PC, the length of the tunnel, is

$$F\cos\theta = -G\frac{Mm}{R^3}r\left(\frac{x}{r}\right) = -G\frac{Mm}{R^3}x.$$

The acceleration of the particle toward C is

$$a = \frac{F\cos\theta}{m} = -\frac{GM}{R^3}x = -\omega^2 x,$$

where $\omega^2 = \frac{GM}{R^3}$. This is the equation of a simple harmonic motion. Hence the particle executes a S.H.M. in the tunnel and its time-period is

$$T = \frac{2\pi}{\omega} = 2\pi\sqrt{\frac{R^3}{GM}}.$$

This result is independent of the length and direction of the tunnel.

Putting the known values of R, G and M, we get

$$T = 2 \times 3.14 \times \sqrt{\frac{\left(6.37\times10^6\,\text{m}\right)^3}{\left(6.67\times10^{-11}\,\text{N-m}^2/\text{kg}^2\right)\left(6.0\times10^{24}\,\text{kg}\right)}}$$

$= 5.05 \times 10^3$ s = 84.2 min.

The time required for crossing the tunnel will be half of this, that is, 42.1 min.

Example 48:

The radius of earth is 6.37 × 10^6 m, its mean density is 55 × 10^3 kg/m^3 and the gravitational constant is 6.67 × 10^{-11} N-m^2/kg^2. Calculate the earth's surface potential.

Solution:

Considering the earth to be a homogeneous sphere, the gravitational potential at its surface is given by

$$V = -G\frac{M}{R}.$$

where M is the mass and R the radius of earth. If ρ be the density of earth, then

$$M = \frac{4}{3}\pi R^3\rho.$$

$$\therefore \quad V = -G\left(\frac{4}{3}\pi R^2 \rho\right).$$

Substituting the given values, we have

$$V = -\left(6.67\times10^{-11}\right)\times\frac{4}{3} \times 3.14 \times (6.37 \times 10^6)^2 \times (5.5 \times 10^3)$$

$$= -6.2 \times 10^7 \text{ J/kg}.$$

Example 49:

What is the potential energy of a mass of 1 kg on the surface of the earth, and at a distance of 10^5 km from the centre of the earth, referred to zero potential energy at infinite distance? ($G = 6.67 \times 10^{-11}$ N-m²/kg²; mass of earth $= 6 \times 10^{24}$ kg and radius $= 6.4 \times 10^6$ m).

Solution:

The gravitational potential energy of a body of mass m at a distance r from the centre of the earth (r > radius R of earth) is

$$U(r) = -\frac{GMm}{r},$$

where M is the mass of earth. For a body on earth's surface, r = R .

$$\therefore \quad U(R) = -\frac{GMm}{R}.$$

Here m = 1 kg.

$$\therefore \quad U(R) = -\frac{\left(6.67\times10^{-11}\right)\times\left(6\times10^{24}\right)\times1}{6.4\times10^6}.$$

$$= -6.2 \times 10^7 \text{ J}.$$

For a body at a distance r (= 10^5 km = 10^8m),

$$U(r) = -\frac{GMm}{r}$$

$$= -\frac{\left(6.67\times10^{-11}\text{Nm}^2/\text{kg}^2\right)\times\left(6\times10^{24}\text{kg}\right)\times1\text{kg}}{10^8\text{ m}}$$

$$= -4.0 \times 10^6 \text{ J}.$$

Example 50:

Calculate the gravitational self-energy of (a) the Sun, (b) the Earth-Sun system. Given : mass of the Sun $= 2 \times 10^{30}$ kg, radius of the Sun $= 7 \times 10^8$ m, mass of Earth $= 6 \times 10^{24}$ kg, mean distance of Earth from the Sun $= 1.5 \times 10^8$ km, $G = 7 \times 10^{-11}$ N-m²/kg².

Solution:

(a) The gravitational self-energy of a uniform sphere of mass M and radius R is given by

$$U_s = -\frac{3}{5}\frac{GM^2}{R}.$$

Putting the given values for Sun, we get

$$U_s = -\frac{3}{5}\frac{\left(7\times10^{-11}\,\text{N-m}^2/\text{kg}^2\right)\left(3\times10^{30}\,\text{kg}\right)^2}{7\times10^{8}\,\text{m}}$$

$$= -2.4 \times 10^{41}\ \text{J}.$$

(b) The self-energy of the Earth-Sun system is the mutual gravitational potential energy of the system. Thus, ignoring the self-energy of Earth and Sun individually, we have

$$U_s = -\frac{GM_sM_E}{R},$$

where R is Earth-Sun distance. Putting the given values, we get

$$U_s = -\frac{\left(7\times10^{-11}\,\text{N-m}^2/\text{kg}^2\right)\left(2\times10^{30}\,\text{kg}\right)\left(6\times10^{24}\,\text{kg}\right)}{\left(1.5\times10^{11}\,\text{m}\right)}$$

$$= -5.6 \times 10^{33}\ \text{J}.$$

Example 51:

Calculate the velocity with which a body must be thrown vertically upward from the surface of the earth so that it may reach a height of 10 R, where R is the radius of the earth and is equal to 6.4 × 10^6 m. (Earth's mass 6 × 10^{24} kg, Gravitational constant G = 6.7 × 10^{-11} N-m^2/kg^2).

Solution:

The gravitational potential energy of a body of mass m on earth's surface is

$$U(R) = -\frac{GMm}{R},$$

where M is the mass of the earth (supposed to be concentrated at its centre) and R is the radius of the earth (distance of the particle from the centre of the earth). The gravitational energy of the same body at a height 10R from earth's surface, *i.e.*, at a distance 11 R from earth's centre is

$$U(11R) = -\frac{GMm}{11R}.$$

∴ change in potential energy

$$U(11R) - U(R) = -\frac{GMm}{11R} - \left(-\frac{GMm}{R}\right) = \frac{10}{11}\frac{GMm}{R}.$$

This difference must come from the initial kinetic energy given to the body in sending it to that height. Now, suppose the body is thrown up with a vertical speed v, so that its initial kinetic energy is $\frac{1}{2}mv^2$. Thus

$$\frac{1}{2}mv^2 = \frac{10}{11}\frac{GMm}{R}$$

or $$v = \sqrt{\frac{20GM}{11R}}.$$

Putting the given values :

$$v = \sqrt{\frac{20\times(6.7\times10^{-11}\,N-m^2/kg^2)\times(6\times10^{24}\,kg)}{11\times(6.4\times10^6\,m)}}$$

$$= 1.07 \times 10^4 \text{ m/s}.$$

Example 52:

The earth may be treated as a symmetrical sphere of radius R = 6500 km with field 9.8 N/kg at its surface. Using the potential energy expression, deduce the vertical speed with which a rocket should be thrown so as to reach upto height 4R from the earth's surface.

Solution:

Proceeding as in last problem, we can show that

$$U(5R) - U(R) = \frac{4}{5}\frac{GMm}{R}.$$

But $GM = gR^2$, where g is acceleration due to gravity (gravitational field) on earth's surface

∴ $$U(5R) - U(R) = \frac{4}{5}mgR.$$

If v be the initial vertical speed of the rocket, then

$$\frac{1}{2}mv^2 = \frac{4}{5}mgR.$$

or $$v = \sqrt{\frac{8gR}{5}}.$$

Putting the given values :

$$v = \sqrt{\frac{8\times(9.8\,\text{N/kg})\times(6.4\times10^6\,\text{m})}{5}}$$

$= 1.0 \times 10^4$ m/s.

Example 53:

If g is the acceleration due to gravity on the earth's surface, calculate the gain in potential energy of an object of mass m raised from die surface of earth to a height h equal to radius·R of the earth.

Solution:

The potential energy of the object m when placed on the surface of earth is

$$U(R) = -\frac{GMm}{R}$$

and when raised to a height h above earth's surface, then

$$U(R+h) = -\frac{GMm}{R+h}.$$

The gain in potential energy is

$$\Delta U = -\frac{GMm}{R+h} - \left(-\frac{GMm}{R}\right) = \frac{GMmh}{(R+h)R}.$$

But $\quad g = \frac{GM}{R^2}.$

$\therefore \quad \Delta U = \frac{mg\,h\,R}{R+h}.$

For $\quad h = R$, we have

$$\Delta U = \frac{mg\,R^2}{R+R} = \frac{1}{2} mg\,R.$$

Example 54:

A satellite of mass 200 kg, initially at rest on the earth, is launched in a circular orbit at a height from the earth, equal to radius of earth. Given $R = 6.37 \times 10^6$ m, $g = 9.8$ m/s^2. Calculate the minimum energy required. **Ans.** 6.24×10^9 J.

Example 55:

Calculate the amount of work done to send a body of mass m from earth's surface to a height R/2 in terms of m, g and R (radius of earth).

Ans. 1/3 mg R.

Example 56:

A projectile is fired vertically from the earth's surface with a velocity of 10 km/s. How far it would go above me earth's surface? The mass and radius of the earth are 6.0 × 10^{24} kg and 6400 km. The air resistance is negligible. (G = 6.67 × 10^{-11} N-m/kg^2).

Solution:

Let m be the mass of the projectile. Its kinetic energy, when fired with velocity v, is

$$K = \frac{1}{2}mv^2.$$

It is this energy which carries it to a height h meter (say) above the earth's surface, and is equal to the work done in carrying the projectile against the gravitation of earth.

Suppose, at any instant the projectile is at a distance r from the centre of the earth (mass M). The gravitational attraction on it is

$$G\frac{Mm}{r^2}.$$

The work required to carry it a distance dr is $G\frac{Mm}{r^2}dr$, and the total work W in carrying it from earth's surface (where r = R, radius of earth) to a height h (where r = R + h) is

$$W = \int_R^{R+h} G\frac{Mm}{r^2}dr$$

$$= GMm\left[-\frac{1}{r}\right]_R^{R+h}$$

$$= GMm\left[-\frac{1}{R+h}+\frac{1}{R}\right] = \frac{GMmh}{R(R+h)}.$$

Equating it to kinetic energy (K = W), we get

$$\frac{1}{2}mv^2 = \frac{GMmh}{R(R+h)}$$

or $\quad Rv^2 (R + h) = 2\,G\,M\,h$

$$\therefore \quad h = \frac{R^2v^2}{2GM - Rv^2}.$$

Now, R = 6400 km = 6.4 × 10^6 m, v = 10 km/s = 10^4 m/s, G = 6.67 × 10^{-11} N-m^2/kg^2 and M = 6.0 × 10^{24} kg.

$$\therefore \quad h = \frac{(6.4\times10^6)^2\times(10^4)^2}{2\times(6.67\times10^{-11})\times(6.0\times10^{24})-(6.4\times10^6)\times(10^4)^2}$$

$$= \frac{6.4\times6.4\times10^7}{2\times6.67\times6.0-6.4\times10} = \frac{6.4\times6.4\times10^7}{16}$$

$$= 2.5 \times 10^7 \text{ meter} = 2.5 \times 10^4 \text{ km.}$$

Example 57:

If a body of mass 10 kg is at rest at a distance R (earth's radius) above the surface of earth, what velocity should be given to it to enable it to escape from the earth's gravitational field? The mass of the earth is 6.0×10^{24} kg and its radius is 6.4×10 m. ($G = 6.67 \times 10^{-11}$ $N\text{-}m^2/kg^2$).

Solution:

The escape velocity of a particle for a planet (say, earth) of mass M is given by

$$v_e = \sqrt{\frac{2GM}{r}},$$

where r is the distance of the particle from the centre of the planet. Here

$$r = R + R = 2R$$

$$= 2 \times (6.4 \times 10^6 \text{ m}) = 1.28 \times 10^7 \text{ m.}$$

$$\therefore \quad v_e = \sqrt{\frac{2\times(6.67\times10^{-11})\times(6.0\times10^{24})}{1.28\times10^7}}$$

$$= 7.9 \times 10^3 \text{ m/s.}$$

Example 58:

A particle is falling from infinity toward the surface of the earth. If the air friction be negligible and initial velocity be zero, then calculate its velocity on reaching the earth's surface. Earth's radius is 6371 km and $g = 98$ m/s^2.

Solution:

A particle projected from earth with escape velocity never returns to the earth, that is, it reaches infinity. Conversely, a particle falling from rest at infinity will reach the earth with velocity of escape, v_e. Now, for earth

$$v_e = \sqrt{2gR}$$

$$= \sqrt{2 \times 9.8 \times \left(6371 \times 10^3\right)}$$

$$= 1.1 \times 10^4 \text{ m/s} = 11 \text{ km/s}.$$

Example 59:

A rocket starts vertically upward with speed v_0. Show that its speed v at height h is given by

$$v_0^2 - v^2 = \frac{2gh}{1+\dfrac{h}{R}},$$

where R is the radius of the earth and g is acceleration due to gravity at earth's surface. Hence deduce an expression for maximum height reached by a rocket fired with speed 0.9 times the escape velocity.

Solution:

The gravitational potential energies of a mass m on earth's surface and at height h are given by

$$U(R) = -\frac{GMm}{R} \text{ and } U(R+h) = -\frac{GMm}{R+h}.$$

$$\therefore \quad U(R+h) - U(R) = -GMm\left(\frac{1}{R+h} - \frac{1}{R}\right)$$

$$= \frac{GMmh}{(R+h)R}$$

$$= \frac{mgh}{1+\dfrac{h}{R}} \qquad [\because GM = gR^2]$$

This increase in potential energy occurs at the cost of kinetic energy which correspondingly decreases. If v is the velocity of the rocket at height h, then the, decrease in kinetic energy is $\frac{1}{2}mv_0^2 - \frac{1}{2}mv^2$. Thus,

$$\frac{1}{2}mv_0^2 - \frac{1}{2}mv^2 = \frac{mgh}{1+\dfrac{h}{R}}$$

or
$$v_0^2 - v^2 = \frac{2gh}{1+\dfrac{h}{R}}.$$

Let h_{max} be the maximum height reached by the rocket, at which its velocity has been reduced to zero. Thus, substituting v = 0 and h = h_{max} in the last expression, we have

$$v_0^2 - \frac{2gh_{max}}{1+\frac{h_{max}}{R}}$$

or $$v_0^2\left(1+\frac{h_{max}}{R}\right) = 2gh_{max}$$

or $$v_0^2 = h_{max}\left(2g - \frac{v_0^2}{R}\right)$$

or $$h_{max} = \frac{v_0^2}{2g - \frac{v_0^2}{R}}$$

Now, it is given that

$$v_0 = 0.9 \times \text{escape velocity} = 0.9 \times \sqrt{2gR}.$$

$$\therefore \quad h_{max} = \frac{(0.9\times0.9)2gR}{2g - \frac{(0.9\times0.9)2gR}{R}}$$

$$= \frac{1.62gR}{2g - 1.62g} = \frac{1.62R}{0.38} = 4.26R.$$

EXERCISES

1. It is said that the first artificial satellite was revolving round the earth at a distance of 560 miles from it. Estimate its velocity and period of revolution, taking the radius of the earth to be 4000 miles and the value of g to be 32 ft/sec^2.

   Ans. 24340 ft/sec; 6207 sec.

   [Hint. Let distance of the satellite from the centre of the earth be R' = R + 560 = 4000 + 560 = 4560 *miles*.

   Then, $mv_0^2/R' = mMG/R'^2$,

   whence, $v_0^2 = MG/R'$, but $MG = gR^2$.

   $\therefore \quad v_0^2 = gR^2/R'$ Or $v_0 = R\sqrt{g/R'}$.

   Hence, *time-period of revolution* or $T = 2\pi R'/v_0$]

2. Show that the time-period of oscillation of a particle dropped in a tunnel right through the earth is the same whether or not

the tunnel passes through the centre of the earth. What is the value of this time- period? **Ans.** 83.72 *min.*

3. Imagine a particle at a point P inside a spherical shell of uniform thickness and density and construct a narrow double cone with apex at P so as to intercept areas A_1 and A_2 on the shell on either side of P. Show that the resultant gravitational force exerted on the particle at P by the intercepted mass elements of the shell is zero and hence show that the gravitational field due to the entire shell is zero at any point inside it.
4. Explain the terms 'gravitational potential' and 'gravitational field'. Obtain expressions for the gravitational potential and gravitational field at a point (i) inside, (ii) outside a hollow spherical shell.
5. Derive expressions for gravitational field and potential at a point inside and outside a thin uniform spherical shell.
6. Define and explain gravitational potential. Calculate the gravitational potential due to a sphere at a point (i) outside the sphere (ii) inside the sphere.

   Show that the potential at the centre of the sphere is one and a half times that on its surface.
7. Calculate the rate of contraction of the sun's radius if the energy released due to its contraction (with no appreciable change in its mass) is radiated and received on the surface of the earth at the rate of 2 calories per sq cm *per minute.* Distance between the sun and the earth 1.5×10^{13} cm. (Mass of the sun = 2×10^{33} gm radius of the sun = 7×10^{10} cm).

   Ans. 2.3 km/year.
8. What it meant by *fundamental lengths and numbers*? Illustrate their importance in Physics by a couple of examples.
9. Construct any three fundamental quantities from the following basic parameters:

   (i) velocity of light (c), (ii) Planck's constant (h), (iii) mass of electron (m), (iv) charge of electron (e), (v) Gravitational constant (G).

   Express the dimensions of the above and the constructed quantities in terms of mass (M), length (L) and time (T).

Ans. *The dimension of the given quantities* are (i) M^0LT^{-1}, (ii) ML^2T^{-1}, (iii) ML^0T^0, (iv) $M^{1/2}L^{3/2}T^{-1}$ and (v) $M^{-1}L^3T^{-2}$.

The *three constructed quantities* are

(i) *Fine structure constant* $\alpha = \dfrac{e^2}{(h/2\pi)c} = \dfrac{2\pi e^2}{hc} = \dfrac{1}{137.04}$

Its *dimensions* are $\dfrac{ML^3T^{-2}}{(ML^2T^{-1})(LT^{-1})} = M^0L^0T^0$, *i.e., no dimension in* M,L,T.

(ii) *Radius of the hydrogen atom,*

$$\alpha_0 = \frac{r_0}{\alpha^2} = \frac{e^2/mc^2}{(2\pi e^2/hc)^2} = \frac{h^2}{4\pi^2 me^2}$$

$= 0.529 \times 10^{-8}$ cm.

Its *dimensions* are $\dfrac{M^2L^4T^{-2}}{(M)(ML^3T^{-2})} = M^0LT^0$.

(iii) *Planck's length,* $l = \left(\dfrac{Gh}{c^3}\right)^{\frac{1}{2}} \approx 10^{-33}$ cm.

Its *dimensions* are $\left[\dfrac{(M^{-1}L^3T^{-2})(ML^2T^{-1})}{L^3T^{-3}}\right]^{\frac{1}{2}} = M^0LT^0$.

10. (a) How may the density of the earth be determined?

(b) The radius of the earth is 6.37×10^8 cm, its mean density. 5.5 gm/c.c. and the gravitational constant, 6.66×10^{-8} C.G.S. units. Calculate the earth's surface potential.

Ans. 6.227×10^{11} *erg/gm.*

11. Obtain an expression for the gravitational attraction at a point (i) outside and (ii) inside a solid sphere and show that in the latter case, it is proportional to the distance from the centre of the sphere.

12. Show that in the case of a hollow sphere (or a thick shell) of density ρ and inner and outer radii, R_1 and R_2, respectively, the gravitational potential at a point inside the hollow sphere is $V = -2\pi rG(R_2^2 - R_1^2)$ and the field at the point, *zero*.

Also show that the field at a point in the material of the hollow sphere (or thick shell) at a distance r from its centre is given by

$$E = -\left(\frac{r^3 - R_1^{\ 3}}{R_2^{\ 3} - R_1^{\ 3}}\right)\frac{M}{R^2}G.$$

13. Show that if a body be projected vertically upward from the surface of the earth so as to reach a height nR above the surface, (i) the increase in its potential energy is [n/ (n+1)] MgR and (ii) the velocity with which it must be projected is $\sqrt{[2n/(n+1)]}\ gR$, where R is the radius of the earth and M, its mass.

    [**Hint:** At the surface of the earth, *i.e.,* at a distance R from its centre, the P.E. of the body = –MmG/R = –mgR²/R = mgR ($\because$ MG = gR²). And. at a distance (nR + R} or (n + l) R from the centre of the earth, its P.E. = –MmG/(n +l) R

    = –mgR²/(n + 1) R= –mgR/ (n +1).

    $\therefore$ increase in P.E. of the body = –mgR/(n + 1)–(–mgR) = [n/(n + l) mgR.

    If v be its velocity of projection, its K.E. $= \frac{1}{2}mv^2 = \left(\frac{n}{n+1}\right)mgR,$

    whence, $\quad V = \sqrt{\left(\frac{2n}{n+1}\right)gR}$.]

14. Deduce an expression for the gravitational potential due to a sphere at an external point. Hence calculate the amount of work required to send a body of mass m from the earth's surface to a height (i) R/2, (ii) 10R and (iii) 1000R, where R is the radius of the earth. Express the result in m, R and g.

    Ans. (i) $\frac{1}{3}$ mgR, (ii) $\frac{10}{11}$mgR, (iii) $\frac{1000}{1001}$mgR.

15. If the density (r) of the earth increases with depth below the surface, show that the value of g may also increase. How should the density vary with depth in order that the value of g may remain unaffected?

Ans. ρ *should be proportional to 1 /r or ρr should remain constant, where r is the distance from the centre of the earth.*

[**Hint:** On the surface of the earth, *i.e.*, at distance R from the centre of the earth, $g = MG/R^2 = 4/3\pi R^3\rho G/R^2 = 4/3\pi GR\rho$, And, at a distance r from the centre of the earth, $g' = 4/3\pi R^3\rho' G/r^2 = 4/3\pi GR\rho'$, where ρ' is now the density. ∴ $g'/g = (r/R)(\rho'/\rho)$. In order that g' = g, we must have $(r/R)(\rho'/\rho) = 1$, or, $\rho/\rho' = r/R$.]

16. In question 17 above, what should be the velocities given to the body to attain the heights R/2, 10R and 1000 R respectively?

Ans. (i) $\sqrt{\frac{2}{3}gR}$, (ii) $\sqrt{\frac{20}{11}gR}$, (iii) $\frac{2000}{1001}gR$.

17. Show by cogent argument why the following two time-periods are the same, *viz.*, 84 minutes: (i) period of oscillation of a particle in a tunnel bored through the earth and (ii) period of revolution of an artificial satellite close to the earth's surface.

18. State the law of gravitational attraction and hence define the gravitational constant G. Describe a method of measuring G.

19. Describe an accurate method for the determination of G.

    A sphere of mass 40 kg is attracted by a second sphere of mass 15 kg, when their centres are 20 cm apart, with a force equal to 1/10 of a milligram weight. Calculate the constant of gravitation. **Ans.** 6.53×10^{-8} C.G.S. units.

20. (a). Describe an accurate method of determining the gravitational constant in the laboratory.

    (b) A smooth tunnel is bored through the earth and a small particle is allowed to move in it from a position of rest. Find the periodic time of one vibration. Given that $G = 6.67 \times 10^{-8}$ *cgs* units and the mean density of the earth = 5.6 gm per c.c.

    If mail were to be delivered through the tunnel, how long would it be between depositing it at one end and its delivery at the other end? **Ans.** 84 min; 42 min.

21. Calculate the mass of the earth from the following data: Radius of the earth = 6×10^8 cm; Acceleration due to gravity = 980 cm/sec^2 and gravitational constant = 6.6×10^{-8} $cm^3\ gm^{-1} sec^{-2}$

22. Calculate the mass of the sun, given that the distance between the sun and the earth is 1.49×10^{13}cm and $G = 6.66 \times 10^{-8}$ C.G.S. units. Take the year to consist of 365 days.

Ans. 19.72×10^{34}gm.

23. Obtain an expression for the limiting velocity required by an artificial satellite for orbiting around the earth. If the radius of the earth be 6.4×10^{8} cm and g = 980 cm/sec^2, calculate the value of this velocity.

Ans. *Limiting velocity* $V_0 = \sqrt{gR} = 7.92 \times 10^5$ cm/sec.

24. Derive the expressions for gravitational potential energy and force inside a sphere of uniform density.

Calculate the time taken by in earth satellite moving in a circular orbit, close to its surface, in completing one round. Take the radius of the earth = 6×10^{8}cm.

Ans. 1 *hr* 21 *min* 56 *sec.*

25. Show that the *escape velocity* from the surface of the earth is $\sqrt{2}$ times the velocity of projection of an artificial satellite orbiting close around the earth.

26. The mean distance of *Mars* from the sun is 1.524 times that of the *Earth* from the sun. How many years would be required for *Mars* to make one revolution around the sun?

Ans. 1.88 *years.*

27. Estimate (i) the value of g, (ii) the escape velocity, on *Mars.* Given, mass of *Mars* ≈ 0.11 of the mass of the earth and its radius, 42/79 that of the earth. (Radius of the earth = 6.37 × 10^8cm.) **Ans.** (i) 381.5 cm/sec^2.; (ii) 5×10^5 cm/sec.

28. Two satellites of equal mass m are moving in the same circular orbit of radius r around the earth, in opposite directions, so as to eventually collide with each other:

(i) Obtain an expression in terms of M, m, r and G (where M is the mass of the earth) for the total mechanical energy of the two satellite-earth system before collision,

(ii) Taking the collision to be perfectly elastic, obtain the total mechanical energy of the system immediately after collision,

(iii) Describe the subsequent motion of the wreckage of the two satellites.

Ans. MmG/r, (ii) –2 MmG/r, (iii) *the wreckage falls down to the earth.*

29. What do you understand by the term 'gravitational self energy' of a body or a system of particles? Show that the gravitational self energy of a system of n particles, each of mass m, at an average distance r from each other is given by $U_s = 1/2Gn(n-1)m^2/r$.

30. Calculate the gravitational self energy of (i) the sun, (ii) the earth-sun system, given that the mass of the sun = 2×10^{30} kg and its radius = 7×10^8 metres, mass of the earth = 6×10^{24} kg and mean earth-sun distance = 1.50×10^8 km Take G = 7×10^{-11} N–m²/kg².

Ans. (i) -2.4×10^{-41} *joules*, (ii) -5.6×10^{33} *joules*.

[**Hint:** Self energy of the sun = $\frac{3}{5}\frac{M_s^{\,2}}{R_s}G$. Self energy of the earth-sun system = $-\frac{M_s M_e}{r_{es}}G$, where M_s, and M_e are the masses of the sun and the earth respectively; R_s, the radius of the sun and r_{es}, the distance between the sun and the earth.]

31. The gravitational self energy of a uniform sphere of mass M and radius R is given by $-\frac{3}{5}G\frac{M^2}{R}$. Explain what is meant by this What happens if the sphere contracts in radius by a small amount a?

Ans. *It results in release of energy* = $\frac{3}{5}G\frac{M^2}{R}a$.

32. Deduce an expression for the electrostatic self energy of a charge q spread uniformly over the surface of a sphere of radius r.

33. 10^5 stars are distributed spherically in a globular cluster, with each star having the same mass as the sun. If the diameter of the cluster be 40 parsec, where 1 parsec = 3×10^{18} cm, calculate:

(i) the number of stars per cubic parsec,

(ii) the gravitational self energy of the cluster (neglecting the self energy of individual stars). (Mass of the sun = 2 × 10^3gm, G = 6.67 × 10^{-8} *C.G.S.* units).

Ans. (i) *number of stars per cubic parsec* = 3, (ii) *nearly* 10^{49} *ergs.*

[Hint: (i) Number of stars per cubic parsec

$$= \frac{\text{total number of stars}}{\text{volume of cluster}} = \frac{10^5}{\frac{4}{3}\pi(20)^3}$$

(ii) Self energy of the cluster,

$$U_s = \frac{3}{5}\frac{M^2}{R}G = \frac{3}{5}\frac{\left(10^5 \times 2 \times 10^{33}\right)^2}{20 \times 3 \times 10^{18}} \times 6.67 \times 10^{-8}$$

3

BENDING OF BEAMS—COLUMNS

INTROCUTION OF BEAM

A rod or a bar of a circular or rectangular cross section, *with its length very much greater than its thickness* (so that there are no shearing stresses over any section of it) is called a *beam*.

Now, a beam may just rest on a support, like a knife-edge or have a small part of it firmly clamped or built into a wall at either end. In the former case, it is called a *supported beam* and in the latter, a *built-in* or an *encastre beam* or usually, simply, a *fixed beam*.

In a supported beam, obviously, the support can merely exert a force on the beam but in a fixed beam, it can also exert a couple on it.

If the beam be *fixed only at one end* and loaded at the other, it is called a *cantilever*.

A beam or a bar of a homogeneous, isotropic material, *subjected to an axial push or compressive stress*, is, in general, called a strut and may be inclined in any direction, including the horizontal and the vertical.

In the *vertical or upright position*, however, a long and slender beam or bar, thus subjected to an-axial push or compressive stress is referred to as a *column, pillar* or *stanchion*.

As will be readily seen, *a column or a pillar is just a vertical strut*. A beam or a bar, *subjected to a pull or a tensile stress*, is called a tie.

BENDING OF A BEAM

Some Definitions

Suppose we have a beam, of a rectangular cross section, say, fixed at one end and loaded at the other (within the elastic limit) so as to be bent a little, as shown in Fig. 3.1, with its upper surface becoming slightly convex and the lower one concave. All the longitudinal filaments in the upper half of the beam thus get extended or lengthened, and therefore under tension, and all those in the lower half get compressed or shortened and therefore under pressure.

These extensions and compressions increase progressively as we proceed away from the axis on either side (as shown in a rough and ready manner by the lengths of the arrows), so that they are the maximum in the uppermost and the lowermost layers of the beam respectively.

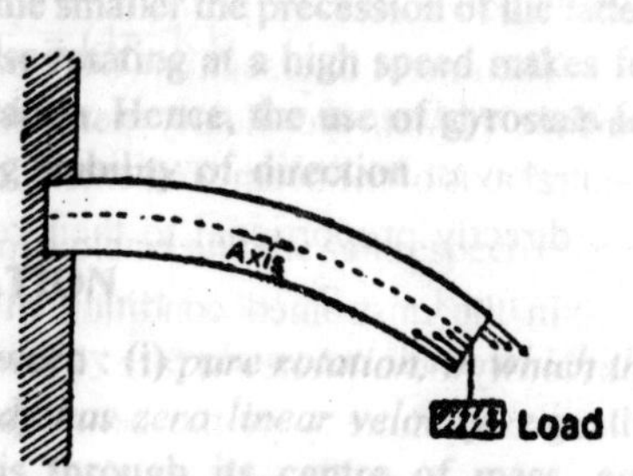

Fig. 3.1

Obviously, there must be a layer between the uppermost and the lowermost layers where the extensions in the upper half change sign to become compressions in the lower half. In this layer or plane, which is *perpendicular to the section of the beam containing the axis*, the filaments neither get extended not compressed, *i.e.*, retain their original lengths. This layer is therefore called the neutral surface of the beam, shown shaded in Fig. 3.2.

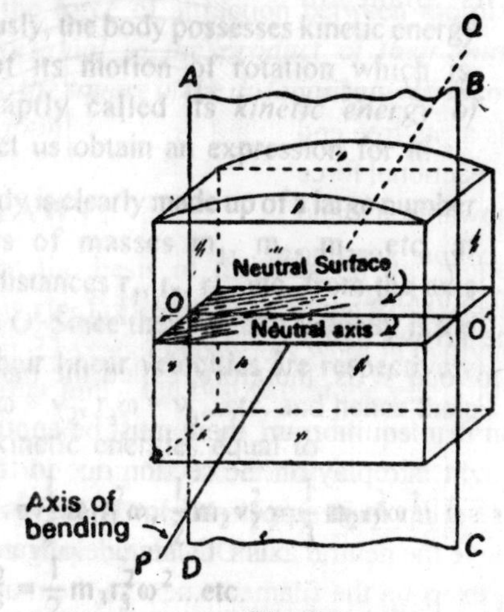

Fig. 3.2

If the material of the beam be homogeneous and isotropic and the bending uniform, the longitudinal filaments all get bent into circular arcs in planes parallel to the *plane of symmetry* ABCD of the beam, which is also, therefore, the *plane of bending*. The centres of curvature of all these arcs lie on a straight line PQ, perpendicular to the plane of bending (or the plane of symmetry), which is referred to as the *axis of bending*.

The line of intersection (OO') of the plane of bending and the neutral surface (which are clearly perpendicular to each other) is called the neutral axis of the beam. The extensions and compressions of the filaments are directly proportional to their respective distances from this axis.

In the unstrained condition of the beam, the neutral surface will obviously be a plane surface. The filament of this surface lying in the plane of symmetry of the bent beam is called the *neutral filament*.

BENDING MOMENT

Let a beam ABCD be fixed at the end AD and loaded (within the elastic limit) with a weight W at the free end BC so as to be bent a little and remaining in equilibrium in the position shown (Fig. 3.3).

Consider a section PBCP′ of the bent beam, cut by a plane PP′ at right angles to its length and its plane of bending (or its plane of symmetry).

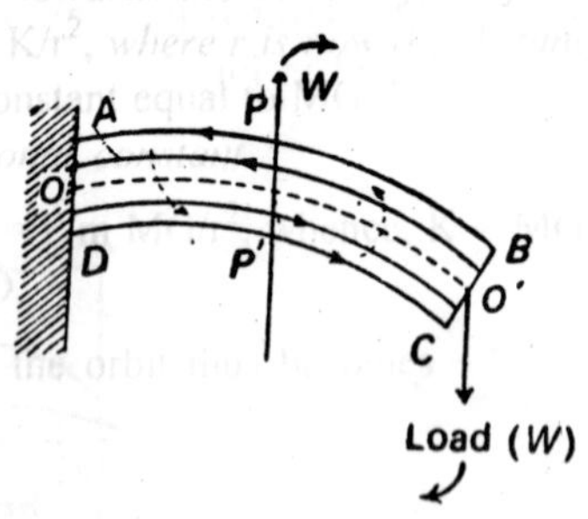

Fig. 3.3

Clearly, the load W acting vertically downwards at its free end gives rise to an equal reactional force W acting vertically upwards at P and the two thus form a *couple*, tending to bend the section *clockwise*, as indicated by full line arrows. This couple applied due to load W is, therefore, called the *bending couple*.

Since the section is in equilibrium, there must be another *equal* and *opposite* couple brought into play on the section due to the tensile and compressive stresses set up in its upper and lower halves respectively. For, the filaments above the neutral axis OO′ being elongated, and hence in a state of tension, exert on the filaments next to them an *inward pull* (towards the fixed end) and the filaments below the neutral axis being shortened and hence in a state of compression, exert on the filaments next to them an *outward pull* (towards the loaded end), as indicated by the arrowheads in the two halves respectively.

These inward and outward pulls thus form a pair of equal and opposite (non-collinear) forces or a *couple*, tending to bend the section in the opposite direction to that due to the applied or the bending couple *i.e., anticlockwise*, as indicated by the dotted arrows. The moment of this couple is called the *moment of the resistance to bending* for the obvious reason that it opposes or resists the bending of the beam due to the applied couple. Since, however, in the equilibrium position of the beam it is equal in magnitude (though opposite in direction) to the bending couple, it is referred to, as the *bending moment* (M).

Expression for bending moment : Consider a portion of the beam to be bent into a circular arc, as shown in Fig. 3.4, subtending an angle θ at its centre of curvature C and let the radius of curvature of the neutral axis OO′ be R.

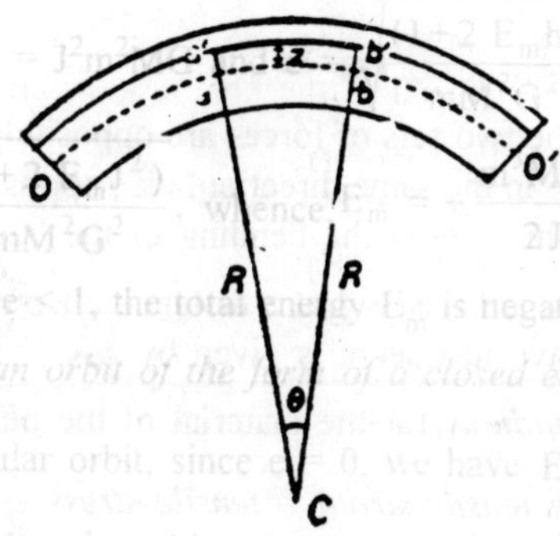

Fig. 3.4

Let the original length of a filament (*i.e.*, its length in the unstrained or unbent condition of the beam) at a distance z from the neutral axis be *ab, the same as that on the neutral axis,* and let its extended length in the strained or bent condition of the beam be a'b'. Then, clearly,

its original length ab $= R\theta$ *and its extended length* $= (R + z)\theta$.

So that, *increase in its lengths* $a'b' - ab = (R + z)\theta - R\theta = z\theta$.

$$\therefore \quad \text{tensile strain} = \frac{\text{increase in length}}{\text{original length}} = \frac{z\theta}{R\theta} = \frac{z}{R},$$

i.e., proportional to z, the distance from the neutral axis.

Taking the beam to be of a rectangular cross-section and considering its section A BCD perpendicular to its length and the plane of bending

(Fig. 3.5) such that the neutral surface meets it along the line EF, we can easily see that the internal forces opposing extension of the filaments all act *inwards* peittfadicular to the upper half ABFE of the section and those opposing contractions of the filaments act *outwards* perpendicula to its lower half EFCD, as indicated in the figure.

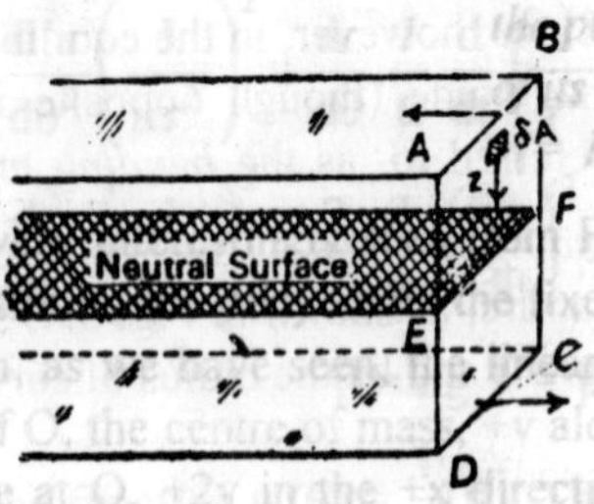

Fig. 3.5

Now, although, the two sets of forces are oppositely directed, their moments about EF are in the same direction, *viz.*, opposite to that of the bending couple and thus, resist the bending of the beam.

If we imagine a small area 8.4 at a distance z from EF, the tensile strain there, as we have just seen, is given by z/R.

Now, *Young's modulus* for the material of the beam is given by

Y = *tensile stress/tensile strain* = *tensile stress*/$\frac{z}{R}$.

$\therefore$ *tensile stress on the small area* $\delta A = \frac{Yz}{R}$, and, therefore,

force on the area δA = *strees* × *area* = $\left(\frac{Yz}{R}\right)\delta A$,

and *moment of this force about EF* $= \frac{Y^2}{R}\delta A \cdot z = \frac{Yz^2}{R}\delta A$.

Since moment of all the forces on elemental areas like δA both in the upper and lower halves of the section act in the same direction, we have

total moment of the forces acting on section ABCD $= \Sigma\frac{Yz^2}{R}\delta A$

$= \frac{Y}{R}\Sigma\delta Az^2$. The quantity $\Sigma\delta Az^2$ is called the *geometrical moment of inertia* of the cross-section of the beam, which we may denote by (to distinguish it from the mechanical moment of inertia for which we use the symbol I). It is also called the *second moment of inertia*.

∴ *total moment of all the forces about EF*, or the bending moment of the beam, $M = \frac{YI_g}{R}$.

The quantity YI_g, which is obviously the external bending moment required to produce a curvature of unit radius in the beam (*i.e.,* when R = 1), is called *flexural rigidity* of the beam and is clearly a measure of the resistance of the beam to bending.

Thus, *bending moment of a beam = flexural rigidity/R.*

For a beam of rectangular cross-section, since $I_g = \frac{bd^3}{12}$, where b is the breadth and d, the depth of the beam, we have

$$\textit{bending moment } (M) = \frac{Ybd^3}{12R},$$

And, *for a beam of a circular cross-section,* since $I_g = \pi R^4/4$, where R is the radius of its cross-section, we have

$$\textit{bending moment } (M) = \frac{Y\pi R^4}{4R}.$$

STIFFNESS OF A BEAM

The stiffness of a beam is taken to be the *ratio between the maximum deflection of its loaded end and its span* and is usually denoted by the symbol 1/n. For steel girders with a large span, n should lie between 1000 and 2000, and for those with a smaller span, between 500 to 700. For timber beams, it should be 360 or above.

ANTICLASTIC CURVATURE

In the above discussion, we have tacitly assumed the bending of the beam to be *pure* or simple, *i.e.,* only longitudinal, with no shearing stresses over any section and hence no transverse bending. In actual practice, however, a longitudinal bending is almost invariably accompanied by a transverse bending.

This is really only to be expected. For, the filaments above the neutral axis, undergoing extension, must suffer a lateral contraction, and those below the neutral axis, undergoing compression, must suffer a lateral extension, σ times as large, where is the *Poisson's ratio* for the material of the beam. As a result, if the crosssection of the beam be rectangular, a layer, initially plane and perpendicular to the plane of bending and containing the neutral filament, assumes the *shape of a saddle*, with the radius or curvature of its longitudinal section in the plane of bending and that of the transverse section in the plane perpendicular to it, so that *the two centres of curvature lie on opposite sides of the beam.*

This may be easily seen by bending a rectangular piece of rubber such that its longitudinal section is concave downwards with its radius of curvature R in the plane of bending (here the plane of the paper). We shall find that its transverse section becomes concave upwards, with its radius of curvature R′ in the plane perpendicular to the plane of bending, the two centres of curvature O and O′ thus lying on opposite sides of the piece, (Fig. 3.6).

Such a surface, with opposite curvatures of its longitudinal and transverse sections is called an anticlastic surface, as opposed to a *synclastic surface* for which the two centres of curvature lie on the same side of the surface as, for example, in the case of a sphere or an ellipsoid.

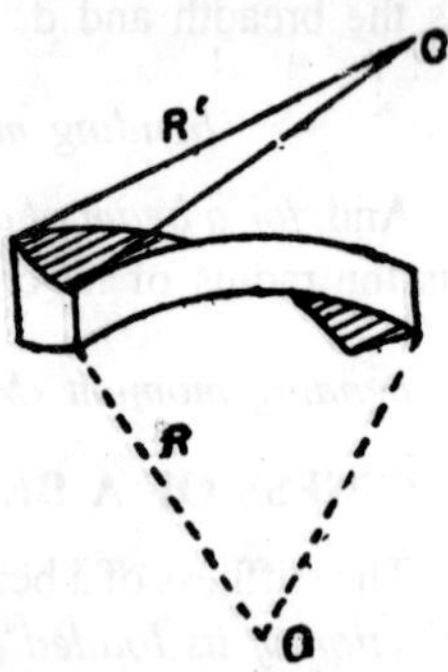

Fig. 3.6

The opposite curvature acquired by the surface in two perpendicular directions (in the plane of bending and in the plane perpendicular to it) is called *anticlastic curvature*.

Incidentally, a measurement of the two radii of curvature, R and R′ (which may easily be done by attaching suitable pointers to the beam and noting the distances and angles transversed by them) enables us to estimate the value of a for the material of the beam. For, as we shall sec in 3.5, the longitudinal and lateral strains produced in the beam at distance z from the neutral axis are respectively given by z/R and z/R′ and, therefore,

$$\sigma = \frac{\text{lateral strain}}{\text{longitudinal strain}} = \frac{\frac{z}{R'}}{\frac{z}{R}} = \frac{R}{R'}$$

It may be pointed out that unless extreme accuracy is called for in any specific case, the transverse bending is hardly ever taken into account and all calculations for the curvature of the beam are based on the theory of simple (or pure) bending, which follows.

THE CANTILEVER—(DEPRESSION OF ITS LOADED END)

A beam rigidly fixed at one end and loaded at the other is called a *cantilever.*

We shall discuss *three cases* of bending here: (i) *When the weight of the cantilever is ineffective i.e.,* produces no bending by itself and (ii) when it is effective, (iii) when the cantilever is uniformly loaded.

Case (i) : (weight of the cantilever ineffective). Let AB be a cantilever of length L, rigidly fixed at the end A and loaded (within the elastic limit) with a weight W at its free end (Fig. 3.7) so as to be bent a little into the dotted position where (the bending being small) end B may be taken to be practically vertically below its earlier position. Clearly, the neutral axis OO' of the beam will now take up the position OO", so that the depression (or deflection) of the loaded end is O'O".

Fig. 3.7

Considering a section PB of the beam by a plane passing through P at right angles to its length and the plane of bending, at a distance x from the fixed end A, we have *external couple acting on the section due to load W*

$$= W \times P'O'' = W(L - x).$$

This, as we know, is, in the position of equilibrium, balanced by the bending moment YI_g/R due to the tensile and compressive stresses set up inside the beam, where Y is the value of *Young's modulus* for the material of the beam, I_g, the geometrical moment of inertia of its cross-section and R, the radius of curvature of the neutral axis of the section at P.

We, therefore, have $\quad W(L-x) = \dfrac{YI_g}{R}.$

Now, the curvature $\frac{1}{R} = \frac{\frac{d^2y}{dx^2}}{\left[1+\left(\frac{dy}{dx}\right)^2\right]^{3/2}}$,

which, when $\frac{dy}{dx}$ is small (as is usually the case), reduces to $\frac{1}{R} = \frac{d^2y}{dx^2}$, the rate of change of slope.

So that, $$W(L-x) = YI_g \frac{d^2y}{dx^2}.$$

Or, $$\frac{d^2y}{dx^2} = \frac{W}{YI_g}(L-x).$$

This, on integration, gives $\frac{dy}{dx} = \frac{W}{YI_g}\left(Lx - \frac{x^2}{2}\right) + C_1$,

where C_1 is a constant of integration.

Since end A is fixed, we have $\frac{dy}{dx} = 0$ at $x = 0$ and, therefore, $C_1 = 0$, and we have

$$\frac{dy}{dx} = \frac{W}{YI_g}\left(Lx - \frac{x^2}{2}\right) \quad ...(i)$$

Integrating once again, we have $y = \frac{W}{YI_g}\left(L\frac{x^2}{2} - \frac{x^3}{6}\right) + C_2$, where C_2 is another constant of integration.

Again, however, $y = 0$ at $x = 0$ and, therefore, $C_2 = 0$.

And so we have $y = \frac{W}{YI_g}\left(\frac{Lx^2}{2} - \frac{x^3}{6}\right)$, ...(ii)

where y gives the depression or deflection of the beam at a distance x from the fixed end A.

Since the free (or the loaded) end B of the beam is at a distance L from A, we have $x = L$ and, therefore,

depression of the loaded end B is given by

$$y = \frac{W}{YI_g}\left(\frac{L^3}{2} - \frac{L^3}{6}\right) = \frac{W}{YI_g} \cdot \frac{2L^3}{6}.$$

Or, $$y = \frac{WL^3}{3YI_g}.$$

And, *slope of the end B relative to A* (or its inclination to the horizontal) is given by $\tan\theta = \theta = \left(\frac{dy}{dx}\right)_{x=L} = \frac{WL^2}{2YI_g}$.

Since for a *beam of rectangular cross-section of breadth b and depth* d, $I_g = \frac{bd^3}{12}$, we have

$$y = \frac{WL^3}{3Y\left(\frac{bd^3}{12}\right)} = \frac{4WL^3}{Ybd^3}$$

And for a *beam of circular cross-section of radius* r, $I_g = \frac{\pi r^4}{4}$

So that, we have $y = \frac{WL^3}{3Y\left(\frac{\pi r^4}{4}\right)} = \frac{4WL^3}{3Y\pi r^4}$.

Case (ii) : (weight of the cantilever effective). Considering again the section PB (Fig. 3.7) at distance x from the fixed end, we have, in addition to the load W acting at B, a weight equal to that of the portion (L – x) of the beam also acting at its mid-point. So that, if w be the weight per unit length of the beam, we have an additional weight w(L – x) acting at a distance (L – x)/2 from P. And, therefore,

total moment of the external couple applied

$$= W(L-x) + w(L-x)(L-x)/2$$

$$= W(L-x) + \frac{w}{2}(L-x)^2.$$

The beam being in equilibrium, this must be balanced by the *bending moment* $\frac{YI_g}{R}$. We, therefore, have

$$W(L-x) + \frac{w}{2}(L-x)^2 = \frac{YI_g}{R} = YI_g\left(\frac{d^2y}{dx^2}\right),$$

which, on integration yields

$$YI_g\frac{dy}{dx} = W\left(Lx - \frac{x^2}{2}\right) + \frac{w}{2}\left(L^2x - 2L\frac{x^2}{2} + \frac{x^3}{3}\right) + C,$$

where C is a constant of integration.

Since at x = 0, $\frac{dy}{dx} = 0$, we have C = 0.

So that,

$$YI_g \int_0^y dy = W\int_0^L \left(Lx - \frac{x^2}{2} \right) dx + \frac{w}{2}\int_0^L \left(L^2x - Lx^2 + \frac{x^3}{3} \right) dx$$

$$= \frac{WL^3}{3} + \frac{wL^4}{8}.$$

Or, since wL= W_0, the *weight of the beam*, we have

$$YI_g y = \frac{WL^3}{3} + \frac{W_0 L^3}{8}, \text{ whence } y = \left(W + \frac{3}{8}W_0 \right) \frac{L^3}{3YI_g}.$$

Thus, *the beam now behaves as though the load (W) at its free end (B) is increased by 3/8 of its own weight.*

Case (iii) : (Cantilever uniformly loaded). In this case, if the load on the cantilever be w per unit length (including its own weight), the bending at the section PB of the beam is produced by a load w(L – x) acting at a distance (L – x)/2 from P or B, with no load suspended at B (Fig. 3.7).

We, therefore, have

$$w(L-x)\frac{(L-x)}{2} = \frac{w}{2}(L-x)^2 = \frac{YI_g}{R} = YI_g\left(\frac{d^2y}{dx^2} \right),$$

which, on integration, gives $YI_g \dfrac{dy}{dx} = \dfrac{w}{2}\left(L^2x - 2L\dfrac{x^2}{2} + \dfrac{x^3}{3} \right) + C,$

where C is a constant of integration.

Since at x = 0, $\dfrac{dy}{dx} = 0$, we have C = 0. And, therefore,

$$YI_g \int_0^y dy = \frac{w}{2}\int_0^L \left(L^2x - Lx^2 + \frac{x^3}{3} \right) dx$$

Or, $$YI_g y = \frac{w}{2}\left(\frac{3L^4}{12} \right) = \frac{wL}{2}\left(\frac{L^3}{4} \right)$$

Since wL = W, the total load on the beam (including its own weight),

we have $$y = \frac{WL^3}{8YI_g},$$

i.e., the bending due to the uniformly distributed load W, instead of the whole of it at B, is 3/8 of that in the latter case and hence the stiffness of the beam increases to 8/3 of its value with the entire load at B.

In case there is *no load placed on the cantilever* and *it bends under its own weight*, we have $W = W_0$. So that, in that case,

$$y = \frac{W_0 L^3}{8YI_g}.$$

THEORY OF SIMPLE BENDING—ASSUMPTIONS

In discussing the cases of bending that follow we shall concern ourselves with only the theory of simple (or pure) bending which makes the following assumptions:

(i) *That Hooke's law is valid for both tensile and compressive stresses and that the value of Young's modulus (Y) for the material of the beam remains the same in either case.*

(ii) *That there are no shearing stresses over any section of the beam when it is bent.* This is more or less ensured if the length of the beam is sufficiently large compared with its thickness.

(iii) *That there is no change in cross-section of the beam on bending,* this is hardly ever true, a rectangular cross-section almost always acquiring an anticlastic curvature and a circular cross-section possibly assuming an oval form. This change in the shape of cross-section may result in a change in its area and hence also in its geometrical moment of inertia I_g.

Any such change is however always much too small and is, in general, ignored.

(iv) *That the radius of curvature of the neutral axis of the bent beam is very much greater than its thickness.* This is invariably true in all cases of bending of beams.

(v) *That the minimum deflection of the beam is small compared with its length.* This too is true, in general, except in cases like that of a clock spring or strongly bent beams.

STRONGLY BENT BEAMS—FLEXIBLE CANTILEVER

In the three cases discussed above, the bending of the beam and hence the inclination of its loaded end with its unbent position being small, its curvature (1/R) has been taken to be equal to the rate of change of slope, d^2y/dx^2. This is obviously not possible if the beam bends strongly so that this angle of inclination is large. In such a case, as we know, the curvature is given by

$$\frac{1}{R} = \frac{\frac{d^2y}{dx^2}}{\left[\left(1+\frac{dy}{dx}\right)^2\right]^{3/2}}$$

This makes the problem complicated and the differential equations obtained do not admit of an easy solution. However, in some cases, like that of a *flexible cantilever, e.g., a clock spring*, the problem becomes simpler and we shall, therefore, consider only this one case here.

If a clock spring be loosely clamped at a point close to its free end and the free end loaded with a weight W, it is found that as more and more of the spring is passed through the clamp, *keeping the load constant*, the loaded end falls progressively down on account of the large bending of the spring until finally it becomes quite vertical, with the horizontal distance between the loaded end and the clamp now the maximum. Any further length of the spring passed out through the clamp merely hangs vertically without any change in this horizontal distance.

Let OAB, in Fig. 3.8, represent the bent position of the spring, clamped at O and loaded at the free end B, such that the tangents to it at A and B make angles θ and ϕ with its unbent position along OX and their horizontal distances from the clamp are x and a respectively. Then, clearly, *moment of the bending couple* at A' = W(a – x) = *bending moment* YI_g/R, where R is the radius of curvature of the spring at A.

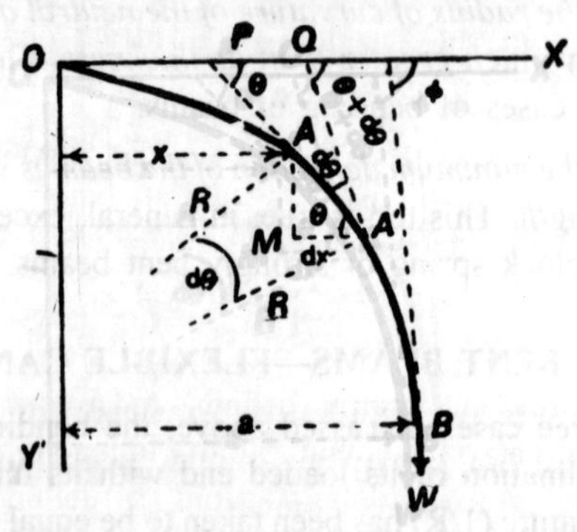

Fig. 3.8

Considering another point infinitely close to A, with AA' = ds, say, so that the radius of curvature of the spring at A' remains the same (R) as at A, if $d\theta$ be the angle that the two radii make at the centre of curvature, we have

$$d\theta = \frac{AA'}{R} = \frac{ds}{R}.$$

Or, $$\frac{1}{R} = \frac{d\theta}{ds}.$$

Draw A'M and AM parallel to the two coordinate axes respectively so as to meet in M. Then, since AA' is sensibly straight and in a line with PA, we have

$$\angle PA'M = \angle AA'M = \angle QPA' = \theta$$

and, therefore. $$\cos\theta = \frac{MA'}{AA'} = \frac{dx}{ds}$$

So that,

$$W(a-x) = \frac{YI_g}{R} = YI_g\frac{d\theta}{ds} = YI_g\frac{d\theta}{dx}\frac{dx}{ds} = YI_g\cos\theta\frac{d\theta}{dx}.$$

Or, $$W(a-x)dx = YI_g\cos\theta d\theta.$$

Integrating between the limits x = 0

and x = a, we have

$$\int_0^a W(a-x)dx = YI_g\int_0^\phi \cos\theta\, d\theta.$$

Or, $$W\left[ax - \frac{x^2}{2}\right]_0^a = YI_g[\sin\theta]_0^\phi,$$

whence, $$\frac{Wa^2}{2} = YI_g\sin\phi. \qquad ...(i)$$

When the loaded end becomes vertical, $\phi = 90°$ and the horizontal distance a, the maximum, say, a_m. Expression (i) therefore becomes

$$\frac{Wa_m^2}{2} = YI_g,\ \text{whence,}\ a_m^2 = \frac{2YI_g}{W}$$

and, therefore, $$a_m = \sqrt{\frac{2YI_g}{W}}. \qquad ...(ii)$$

As will be readily seen, either of the two expressions may be used to obtain the value of Y for the material of the spring, if we know and a in the first, and a_m in the second, case.

DETERMINATION OF Y FOR THE MATERIAL OF A CANTILEVER

Statical Method

This consists in noting the depression or deflection (y) of the loaded end of the cantilever for different loads (W) suspended from its free end and plotting a graph between W and y when we obtain a straight line OA passing through the' origin, as shown in Fig. 3.9.

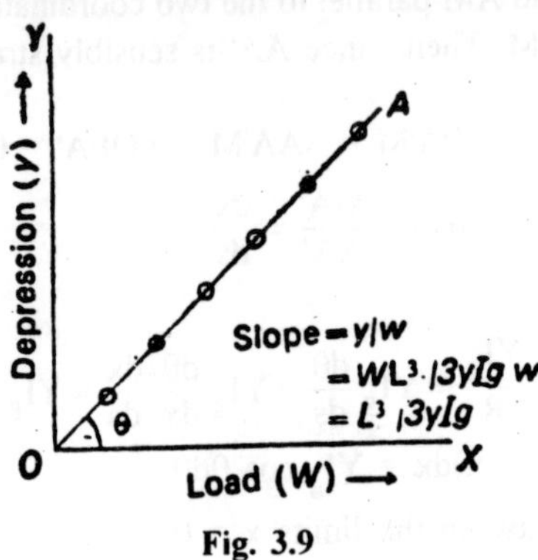

Fig. 3.9

The slope of the straight graph is clearly

$$\tan\theta = \frac{y}{W} = \frac{WL^3}{3YI_g}.W = \frac{L^3}{3YI_g},$$

whence, knowing L and I_g, the value of Y for the material of the cantilever may be easily obtained.

Dynamical Method (Transverse Vibration of the Cantilever)

As we know, in the position of equilibrium of a cantilever loaded with a weight W at its free end, we have *displacement of the loaded end, i.e.,*

$$y = \frac{WL^3}{3YI_g}, \text{ whence, } W = \left(\frac{3YI_g}{L^3}\right)y. \qquad ...(i)$$

This must obviously be equal to the inertial reaction of the cantilever balancing it and hence directed oppositely to it.

But if m be the mass of the weight W and $\frac{d^2y}{dt^2}$, the acceleration (upwards), we also have *inertial reaction* $= -\frac{md^2y}{dt^2}$, the — ve sign indicating its direction, opposite to that of the displacement of *mass* m or the free end of the cantilever.

We, therefore, have $-m\frac{d^2y}{dt^2} = \frac{3YI_g}{L^3}Y.$

Or, $\frac{d^2y}{dt^2} = -\frac{3YI_g}{mL^3}y,$

i.e., $\frac{d^2y}{dt^2} = -\mu y,$

where $\frac{3YI_g}{mL^3} = \mu$, a constant.

The acceleration of mass m or the free end of the cantilever is thus proportional to its displacement and is directed oppositely to it. It therefore executes a S.H.M. of time-period T, given by

$$T = 2\pi\sqrt{\frac{1}{\mu}} = 2\pi\sqrt{\frac{mL^3}{3YI_g}}. \quad \text{...(i)}$$

Or, since $y = \frac{WL^3}{3YI_g} = \frac{mgL^3}{3YI_g}$,

we have $\frac{mL^3}{3YI_g} = \frac{y}{g}$.

And, therefore $T = 2\pi\sqrt{\frac{y}{g}}$. ...(ii)

This gives us a dynamical method for the determination of Y for the material of the cantilever, particularly a light one whose weight is ineffective, *e.g.*, a *metre stick*. For, plotting a graph between m and T^2, we obtain a straight line passing though the origin, whose slope is given by $\frac{4\pi^2L^3}{3YI_g}$, whence the value of Y may be easily calculated.

DEPRESSION OF SUPPORTED AND FIXED BEAMS

We shall deal with the two cases separately, restricting ourselves to beams which are supported or fixed in a horizontal plane.

Supported Beam, Centrally Loaded

(a) *When the weight of the beam is ineffective* : Let a beam be supported on two knife-edges A and B, distance L apart and loaded in the middle at O with a weight W, as shown in Fig. 3.10.

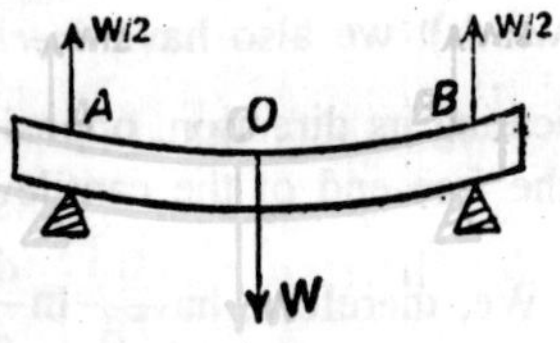

Fig. 3.10

The upward reaction at each knife-edge being W/2 and the middle part of the beam being sensibly horizontal, it may be taken to be a combination of two inverted cantilevers OA and OB, each of effective length L/2, fixed at O and bending upwards under a load W/2 acting at A and B. Clearly, then, the elevation of A or B above O, or the depression of O below A and B is given by

$$y = \frac{\left(\frac{W}{2}\right)\left(\frac{L}{2}\right)^3}{3YI_g} = \frac{WL^3}{48YI_g}. \quad ...(i)$$

If, therefore, the beam be of a *circular cross-section*, of *radius* r, so that $I_g = \frac{\pi r^4}{4}$, we have

$$y = \frac{WL^3}{48Y}\left(\frac{4}{\pi r^4}\right) = \frac{WL^3}{12Y\pi r^4} \quad ...(ii)$$

And, if the cross-section of the beam be *rectangular*, of *breadth b* and *depth* d, so that $I_g = \frac{bd^3}{12}$, we have

$$y = \frac{WL^3}{48Y}\left(\frac{12}{bd^3}\right) = \frac{WL^3}{4Ybd^3}. \quad ...(iii)$$

Alternatively, we could obtain expression (i) for y directly as follow:

Considering the section PB of the cantilever OB, say, distant x from its fixed end O, (Fig. 3.11) we have *moment of bending couple due to load W/2*

$$= \frac{W}{2}\left(\frac{L}{2} - x\right)$$

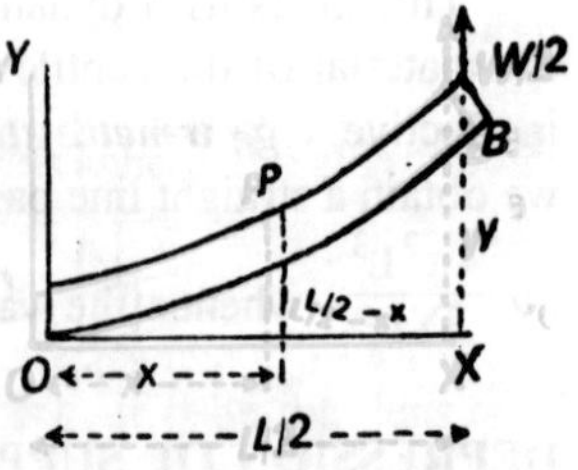

Fig. 3.11

The beam being in equilibrium, this must be just balanced by the *bending moment* (or the moment of the resistance to bending), YI_g/R, where R is the radius of curvature of the section at P, (3.5).

We, therefore, have $\frac{YI_g}{R} = YI_g \frac{d^2y}{dx^2} = \frac{W}{2}\left(\frac{L}{2} - x\right)$,

which on integration gives $\frac{dy}{dx} = \frac{W}{2YI_g}\left(L\frac{x}{2} - \frac{x^2}{2}\right) + C$.

Since at $x = 0$, $\frac{dy}{dx} = 0$, we have $C = 0$ and, therefore,

$$\frac{dy}{dx} = \frac{W}{2YI_g}\left(L\frac{x}{2} - \frac{x^2}{2}\right).$$

Or, $$dy = \frac{W}{2YI_g}\left(L\frac{x}{2} - \frac{x^2}{2}\right)dx,$$

which, on further integration between the limits $x = 0$ and $x = L/2$, gives

$$y = \frac{W}{2YI_g}\left[\frac{L^3}{16} - \frac{L^3}{48}\right] = \frac{WL^3}{48YI_g},$$

the same expression as (i) above.

And, clearly, slope of B, and hence also that of A, relative to O (or their angle of incination θ to the horizontal) is given by

$$\tan\theta = \theta = \left(\frac{dy}{dx}\right)\frac{L}{2} = \frac{W}{2YI_g}\left(\frac{L^2}{4} - \frac{L^2}{8}\right) = \frac{WL^2}{16YI_g}$$

(b) *When the weight of the beam is effective*: Let the beam, of weight w per unit length be resting in a horizontal plane on two knife-adges A and B distance L apart and let a load W be applied to it at its mid-point O, (Fig. 3.12). Then, clearly, *total load acting on the beam at O = W + weight of the beam = W + wL = (W + W_0), where wL = W_0, the weight of the beam.*

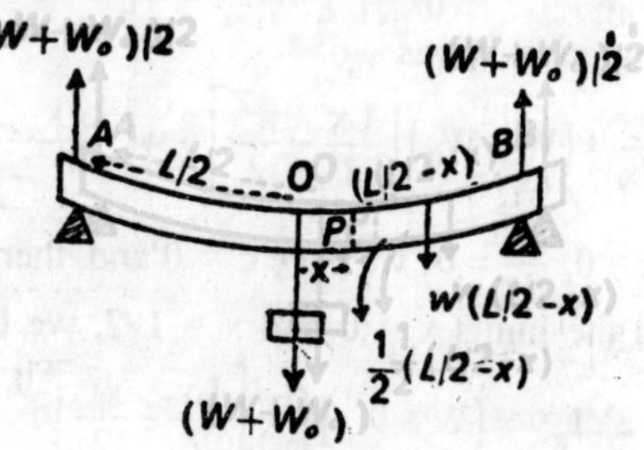

Fig. 3.12

$\therefore$ *upward reaction at each knife-edge* = $(W + W_0)/2$, as shown.

Now, as in case (a), so also here, the beam behaves as a combination of two cantilevers OA and OB, each of length L/2 and fixed at O, with a load $(W + W_0)/2$ acting upwards at their free ends A and B. The elevation of A or B above O thus gives the depression y of the centre O of the beam below A or B.

Considering a section PB of the cantilever OB, distant x from the fixed end O, we have *weight of section PB of the beam* $= w\left(\frac{L}{2} - x\right)$, acting vertically downwards midway between P and B, *i.e.*, at a distance $\frac{1}{2}\left(\frac{L}{2} - x\right)$ from P or B, as indicated.

Taking moments about P, we have *moment of* $\frac{(W+W_0)}{2}$ *about* $P = \left(\frac{W+W_0}{2}\right)\left(\frac{L}{2} - x\right)$ *anticlockwise.*

And *moment of* $w\left[\frac{L}{2} - x\right]$ *about* $P = \frac{1}{2}w\left[\frac{L}{2} - x\right]\left[\frac{L}{2} - x\right]$

$$= \frac{w}{2}\left[\frac{L}{2} - x\right]^2, \text{ clockwise.}$$

$\therefore$ *resultant moment about* $P = \frac{W+W_0}{2}\left[\frac{L}{2} - x\right] - \frac{w}{2}\left[\frac{L}{2} - x\right]^2$, tending to bend the beam in the anticlockwise direction.

Since the beam is in equilibrium, this must be just balanced by the *bending moment* $\frac{YI_g}{R}$, where R is the radius of curvature of the section at P. We, therefore, have

$$\frac{YI_g}{R} = YI_g\frac{d^2y}{dx^2} = \left[\frac{W+W_0}{2}\right]\left[\frac{L}{2} - x\right] - \frac{w}{2}\left[\frac{L}{2} - x\right]^2,$$

which, on integration, gives

$$2YI_g\frac{dy}{dx} = [W + W_0]\left[\frac{Lx}{2} - \frac{x^2}{2}\right] - w\left[\frac{L^2x}{4} - \frac{Lx^2}{2} + \frac{x^3}{3}\right] + C.$$

Since at $x = 0$, $\frac{dy}{dx} = 0$, we have $C = 0$ and, therefore, on integrating again between the limits $x = 0$ and $x = L/2$, we have

$$2YI_g y = [W + W_0]\int_0^{L/2}\left[\frac{Lx}{2} - \frac{x^2}{2}\right]dx$$

$$-w\int_0^{L/2}\left[\frac{L^2x}{4}-\frac{Lx^2}{2}+\frac{x^3}{3}\right]dx$$

$$=\left[W+W_0\right]\left[\frac{Lx^2}{4}-\frac{x^3}{6}\right]_0^{L/2}-w\left[\frac{L^2x^2}{8}-\frac{Lx^3}{6}+\frac{x^4}{12}\right]_0^{L/2}$$

$$=\left[W+W_0\right]\left[\frac{L^3}{24}\right]-wL\left[\frac{3L^3}{192}\right]=\frac{L^3}{24}\left[W+W_0-\frac{3}{8}W_0\right].$$

$$[\because wL = W_0]$$

Or, $$y=\frac{L^3}{48YI_g}\left[W+\frac{5}{8}W_0\right]$$

Thus, the *weight of the beam has the effect of increasing the effective load on the beam by 5/8 of its own value.*

(ii) *Supported beam uniformly loaded* : As before, let the beam be resting in a horizontal plane on two knife-edges A and B, distance L apart, (Fig. 3.13). Then, if w be the load per unit length in the beam (including its own weight), the total load on it is wL = W, acting at its mid-point O and, therefore, upward reaction at each knife-edge is W/2, as shown,

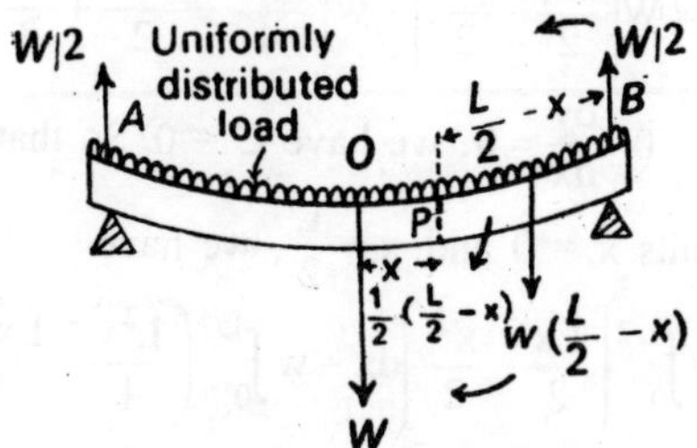

Fig. 3.13

As in cases (i) (a) and (b), the beam behaves as a combination of two cantilevers OA and OB, each of length L/2, with its end O fixed and with a load W/2 acting upwards at its other end A or B. So that, elevation of the end A as B above the fixed end O gives the depression y of the mid-point O of the beam below A or B.

Considering again a section PB of the cantilever OB, distant x from O, we have weight of the section PB, equal to $w\left[\frac{L}{2}-x\right]$ acting vertically downwards at its mid-point, *i.e.*, at a distance $\frac{1}{2}\left(\frac{L}{2}-x\right)$ from P or B.

We, therefore, have

moment of $\frac{W}{2}$ *(at B) about* $P = \frac{W}{2}\left[\frac{L}{2} - x\right]$, *anticlockwise,* and *moment of* $w\left[\frac{L}{2} - x\right]$ about $P = \frac{1}{2}w\left[\frac{L}{2} - x\right]\left[\frac{L}{2} - x\right]$

$$= \frac{w}{2}\left[\frac{L}{2} - x\right]^2, \text{ clockwise.}$$

∴ *resultant moment about* P, tending to bend the cantilever anticlockwise $= \frac{W}{2}\left[\frac{L}{2} - x\right] - \frac{w}{2}\left[\frac{L}{2} - x\right]^2$

Since the cantilever is in equilibrium in its bent position, this must just be balanced by the bending moment (or the moment of the resistance to bending) YI_g/R, where R is the radius of curvature of the section at P. Thus,

$$\frac{YI_g}{R} = YI_g\frac{d^2y}{dx^2} = \frac{W}{2}\left[\frac{L}{2} - x\right] - \frac{w}{2}\left[\frac{L}{2} - x\right]^2,$$ which, on integration gives $2YI_g\frac{dy}{dx} = W\left[\frac{Lx}{2} - \frac{x^2}{2}\right] - w\left[\frac{L^2x}{4} - \frac{Lx^2}{2} + \frac{x^3}{3}\right] + C.$

Since at x = 0, $\frac{dy}{dx} = 0$, we have C = 0. So that, integrating again between the limits x = 0 and $x = \frac{L}{2}$, we have

$$2YI_g y = W\int_0^{L/2}\left[\frac{Lx}{2} - \frac{x^2}{2}\right]dx - w\int_0^{L/2}\left(\frac{L^2x}{4} - \frac{Lx^2}{2} + \frac{x^3}{3}\right)dx$$

$$= W\left[\frac{L^3}{24}\right] - W\left[\frac{3L^2}{192}\right] = \frac{WL^3}{24}\left[1 - \frac{3}{8}\right], \qquad [\because wL = W]$$

whence, $$y = \frac{5}{8}\frac{WL^3}{48YI_g},$$

showing that *the depression is 5/8 of that when the same load is applied centrally.* In other words, *the stiffness of the beam now increases to 8/5 or 1.6 times its value in the former case.*

If there be *no load placed on the beam,* we have $W = W_0$, the weight of the beam itself and, therefore, $y = \frac{5}{8}\frac{W_0L^3}{48YI_g}$, which thus gives the *depression of the beam when bending under its own weight.* As will be readily seen, we obtain the same expression for y by putting W = 0 in case (i) (b) above.

(iii) *Fixed (or encastre) beam, centrally loaded :* As already pointed out (3.1), each end, in the case of a fixed beam, is securely clamped or built into a wall, so that it remains horizontal all through, *i.e.,* its slope remains zero both before and after the application of the load. To ensure, however, that the fixing of the ends does not hinder the horizontal movement of the beam as a whole, one of the ends is usually held between frictionless rollers, as shown in Fig. 3.14. This, while it prevents any angular or vertical movement of the beam, allows it to move longitudinally.

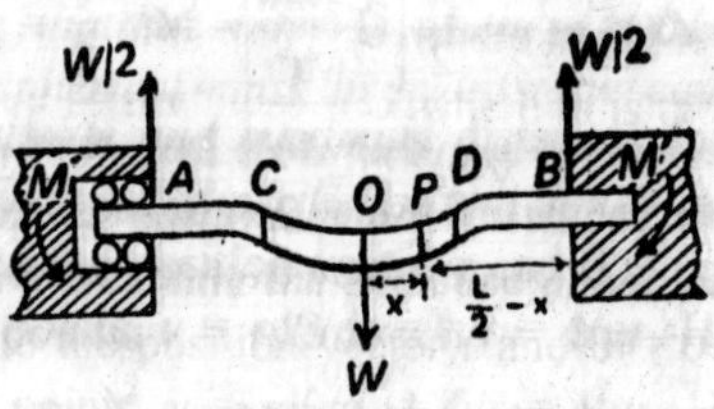

Fig. 3.14

If the beam were simply supported at the ends, the centrally applied load would bend the entire beam, *including the ends*, so as to be concave upwards. In order to keep the ends horizontal, as here, therefore, a couple must be applied in the opposite direction to that due to the load. This couple is provided by the clamp or the built-in support and is called the *fixing* moment. The beam being symmetrical, its magnitude is the same at either end, say, M'.

Now, as the load W is applied to the centre O of the beam, it tends to bend the beam into a curve, *concave upwards*, but the fixing moment M', which is effective all over the beam, tends to bend it into a curve *concave downwards*. As a result, the beam assumes a shape such that *its middle part CD is concave upwards and its parts AC and BD between it and the two ends are concave downwards, with well-marked points of inflexion at C and D on either side of, and equidistant from, its centre O and where, obviously, the slope (dy/dx) is the maximum* (because the curvature of the beam changes sign there).

Again, considering a section PB of the right half of the beam, distant x from its mid-point O and remembering that there is an upward reaction W/2 at A and B, we have *moment of the bending couple due to the applied load about P*

$$= \frac{W}{2}\left(\frac{L}{2} - x\right).$$

Since the fixing moment M′ is directed oppositely to it, we have *resultant moment of the bending couple about* $P = \frac{W}{2}\left[\frac{L}{2} - x\right] - M.$

For equilibrium of the beam in its bent position, this must just be balanced by the bending moment (or the moment of the resistance to bending) YI_g/R, where R is the radius of curvature of the section at P. We, therefore, have

$$\frac{YI_g}{R} = YI_g \frac{d^2y}{dx^2} = \frac{W}{2}\left[\frac{L}{2} - x\right] - M', \qquad \text{...(i)}$$

which, *on integration*, gives $YI_g \frac{dy}{dx} = \frac{W}{2}\left[\frac{Lx}{2} - \frac{x^2}{2}\right] - M'x + C$

Since at $x = 0$, $\frac{dy}{dx} = 0$, we have $C = 0$, and therefore,

$$YI_g \frac{dy}{dx} = \frac{W}{2}\left[\frac{Lx}{2} - \frac{x^2}{2}\right] - M'x \qquad \text{...(ii)}$$

Now, at $x = \frac{L}{2}$, *i.e.*, at the end B of the beam also, $\frac{dy}{dx} = 0$. So that,

$$0 = \frac{W}{2}\left[\frac{L}{2}\left(\frac{L}{2}\right) - \frac{(L/2)^2}{2}\right] - \frac{M'L}{2}$$

Or,

$$\frac{M'L}{2} = \frac{W}{2}\left[\frac{L^2}{8}\right],$$

whence,

$$M' = \frac{WL}{8}.$$

Substituting this value of M' in relation (ii), we have

$$YI_g \frac{dy}{dx} = \frac{W}{2}\left[\frac{Lx}{2} - \frac{x^2}{2}\right] - \frac{WLx}{8}. \qquad \text{(iii)}$$

Therefore, maximum elevation or the elevation of end B above O, or what is the same thing, *depression y of mid-point O of the beam below the ends A and B*, is obtained by integrating expression (iii) for the limits $x = 0$ and $x = L/2$. Thus,

$$YI_g y = \frac{W}{2}\int_0^{L/2}\left(\frac{Lx}{2} - \frac{x^2}{2}\right)dx - wL\int_0^{L/2}\frac{x}{8}dx$$

$$= \frac{W}{2}\left[\frac{Lx^2}{4} - \frac{x^3}{6}\right]_0^{L/2} - WL\left[\frac{x^2}{16}\right]_0^{L/2} = WL^3\left(\frac{1}{32} - \frac{1}{96} - \frac{1}{64}\right) = \frac{WL^3}{192};$$

$$\therefore y = \frac{WL^3}{192YI_g} = \frac{1}{4}\frac{WL^3}{48YI_g},$$

i.e., the depression of the mid-point of the beam is one-fourth of that when it is simply supported at the ends.

This means, in other words, *that the stiffness of the beam, when its ends are fixed, is four times that when the ends are simply supported.*

Further, as we have just seen, *moment of bending couple about P distant x from 0*

$$= \frac{W}{2}\left(\frac{L}{2} - x\right) - M'.$$

Now, $M' = \frac{WL}{8}$ and for the mid-point O of the beam, $x = 0$.

$\therefore$ *moment of bending couple about the mid-point of the beam*

$$= \frac{W}{2}\left(\frac{L}{2} - 0\right) - \frac{WL}{8} = \frac{WL}{8}$$

On the other hand, *when the beam is simply supported at the ends, the moment of the bending couple about the mid-point of the beam*

$$= \frac{W}{2}\left(\frac{L}{2} - 0\right) = \frac{WL}{4}.$$

Thus, *fixing the ends of the beam reduces the moment of the bending couple to half the value it would have if the ends are simply supported.* So that, for the same value of the bending couple (WL/4), the fixed beam can take twice the load that can be taken by the supported beam. In other words, *the strength of a beam, when fixed*, is twice *that when if is merely supported.*

Positions of points of inflexion: The positions of the points of inflexion C and D may be easily obtained from the fact that because of the curvature changing sign at these points, the slope dy/dx is the maximum there and hence the rate of change of slope, $d^2y/dx^2 = 0$.

Thus, putting $d^2y/dx^2 = 0$ in relation (i) above and substituting the value of $M' = WL/8$, we have

$$0 = \frac{W}{2}\left(\frac{L}{2} - x\right) - \frac{WL}{8}.$$

Or, $\frac{WL}{4}-\frac{WL}{8}=\frac{Wx}{2}$, whence, $x=\frac{L}{4}$.

This gives the point D at a distance L/4 from O and hence also L/4 from B.

Similarly, point C lies at a distance L/4 from O to the left or L/4 from A.

The beam is thus, in effect, a combination of a central beam of length L/2, with a pair of cantilevers, each of length L/4, on either side, with the loading as shown in Fig. 3.15.

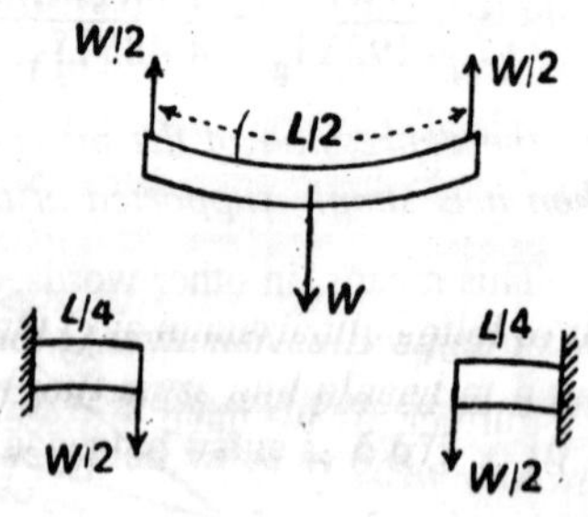

Fig. 3.15

DETERMINATION OF ELASTIC CONSTANTS

(i) *Searle's method for the material of a wire* : G.F.C. Searle devised, in the year 1900, a simple method for the determination of elastic constants of an *isotropic material in the form of a wire.*

A short length of the given wire (w) is fastened to the mid-points P_1 and P_2 of two identical metal bars AB and CD, suspended from points directly above P_1 and P_2 by means of two equal lengths of *cotton* threads, each about a metre long, (so that the torsional couple set up in the suspension threads may be negligibly small).

Initially the bars, which are chosen to be of a circular or rectangular cross-section, (so that their M.I. about their respective suspension threads may be easily obtained from their mass and physical dimensions), lie parallel to each other, with the wire w quite straight and horizontal, as shown in Fig. 3.16 (a).

To determine the clastic constants for the material of the wire, we then proceed as follows:

(a) *Determination of Y* : The ends A and C of the two bars respectively are pulled symmetrically inwards through small equal distances so that the wire gets bent into a circular are of radius R, as shown in Fig. 3.16(b), and then released.

Due to the torque exerted by the wire, the two bars are thus set oscillating in a horizontal plane about their respective suspension threads, with the wire forming an are first facing one side and then the other, the mid-points of the bars remaining practically at rest and the bending moment at all transverse sections of the wire constant.

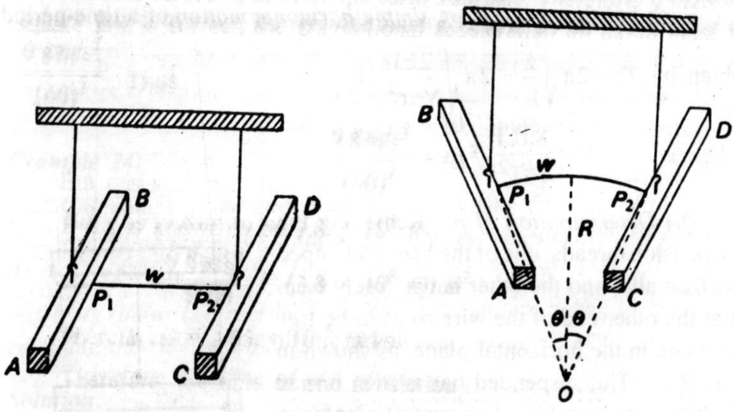

Fig. 3.16

If L be the length of the wire and θ, the angle through which each bar has been pulled in from its initial position, clearly, *angle subtended by the circular arc at its centre O* = 2θ,

and we, therefore, have L = R.2θ, whence, $R = \frac{L}{2\theta}$.

Now, *bending moment of the wire*, $M = \frac{YI_g}{R}$,

where I_g, is the *geometrical M.I.* of the cross-section of the wire, equal to $\frac{\pi r^4}{4}$, (r being the *radius of the wire*).

Substituting the values of I_g and R, therefore, we have

$$M = \frac{Y\pi r^4}{4} \cdot \frac{2\theta}{L} = \frac{Y\pi r^4}{2L}\theta.$$

This being equal to the *restoring couple* acting on each bar, its *equation of motion is* $M + \frac{Id\omega}{dt} = 0$,

where dω/dt is the angular acceleration produced in each bar and I, its M.I. about the axis through its mid-point and perpendicular to its length, *i.e.,* about its suspension thread.

So that, $$\frac{Id\omega}{dt} = -M = -\frac{Y\pi r^4}{2L}\theta = -\mu\theta$$

Or, $$\frac{d\omega}{dt} \alpha\, \theta \qquad \left[\text{where } \frac{Y\pi r^4}{2LI} = \mu\right]$$

Each bar thus executes a *simple harmonic motion* of a time-period given by $T_1 = 2\pi\sqrt{\frac{I}{\mu}} = 2\pi\sqrt{\frac{2LI}{Y\pi r^4}}$,

whence, $$Y = \frac{8\pi LI}{r^4 T_1^2}.$$

(b) *Determination of n :* Removing the suspension threads, one of the bars is clamped horizontally, and the other hangs freely from it at the other end of the wire so as to be free to rotate in the horizontal plane, as shown in Fig. 3.17. The suspended bar is then turned a little in its own (*i.e.,* horizontal) plane and released, when it executes torsional oscillations about the wire, with its time-period given by

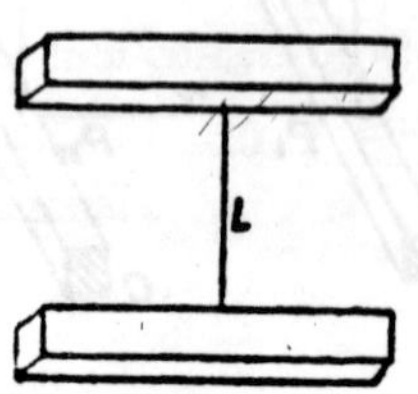

Fig. 3.17

$$T_2 = 2\pi\sqrt{\frac{I}{C}},$$

where C is the *torsional couple per unit twist* of the wire, equal to $\frac{n\pi r^4}{2L}$.

So that, $T_2 = 2\pi\sqrt{\frac{I.2L}{n\pi r^4}}$, whence, $n = \frac{8\pi LI}{r^4 T_2^2}$ and may be easily evaluated.

(c) *Determination of σ :* As we know, *Poisson's ratio* $\sigma = \left(\frac{Y}{2n}\right) - 1$.

Substituting for Y and n, therefore, we have

$$\sigma = \frac{T_2^2}{2T_1^2} - 1 = \frac{T_2^2 - 2T_1^2}{2T_1^2}$$

(d) *Determination of K :* The value of *bulk modulus* K for the material of the wire may either be obtained from the relation

$$\frac{9}{Y} = \frac{3}{n} + \frac{1}{K},$$

whence, $K = \frac{\eta Y}{(9n - 3Y)}$ or from the relation $K = \frac{Y}{3(1-2\sigma)}$, whence, we have $K = \dfrac{8\pi LI}{3r^4\left(3T_1^2 - T_2^2\right)}$.

Thus, all the four elastic constants, Y, n, K and σ, for the material of the wire may be easily obtained.

It will be noted that the method has the merit of (i) *requiring only a short length (just a few centimetres) of the wire and (ii) giving the value of s in terms of two accurately measurable quantities* T_1 and T_2, eliminating altogether the chief source of error, namely, the measurement of the radius (r) of the wire.

(ii) *Cornu's optical method for a glass plate or a metal beam* : A sensitive optical method, based on the phenomenon of interference of light, was devised by *Cornu* in the year 1869 for the determination of the elastic constants of a material in the form of a beam. The method is obviously applicable to substances like glass or metals which can take a high polish. It is particularly suited for glass which can be subjected to only small stresses with safety.

The glass plate or the metal beam, with its upper surface highly polished is supported *symmetrically* on two knife-edges K_1 and K_2 in a horizontal plane, with the knife-edges normal to the axis of the plate (or the beam) and an *optically plane* cover glass G placed on it at its mid-point O, [Fig. 3.18 (b)].

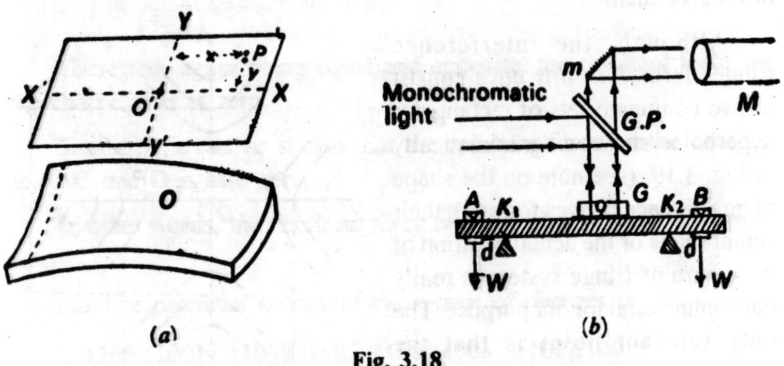

Fig. 3.18

The beam (or the plate) is then subjected to a *uniform bending couple* Wd by suspending equal weights W and W from it at A and B, a little outside the two knife-edges at an equal distance d from them, as shown. Thus loaded, the beam experiences no shear anywhere between the two knife-edges and the bending couple is uniform all over in this portion. It thus undergoes a pure or simple bending about the mid-point O, with its longitudinal axis concave downwards, having a radius of curvature R_1, say.

This, as we know, is accompanied by the transverse axis of the beam acquiring an anticlastic curvature or a curvature in the opposite direction to that of the longitudinal axis, *i.e.,* it becomes concave upwards, as shown in Fig. 3.18 (a), having a radius of curvature R_2.

So that, although the mid-point O' of the lower surface of the cover glass still lies vertically above the mid-point O of the upper surface of the beam, the two no longer remain in contact and a thin film of air is formed between them.

If, therefore, a parallel beam of monochromatic light from a suitable source like a sodium lamp is allowed to fall normally on the film by reflecting it downwards by means of a glass plate G.P., interference fringes are formed due to reflection of light from the upper and lower surfaces of the film or, what is the same thing from the lower surface of the cover glass and the upper surface of the beam or the glass plate). These fringes are localised in the film itself and may be observed by means of a long-focus travelling microscope M arranged vertically above or, more conveniently, by deflecting the rays through 90° by means of a plane mirror m and arranging the microscope horizontally to receive them.

Although, the interference fringes formed *in this case* consist of two conjugate sets of rectangular hyperbolae, shown diagrammatically in Fig. 3.19, (see note on the shape of interference fringes formed), their actual shape or the actual position of the centre of fringe system is really quite immaterial for our purpose. The only relevant point is that the distances between any two pairs of conjugate fringes along and across the beam, measured by a horizontal and a vertical travel of the microscope respectively, enable us to obtain the radii R_1 and R_2 of the longitudinal and transverse sections of the

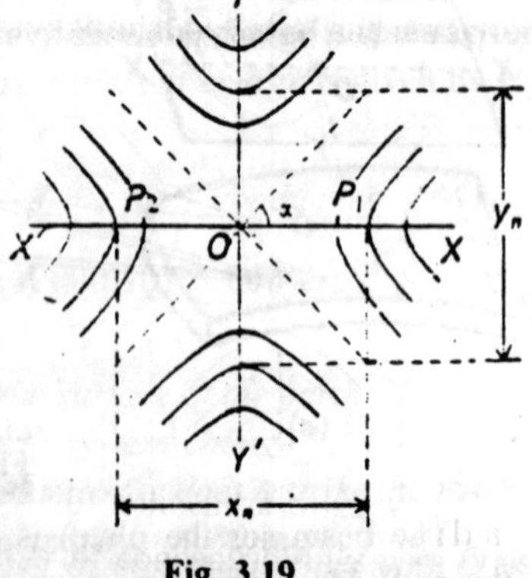

Fig. 3.19

beam respectively, and hence to calculate the values of the elastic constants of the material of the beam, as explained hereunder.

Calculations

(i) *For Poisson's ratio, σ:* Considering the conjugate set of hyperbolic fringes formed along the direction of the longitudinal axis of the beam, (Figs. 3.19 and 3.20) and concentrating on the dark fringes, which are the more prominent when seen by reflected light, if P_1 and P_2 be the positions of the nth pair of them, we have *thickness of the air film between the beam and the cover glass*

= O'C = OO' + OC.

Or, putting the distance OO' between the centre of the cover glass and the mid-point of the beam surface equal to Δ and OC = p, we have *thickness of the air film* = Δ + p.

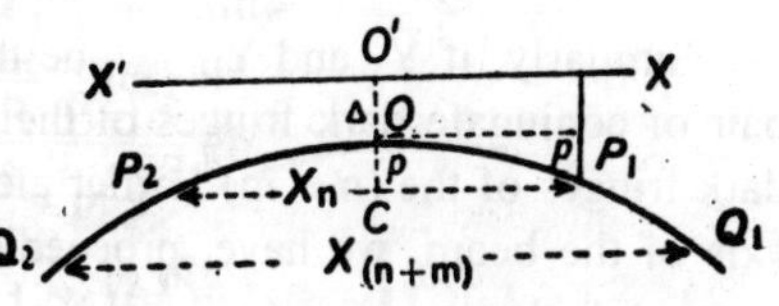

Fig. 3.20

Now, clearly, $OC(2R_1 - OC) = CP_1 \times CP_2$. Or, $p(2R_1 - p) = (X_n/2)^2$ where R_1 is the radius of curvature of the bent beam, or rather of the neutral axis of its longitudinal section, and $P_1P_2 = X_n$.

Or, $2R_1p - p^2 = X_n^2/4$. Or, neglecting p^2 compared with 2Rp, we have

$$p = \frac{X_n^2}{8R_1}.$$

So that, *thickness of the enclosed air film*

$$= \Delta + p = \Delta + \frac{X_n^2}{8R_1}.$$

$\therefore$ *path difference for the interfering rays* = *2(thickness of the air film)*

$$= 2\left(\Delta + \frac{X_n^2}{8R_1}\right) = 2\Delta + \frac{X_n^2}{4R_1}. \quad ...(i)$$

The case being similar to that of Newton's rings, this path difference for a dark fringe must be equal to a whole number of wave-lengths (λ) of the monochromatic light used (usually sodium light).

So that, $$2\Delta + \frac{X_n^2}{4R_1} = n\lambda.$$

Again, if $Q_1Q_2 = X_{(n+m)}$ be the distance between another conjugate pair of dark fringes of the $(n + m)^{th}$ order along the same direction (*i.e.*, along the direction of the longitudinal axis of the beam), we have

$$2\Delta + \frac{X_{(n+m)}^2}{4R_1} = (n+m)\lambda. \qquad \text{...(ii)}$$

Subtracting relation (i) from (ii) we have

$$\frac{X_{(n+m)}^2 - X_n^2}{4R_1} = m\lambda,$$

whence, $$R_1 = \frac{X_{(n+m)}^2 - X_n^2}{4m\lambda}. \qquad \text{...(iii)}$$

Similarly, if Y_n and $Y_{(n+m)}$ be the respective distances between a pair of conjugate dark fringes of the nth order and a pair of conjugate dark fringes of the $(n + m)^{th}$ order along the direction of the transverse axis of the beam, we have, proceeding exactly as above,

$$\frac{Y_{(n+m)}^2 - Y_n^2}{4R_2} = m\lambda,$$

whence, $$R_2 = \frac{Y_{(n+m)}^2 - Y_n^2}{4m\lambda}. \qquad \text{..(iv)}$$

So that, $$\frac{R_1}{R_2} = \frac{X_{(n+m)}^2 - X_n^2}{Y_{(n+m)}^2 - Y_n^2}.$$

Thus, since $\frac{R_1}{R_2} = \sigma$, *Poisson's ratio* for the material *of the beam,* we have

$$\sigma = \frac{X_{(n+m)}^2 - X_n^2}{Y_{(n+m)}^2 - Y_n^2}. \qquad \text{...(v)}$$

Hence, measuring distances X_n and $X_{(n+m)}$ by moving the microscope horizontally and distances Y_n and $Y_{(n+m)}$ by moving it vertically, we can easily determine the value of *Poisson's ratio* for the material of the beam (or the glass plate).

We may also obtain the value of σ from the angle a that the asymptotes in Fig. 3.21 make with the x-axis. For, $\sigma = \frac{R_1}{R_2} = \cot^2 \alpha$, where α can be measured by means of a goniometer eye-piece.

(ii) *For Young's modulus Y* : We know that the bending couple applied to the beam is uniform and equal to Wd. For the equilibrium of the beam in its bent position, therefore, this must be balanced by the bending moment or the moment of the resistance to bending. We, therefore, have for the longitudinal section of the beam,

$$Wd = YI_g\left(\frac{1}{R_1} - \frac{1}{R_0}\right), \tag{vi}$$

where $1/R_0$ is any slight curvature the beam may possibly have along this section on account of its own weight.

So that $\dfrac{YI_g}{R_1} = \dfrac{Wd + YI_g}{R_0}$.

Or, putting $\dfrac{YI_g}{R_0} = \beta$ and substituting the value of R_1 from relation (iii) above, we have

$$YI_g \frac{4m\lambda}{X^2_{(n+m)} - X^2_n} = Wd + \beta$$

Or $$\frac{4mYI_g\lambda}{d}\left(\frac{1}{X^2_{(n+m)} - X^2_n}\right) = W + \frac{\beta}{d}.$$

Again, putting $\dfrac{1}{\left(X^2_{(n+m)} - X^2_n\right)} = x$ and $W = y$, we have

$$y = \frac{4mYI_0\lambda}{d}x - \frac{\beta}{d},$$

which is the equation to a straight line, of slope $\dfrac{4mYI_g\lambda}{d}$ and intercepting the y-axis at $-\dfrac{\beta}{d}$.

If, therefore, we change the values of W in equal steps, obtain the corresponding values of $\dfrac{1}{\left(X^2_{(n+m)} - X^2_n\right)}$ and plot.

$$y = W \text{ against } x = \frac{1}{\left(X^2_{(n+m)} - X^2_n\right)},$$

we obtain a straight line A (Fig. 3.23) intercepting the y-axis at $\dfrac{\beta}{d}$ below the origin and having a slope $\dfrac{4mYI_g\lambda}{d}$, whence the value of Y for the material of the beam can be easily calculated.

Now, since $R_1 = \sigma R_2$ we may put expression (vi) for Wd in terms of R_2 as

$$Wd = YI_g\left(\frac{1}{\sigma R_2} - \frac{1}{R_0}\right)$$

Or, $$\frac{YI_g}{\sigma}\left(\frac{1}{R_2}\right) = Wd + \frac{YI_g}{R_0} = Wd + \beta.$$

Or, substituting the value of R_2 from relation (iv) above, we have

$$\frac{YI_g}{\sigma}\left(\frac{4m\lambda}{Y^2_{(n+m)} - Y^2_n}\right) = Wd + \beta$$

Or $$\frac{4mYI_g\lambda}{\sigma d}\left(\frac{1}{Y^2_{(n+m)} - Y^2_n}\right) = W + \frac{\beta}{d}.$$

Again, putting $\frac{1}{\left(Y^2_{(n+m)} - Y^2_n\right)} = x$ and $W = y$, we have

$$y = \frac{4mYI_g\lambda}{\sigma d}x - \frac{\beta}{d},$$

which is also the equation to a straight line of slope $\frac{4mYI_g\lambda}{\sigma d}$ and intercepting the y-axis at $-\frac{\beta}{d}$.

So that plotting $y = W$ against $x = \frac{1}{\left(Y^2_{(n+m)} - Y^2_n\right)}$, we obtain a straight line B (Fig. 3.21), intercepting the y-axis at β/d below the origin, *i.e.*, at the same point as A, and having a slope $= 4mYI_g\lambda/\sigma d$.

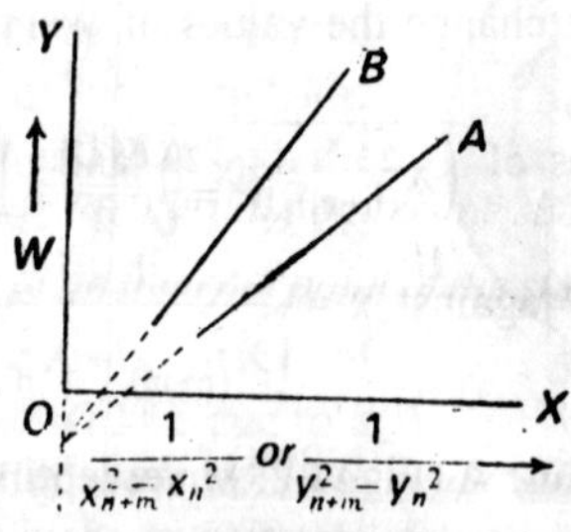

Fig. 3.21

$$\text{Clearly, therefore } \frac{\text{slope of A}}{\text{slope of B}} = \frac{\dfrac{4mYI_g\lambda}{d}}{\dfrac{4mYI_g\lambda}{\sigma d}} = \sigma$$

Thus, the value σ for the material of the beam may also be obtained from the slopes of the two curves A and B.

Now, knowing σ and Y for the material of the beam, we can obtain the other two elastic constants K and n for it from the relations

$$K = \frac{Y}{3(1-2\sigma)} \text{ and } n = \frac{Y}{2(1+\sigma)}.$$

Shape of interference fringes formed : As mentioned already, the actual shape of the interference fringes formed is of no consequence to us. Nevertheless, it may interest the student to see why they assume a hyperbolic shape in the case discussed above and how they may have other shapes as well.

In Fig. 3.22, let EF and GH represent the longitudinal and transverse axes of the bent portion of the beam (or the glass plate) between the two knife edges, concave downwards and upwards respectively, with the tangent planes to them (shown dotted) touching the surface of the beam at its mid-point O. And, let XO'X' and YO'Y' represent the longitudinal and the transverse axes respectively of the under surface of the cover glass.

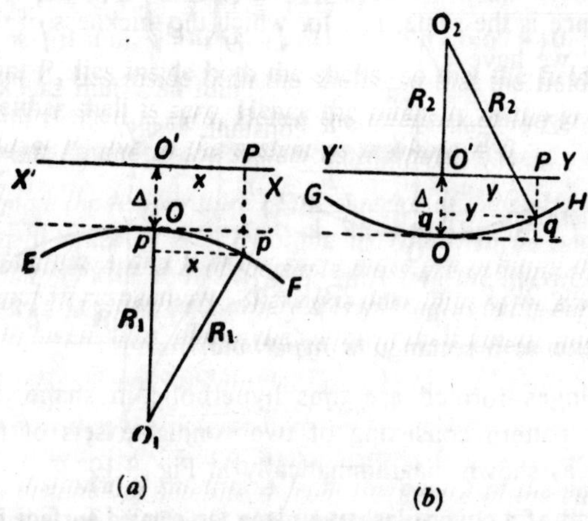

Fig. 3.22

As already pointed out, the mid-point O' of the cover glass lies vertically above the mid-point O of the beam but, in view of the anticlastic curvature of the latter, does not actually touch it. So that, the thickness of the *air film* at O or O' is not zero. Let it be equal to A.

And, at a point such as P (coordinates x, y), the thickness of the air film is increased by an amount p on account of the longitudinal curvature of the beam [Fig. 3.24(a)] and decreased by an amount q on account of its transverse curvature [Fig. 3.24(b)]. Its thickness at P is thus $(\Delta + p - q)$.

Now, from the geometry of Figure (a) we have

$$p(2R_1 - p) = x^2.$$

Or, $$2R_1p - p^2 = x^2.$$

p^3 being negligibly small compared with R_1, we $p = \frac{x^2}{2R_1}$. Similarly, from Fig. 3.242 (b), we have $q = \frac{y^2}{2R_2}$. where R_1 and R_2, as we know, are the radii of curvature of the longitudinal and transverse sections of the beam respectively.

The thickness of the air film at P is thus $\Delta + \left(\frac{x^2}{2R_1}\right) - \left(\frac{y^2}{2R_2}\right)$. Since an interference fringe is really the locus of points for which the optical path difference is the same, *i.e.*, for which the thickness of the air film is the same, we have

$$\Delta + \frac{x^2}{2R_1} - \frac{y^2}{2R_2} = \text{a constant, k say.} \qquad ...(I)$$

And, therefore, $\frac{x^2}{2R_1k} - \frac{y^2}{2R_2k} = 1$.

Or putting $2R_1k = a^2$ and $2R_2k = b^2$, we have $\frac{x^2}{a^2} - \frac{y^2}{b^2} = 1$, which is the equation to a rectangular *hyperbola.*

The fringes formed are thus hyperbolic in shape, the whole interference pattern consisting of two conjugate sets of rectangular hyperbolae, as shown diagrammatically in Fig. 3.19.

If instead of a cover glass, we place the curved surface of a plano-convex lens of a large radius of curvature r on the surface of the beam, we obtain *Newton's rings* in the unbent position of the beam (as expected) which, on the bending of the beam, assume an elliptical or a *hyperbolic* shape according as R_2 > or < r. Since, generally, $R_2 > r$, the fringes are elliptical in shape. Further, in the event of r being equal to either R_1 or R_2, we obtain straight or linear fringes.

COLUMNS OR PILLARS—EULER'S THEORY

A column or a pillar, as mentioned earlier (3.1), is just a long, straight and vertical beam or a bar of uniform cross-section, subjected to compressive axial loads. There are three modes of fixation of its ends:

(i) *Rounded, hinged or pin-jointed* : The ends of a column are said to be rounded, hinged or pin-jointed if they are fitted into a metal socket such as shown in Fig. 3.23 (a), so that they are *fixed in position but not in direction*. Thus, although the deflection at the ends is always zero, the column is free to acquire or change its slope all along its length.

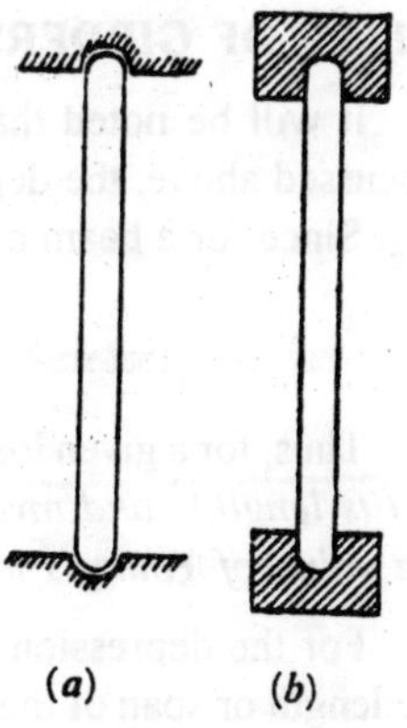

Fig. 3.23

(ii) *Fixed* : The ends of a column are said to be fixed [Fig. 3.23 (b)] if they are clamped tightly or built into a structure (as in the case of an encastre beam), so that *they can change neither their position nor direction*. The deflection at the ends is zero under all conditions of loading and there are well-marked points of inflexion along the length of the column.

(iii) *Free* : If the end of a column is neither hinged (or pin-jointed) nor fixed, it is said to be free. The column can, in this case, be deflected as well as acquire slope, *i.e., the end can change us position as well as direction.*

The mode of fixation of the ends of a column has a great bearing on its strength, *i.e.*, its capacity to withstand or resist the thrust due to the axial load placed on it.

Euler's theory. The noted Swiss mathematician *Leonard Euler* propounded, in the year 1757, his *theory of long columns* on the basis of the following assumptions:

(i) *The material of the column is homogeneous and isotropic,*

(ii) *its cross-section is uniform,*

(iii) *its length very much (i.e., at least 10 times) greater than its cross-sectional or lateral dimensions,*

(iv) *if pin-jointed, the pin-joints are frictionless and if fixed, the fixed ends are rigid,*

(v) *the column is initially straight,*

(vi) *it is laded axially and*

(vii) *its own weight is neglected.*

In the articles below, we shall follow Euler's treatment of long columns in accordance with his theory.

SHAPE OF GIRDERS

It will be noted that in all the various cases of bending of beams discussed above, the depression of the beam (y) is proportional to WL^3/YI_g. Since for a beam of rectangular cross-section, $I_g = bd^3/12$, we have

$$y \propto \frac{WL^3}{Ybd^3}.$$

Thus, for a given load (W), the depression (y) is *directly proportional to (its length)^3 and inversely proportional to its breadth, (depth)^3 and the value of Young's modulus (Y) for its material.*

For the depression (y) to be small for a given load (W), therefore, the length or span of the girder should be small and its breadth and depth large, as also the Young's modulus for its material.

Since in a supported or fixed beam, the middle portion gets depressed, its upper and lower halves, (above and below the neutral surface) get compressed and extended respectively. These compressions and extensions, and hence the corresponding stresses, are, as we know, the maximum at the upper and the lower surfaces' and progressively decrease to zero as we approach the neutral surface from either face. Obviously, therefore, the upper and the lower surfaces of the beam must be stronger than the intervening part. That is why the two surfaces of a girder or iron rails (for railway tracks etc.) are made much broader than the rest of it, thus giving its cross-section the shape of the letter I.

This naturally effects a good deal of saving in the material of the girder, without appreciably impairing its strength.

DETERMINATION OF YOUNG'S MODULUS FOR THE MATERIAL OF A BEAM—METHOD OF BENDING

As can be readily seen, any one of the expressions for the depression (y) of a supported or a fixed been under an applied load may be used to obtain the value of Young's modulus (Y) for its material, if we can accurately measure y for a known load W and if the physical dimensions of the beam (*i.e.*, its length, breadth and depth) be known.

Konig's Method

Konig devised, in the year 1885, a method in which the depression of the beam is not measured directly but obtained from the slope of its ends or their inclination to the horizontal. This angle of inclination can be measured far more accurately by the scale and telescope method than the depression of the beam. Also much smaller loads can be used, thereby eliminating the possibility of the elastic limit being exceeded or of the knife-edges biting into the beam.

The beam, mounted on two knife-edges K_1 and K_2, distance L apart, carries two plane-mirror pieces m_1 and m_2 near its two ends, a *small equal* distance beyond the knife-edges, such that they are inclined slightly outwards from the vertical. (Fig. 3.24).

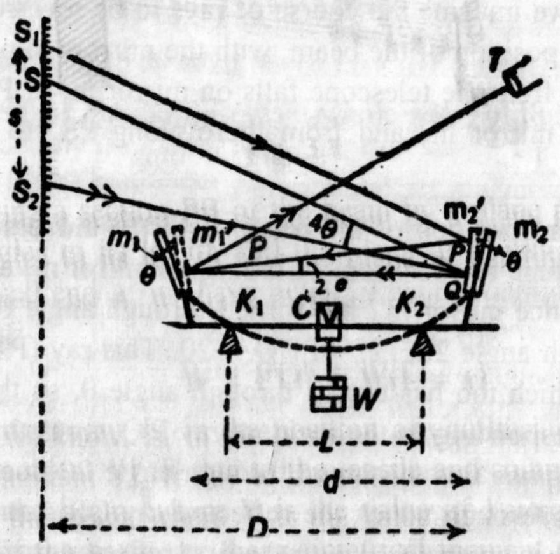

Fig. 3.24

A suitable load W can be placed on a hanger pivoted on the beam at its mid-point C (*i.e.*, midway between the two knife edges).

An illuminated translucent scale and a telescope (T) are arranged, as shown, so that a ray of light proceeding from a division on the scale first gets reflected at mirror m_2 and then at m_1 and from there on to the telescope where the image of the scale division is brought into sharp focus on a cross wire of the eye-piece.

To start with, *i.e.,* with no load yet applied to the beam, let a ray S_1P from a scale division S_1 fall on mirror m_2 at P, get reflected along PP' to fall on mirror m_1 at P' and from there, along P'T on to the telescope, so that the image of S_1 is formed on the cross-wire.

Let a small, suitable load W be now applied to the mid-point C of the beam so that the beam gets depressed a little, say, through a distance y at C into the dotted position shown. The two mirrors then turn inwards through a small angle θ into their dotted positions m_1' and m_2', where θ is obviously the slope of the ends of the beam relative to its mid-point C or their angle of inclination with the horizontal.

Obviously, the image of division S_1 of the scale will not now fall on the cross-wire of the eye-piece of the telescope but that of some other division S_2. For, a ray S_2Q from S_2 falling on mirror m_2, at Q will get reflected along QP' to mirror m_1 and from there along P'T (the same path as previously taken by the ray from S_1) to the telescope.

Now, if we imagine the course of rays to be reversed, we find that in the unbent position of the beam, with the mirrors in positions m_1 and m_2, a ray TP' from the telescope falls on mirror m_1 at P', gets reflected along P'P to mirror m_2 and from there along PS_1 to meet the scale in S_1.

In the bent position of the beam, with the mirrors in position m'_1 and m'_2, the same ray TP' from T, falling on mirror m_1 at P', is reflected along P'Q. Since mirror m'_1 has turned through angle θ, the ray P'Q is turned through angle 2θ, *i.e.,* $<PP'Q = 2\theta$. This ray (P'Q) now falls on mirror m'_2 which too has turned through angle θ, so that the reflected ray QS_2 corresponding to the incident ray P'Q, turns through angle 2θ. Since ray P'Q has already turned through angle 2θ due to reflection at m_1, the angle through which the reflected ray QS_2 turns in $2\theta + 2\theta = 4\theta$, *i.e.,* it makes an angle 4θ with the first reflected ray PS_1 or with QS, where QS is drawn parallel to PS_1. Thus,

$$<SQS_2 = 4\theta \text{ and}$$

$$\therefore \quad \text{distance } SS_2 = 4\theta.\ (D),$$

where D is the distance between mirror m_2 and the scale.

And, since $\angle PP'Q = 2\theta$, we have $PQ = 2\theta\,(d) = S_1S$ [$\because$ $QS \| PS_1$] where d is the distance between the two mirrors.

$\therefore$ distance $S_1S_2 = s$, say, between the two scale divisions S_1 and S_2, whose images are received on the cross wire of the telescope eye-piece before and after the bending of the beam respectively, is given by

$$s = S_1S_2 = S_1S + SS_2 = 2\theta(d) + 4\theta(D) = 2\theta(d + 2D),$$

whence, $$\theta = \frac{s}{2(d+2D)}$$

But, as we have seen in 3.10, $\theta = WL^2/16YI_g$, where the symbols have their usual meanings. We therefore have

$$\frac{s}{2(d+2D)} = \frac{WL^2}{16YI_g},$$

whence, $$Y = \frac{(d+2D)L^2}{8I_g}\left(\frac{W}{s}\right) \qquad ...(I)$$

As we know, in the case of a beam of *rectangular cross-section,*

$$I_g = \frac{bd^3}{12} \text{ and } \therefore\ Y = \frac{3}{2}\frac{(d+2D)L^2}{bd^3}\left(\frac{W}{s}\right). \qquad ...(i)$$

And, in the case of a beam of *circular cross-section,*

$$I_g = \frac{\pi R^4}{4}.$$

So that; in this case, $$Y = \frac{(d+2D)L^2}{2\pi r^4}\left(\frac{W}{s}\right). \qquad ...(ii)$$

In actual practice, we gradually increase the load (W) and note the corresponding values of s and then plot s against W, when we obtain a straight line graph passing through the origin. The slope of the straight line gives the mean value of (W/s) and this is substituted in relation (i) or (ii) for Y above, as the case may be.

Simple Laboratory Method

We use a light beam of a rectangular cross-section for the purpose, supported at its two ends and loaded centrally, when, measuring y, we may use relation (iii) to calculate Y for the material of the beam.

The arrangement will be easily understood which shows the beam supported symmetrically in the horizontal plane on two knife-edges, placed parallel to each other, a known distance L apart, with the load applied at its mid-point by placing suitable weights in a scale pan, also supported on a knife-edge. The depression y of the mid-point is noted directly on a micrometer screw placed on the beam, with greater precision, with the help of a microscope fitted with a micrometer eye-piece, carrying cross wires. (Fig. 3.25).

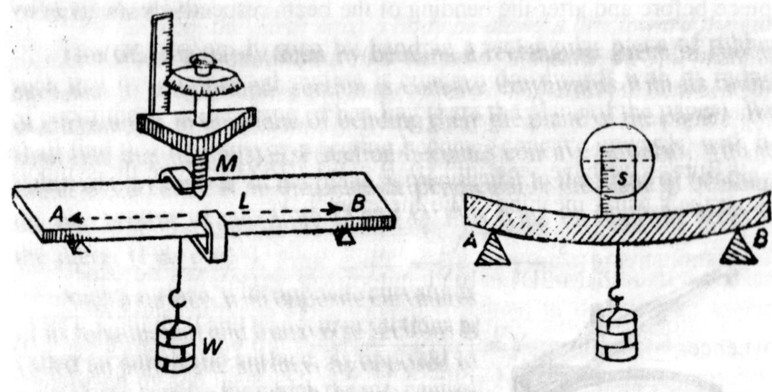

Fig. 3.25

The procedure consists in first increasing the load in equal steps W and then decreasing it in the same equal steps and noting the values of y in each case. The mean of all these readings is then taken to be the correct value of y for a load W. Then, from the relation $y = WL^3/4\,Ybd^3$ we have $Y = WL^3/4ybd^3$ and the value of Y for the material of the beam thus easily obtained.

The method, though simple enough, suffers from the following *drawbacks*:

(i) *The depression of the beam cannot be measured sufficiently accurately.*

(ii) *Fairly large loads are required to be used to produce a measurable depression and the elastic limit may possibly get exceeded, resulting in permanent damage to the beam..*

(iii) *With increasing load applied to the beam, the knife-edges may bite into it, thereby seriously affecting the readings taken.*

All these defects have been removed in *Konig's method* which we shall now proceed to discuss.

CRITICAL LOAD FOR COLUMN

Let a *long, straight* and vertical column of length L be represented by a strip AB of wood or metal, with its ends rounded or hinged (Fig. 3.26). Let a cylinder C, free to move only vertically between two parallel guides G, G, be arranged at A, such that varying amounts of lead shot or mercury may be placed inside it and any desired load thus applied axially to the column or the strip AB.

Initially, with no load at A or a load insufficient to bend the column or the strip, if a lateral force f be applied to its mid-point O so as to deflect it through a distance y, it is found that on removal of this force, the column or the strip straightens itself out. On gradually increasing the load at A, however, we arrive at a certain load P, fixed for the given column, such that on applying the lateral force f, as before, and then removing it, the column no longer straightens out but remains in its bent position with its mid-point O deflected through y. The same is true for any other deflection that may be given to the mid-point of the column (within the elastic limit) by some other lateral force, so long as the load at A remains P

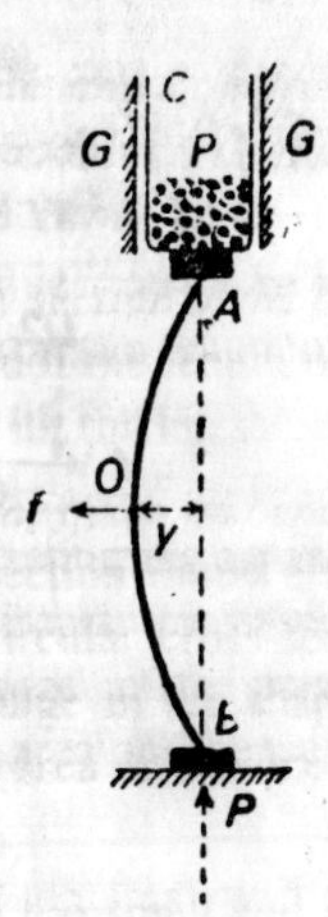

Fig. 3.26

This maximum load P which a column can support such that, if initially straight, it still remains straight and, if given a slight bend, it remains bent in that very position without tending to bend further, is called the critical, buckling or crippling load for the column.

For a load less than the critical load, the column, if bent, will tend to straighten out and for a load greater than the critical load, it will collapse due to buckling, (*i.e.,* bending further or bulging out).

Only under the critical load will the column remain in equilibrium whatever the deflection we may choose to give it within its elastic limit. This may be seen from the following:

Let us consider a column AB, of length L, under a vertical load P_1 (less than its critical load P) and a lateral force/at its mid-point O deflecting it through a distance y [Fig. 3.27 (a)].

For equilibrium of the column, obviously, the lateral force f must be balanced by two horizontal forces each equal to half its value and directed oppositely to it at A and B, as shown, with the total bending couple about O equal to zero. We, therefore, have

$$P_1y+\left(\frac{f}{2}\right)\left(\frac{L}{2}\right)+M=0,$$

where M is the *bending mo*ment or the moment of the resistance to bending set up m the column at O.

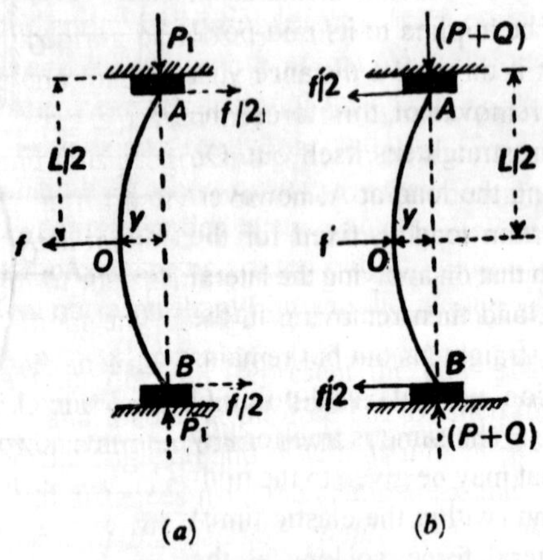

Fig. 3.27

If R be the *radius of curvature* of the column at O, we have $M=\frac{YI_g}{R}$, where Y is the *Young's modulus* for the material of the column and I_g, the *geometrical moment of inertia* of its cross-section.

So that, $P_1y+\frac{fL}{4}+\frac{YI_g}{R}=0.$

Now, if we gradually decrease the lateral force f to zero and increase the load P_1 to P, such that the deflection at the mid-point O of the column remains unchanged at y, we have $Py+YI_g/R=0$, where P is the *critical load* for the column, showing clearly that, deflection y being proportional to curvature 1/R, the column will continue to be in equilibrium under the critical load for it whatever the deflection given to it within its elastic limit.

If, however, we increase the load on the column beyond its critical value P to, say, (P + Q), as shown in Fig. 3.27 (b), we should have for equilibrium of the column

$$(P+Q)+\frac{YI_g}{R}=0.$$

Or, $$Py+Qy+\frac{YI_g}{R}=0.$$

This is obviously, *not possible*, since as we have seen above $Py + YI_g/R = 0$, indicating that the bending moment YI_g/R can balance only the part Py of the bending couple (Py + Qy). The part Qy of the bending couple thus. remains unbalanced and bends the column further beyond y.

In order to restrict the deflection of the column to y, therefore, we shall have to apply a lateral force f', say, at O in a direction opposite to that of f in the first case to balance the part Qy of the bending couple.

Since, again, this lateral force f' may be supposed to be balanced by two horizontal forces, each equal to f'/2 directed oppositely to f' at A and B, we shall have

$$Qy+\left(\frac{f'}{2}\right)\left(\frac{L}{2}\right)=0.$$

Or, $$Qy+f'\frac{L}{4}=0.$$

Now, the bending moment (or the moment of the resistance to bending) is proportional to the deflection of the column only within the elastic limit. Once the elastic limit is crossed, the column acquires a permanent set, though, quite possibly, the bending moment increasing more rapidly beyond the elastic limit, the column may attain a new position of equilibrium under the load (P + Q). If this does not come about, however, the column simply collapses due to buckling.

CALCULATION OF CRITICAL LOAD—EULER'S FORMULAE FOR LONG COLUMNS

We shall now proceed to obtain expressions for the critical load in four possible cases of fixation of columns, viz.,

(i) *both ends hinged or pin-jointed,*

(ii) *one end fixed and the other pin-jointed,*

(iii) *both ends fixed and*

(iv) *one end fixed and the other free.*

The expressions for the critical load (P) in all these cases are referred to as *Euler's formulae.*

(i) *When both ends of the column are rounded or hinged* : Let AB (Fig. 3.28) represent a long and initially straight and upright column of an isotropic material,, of length L and having a uniform cross-section and uniform elasticity, with both its ends rounded or hinged so as to be free to bend all along its length. And, let the critical load P for it be applied to it *axially* (*i.e.,* with its line of action coinciding with the axis of the column in its unbent position AB) and a small bend given to it with the help of some lateral force applied to it for an instant.

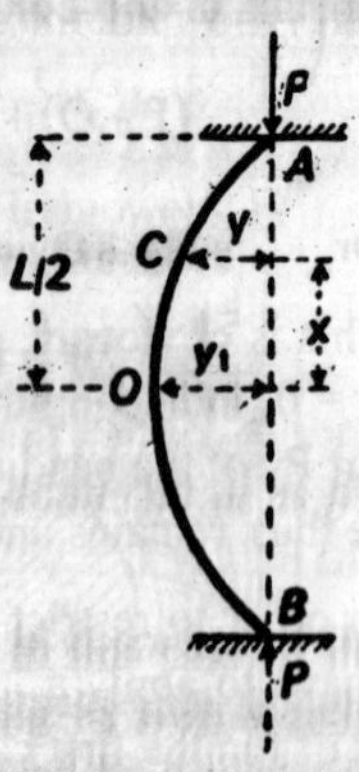

Fig. 3.28

The column being in equilibrium, the bending couple at every section of it is balanced by the bending moment (or the moment of the resistance to bending) there.

Consider the equilibrium of the portion CA of the column, where the point C lies a vertical distance x from its mid-point O and is deflected through a distance y from its initial position.

If the radius of curvature of the column at C be R, we have, for equilibrium,

$$Py + \frac{YI_g}{R} = 0,$$

where Y is the value of Young's modulus for the material of the column and I_g, the geometrical M.I. of its cross-section.

Since the curvature 1/R may, as we know, be taken to be equal to d^2y/dx^2, we have

$$Py + YI_g \frac{d^2y}{dx^2} = 0.$$

$$\text{Or,} \quad \frac{d^2y}{dx^2} + \frac{P}{YI_g} y = 0.$$

The solution to this equation is $y = A \sin \lambda x + B \cos \lambda x$, ...(i)

where $\lambda = \sqrt{\frac{P}{YI_g}}$ and A and B are constants.

On differentiating with respect to x, we have

$$\frac{dy}{dx} = A\lambda \cos\lambda x - B\lambda \sin\lambda x. \qquad ...(ii)$$

So that, (a) at x = 0, *i.e., at the mid-point of the column*, dy/dx=0, $\lambda x = 0$ and, therefore, $\sin \lambda x = 0$ and $\cos \lambda x = 1$. Hence, A = 0. And if the deflection there be y_1, we have, from relation (i) above, $y_1 = B$. And (b) with x = L/2, *i.e.*, at the end A or B of the column, y = 0 and, therefore, from relation (i), $0 = B \cos \lambda L/2$. ($\because A \sin \lambda x = 0$.)

This clearly means that either B or $\cos \lambda L/2 = 0$. Since, as we have just seen, $B \neq 0$ but equal to y_1, we have

$$\cos \lambda \frac{L}{2} = 0,$$

Or, $$\frac{\lambda L}{2} = \frac{\pi}{2}, \frac{3\pi}{2}, \frac{5\pi}{2} \text{ etc.}$$

Ignoring all other values except the first (since they indicate points of inflexion in the column), we have

$$\sqrt{\frac{P}{YI_g}} \cdot \frac{L}{2} = \frac{\pi}{2}.$$

Or, $$\frac{P}{YI_g} = \frac{\pi^2}{L^2},$$

whence, the critical load $P = \frac{\pi^2 YI_g}{L^2} = 9.87 \frac{YI_g}{L^2}$.

Thus, for given values of Y and I_g, the smaller the value of L, *i.e., the shorter the column*, the greater the critical load for it.

The critical loads in all other cases can be expressed in terms of the critical load for this type of hinged column. It is, therefore, taken to be a *fundamental case* and the column is referred to as the *fundamental hinged column.*

The length of such a column, for which the critical load is the same as that for any other type of column of length L, is called its *equivalent or effective length* corresponding to the latter and is denoted by the symbol L_e. This enables us to easily compare the strengths of columns with different types of end-fixation with that of this fundamental hinged column of the same length

(ii) ***When one end of the column is fixed and the other hinged*** : Fig. 3.29 represents such a column with its lower end fixed and the upper one hinged and loaded. In order, however, to keep its two ends in the same vertical line, the upper hinged end is constrained to move between two parallel guides G, G.

It can be shown that such a column of length L is equivalent to a column of length 0.7 L of the fundamental hinged type [*i.e.*, case (i)]. So that, $L_e = 0.7L$. All we have to do to obtain the value of the critical load, therefore, is to substitute L_e for L in the expression for P in case (i). When we obtain

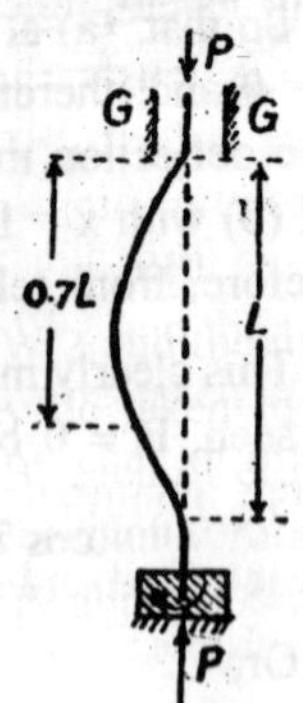

$$\textit{critical Load } P = \frac{\pi^2 Y I_g}{L_e^2} = \frac{\pi^2 Y I_g}{(0.7L)^2} \approx \frac{2\pi^2 Y I_g}{L^2}$$

which is twice the critical load for a fundamental type of hinged column of the same length (L).

Fig. 3.29

Thus, a column fixed at one end and hinged at the other has twice the strength it will have when hinged at both ends.

(iii) *When both ends of the column are fixed* : Let the ends A and B of a column of length L be fixed as shown in Fig. 3.30 so that when bent, the tangents at the ends A and B, as also at the mid-point O, are all vertical. The line of action of the resultant load P now no longer coincides with the axis of the column but lies in between the axis (AB) and the mid-point O, cutting the bent column in points C and D, which due to the absence of any bending couple there, are points of opposite flexure.

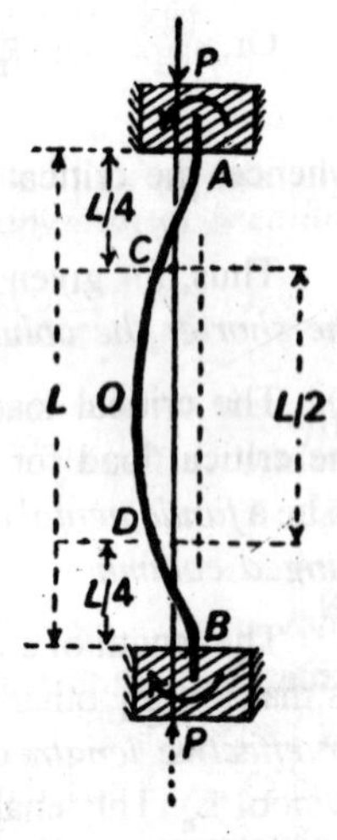

Clearly, at all those points in curves CA and CO of the bent column where the deflections (as measured from the vertical line through C) are equal, the bending couples must also be equal and, since the column has a uniform cross-section, the radii of curvature too at these points must be the same.

Fig. 3.30

Thus, since the slope of the curve at C, as also at A and O, is the same (the tangents there being all vertical), it follows that curves CA and CO are similar and equal. The same being true of curves DB and

DO, it is clear that points C, O and D divide the column into four equal parts, so that the length of the bent portion COD is equal to half the length of the entire column, *i.e.*, equal to L/2.

Obviously, therefore, the portion COD of the column, the whole of which is bent, behaves like a column of length L/2, with its two ends rounded or hinged (*i.e.*, the fundamental hinged type), carrying a load P at C. So that, the *equivalent or effective length* $L_e = \frac{L}{2}$.

Hence, proceeding exactly as in case (i) above, we have

$$\textit{critical load } P = \frac{\pi^2 YI_g}{L_e^2} = \frac{\pi^2 YI_g}{\left(\frac{L}{2}\right)^2} = \frac{4\pi^2 YI_g}{L^2},$$

showing that *the critical load (P) is now four times that for the fundamental hinged type of the same length.*

Thus, a *column, with its two ends fixed, has four times as much strength as when its two ends are rounded or hinged.*

(iv) *When one end of the column is fixed and the other free* : The critical load for the column in this case may be deduced directly from case (i) of a column hinged at both ends (the fundamental hinged column) as will be seen from the following ·

Let AB be a column of length L', with both its ends rounded or hinged, and P, the critical load on it [Fig. 3.31 (a)]. The tangent to the bent column at its mid-point O being vertical, if we clamp it tightly, *i.e.*, fix it, at O without in any way disturbing the position of the tangent, we have, for all practical purposes, a column OA, of length L'/2 = L_e with its lower end O fixed and the upper and A behaving as a free end (because it does not lie on the vertical line through the fixed end (O) and thus suffers lateral displacement), as shown in Fig. 3.31 (b). The lower half OB of the column now plays no part and may as well be removed.

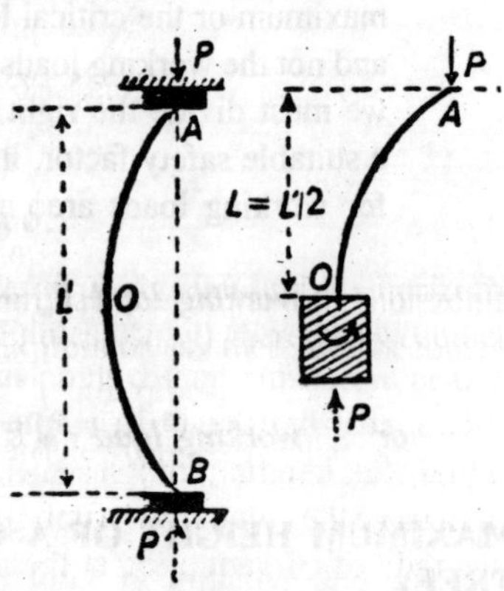

Fig. 3.31

We may thus take a column of length L, fixed at one end and free at the other, to be equivalent to a column of length L' = 2L, with its two ends rounded or hinged, *i.e.*, the fundamental hinged column. So that, we have *effective or equivalent length,* $L_e = 2L$.

$\therefore$ *for a column fixed at one end and free at the other we have*

$$\textbf{Critical load } P = \frac{\pi^2 Y I_g}{L_e^2} = \frac{\pi^2 Y I_g}{(2L)^2} = \frac{\pi^2 Y I_g}{4L^2},$$

which is clearly one-fourth of the critical load for the fundamental hinged column of the same length.

Thus, *the strength of a column fixed at one end and free at the other in one-fourth of that of the fundamental hinged column (or a column hinged at both ends) of the same length.*

Notes:

1. The expression for the critical load (P) in all the different cases of end fixation may be put in either the general form $P = \frac{N\pi^2 Y I_g}{L^2}$ or $P = \frac{\pi^2 Y I_g}{L_e^2}$, where N is equal to 1, 2, 4 and 1/4 and L_e equal to L, 0.7 L, 1/2 L and 2L in the four cases respectively.
2. In all the cases discussed above, Euler's formulae for P give the maximum or the critical load for the different types of columns and not the working loads for them. To obtain the working loads we must divide the right hand side of the expression for P by a suitable safety factor, n, say. So that, the general expressions for working loads are

$$\textit{working load } P_w = \frac{N\pi^2 Y I_g}{nL^2}$$

$$\textit{or} \quad \textit{working load } P_w = \frac{\pi^2 Y I_g}{nL_e^2}.$$

MAXIMUM HEIGHT OF A COLUMN OR A PILLAR (OR A TREE)

We have just seen in case (iv) of the previous article that the critical load for a column or a pillar, fixed at one end and loaded at the other free end is given by $P = \frac{\pi^2 Y I_g}{4L^2}$.

Now, when there is no external load placed on the upper free end of the column, the only load on it is its own weight which acts at its centre of gravity. Since the centre of gravity of a column of uniform diameter and homogeneous composition lies at its mid-point, we have, in effect, a column of length L/2, fixed at its lower end and carrying a load equal to its own weight at the upper free end. So that, putting L/2 for L in the expression for P in case (iv), we have

$$P = \frac{\pi^2 Y I_g}{4(L/2)^2} = \frac{\pi^2 Y I_g}{L^2}$$

All that we have to do, therefore, to determine the maximum length or height of a column or a pillar is to find the value of L corresponding to the critical load P, equal to its own weight.

Thus, P = *weight of the column* = $\pi r^2 L\rho g$. where r is its radius and ρ, the *density* of its material, and $I_g = \frac{\pi r^4}{4}$. And, therefore,

$$\pi r^2 L\rho g = \frac{\pi^2 Y I_g}{L^2} = \frac{\pi^2 Y \pi r^4}{4L^2} = \frac{\pi^2 Y r^4}{4L^2}$$

Or, $$L^3 = \frac{\pi^3 Y r^4}{4\pi r^2 \rho g} = \frac{\pi^2 Y r^2}{4\rho g},$$

whence, $L = \left(\frac{\pi^2 Y r^2}{4\rho g}\right)^{1/2}$ and can be easily evaluated.

This value of L then gives the maximum length or height of a column which can just support its own weight.

Similarly, if we can imagine a tree to have a uniform area of cross-section all along its length, we can obtain the value of L for it from a knowledge of its diameter (or radius) and the values of ρ and Y for its material. And this gives the maximum height it can attain without collapsing under its own weight.

SOLVED EXAMPLES

Example 1:

A wooden bar, 100 cm long and 2.0 cm × 2.0 cm in section, rests horizontally and symmetrically on two knife-edges 60 cm apart. When a weight, of 2 kg is suspended from each end of the bar, the centre is elevated through a distance 0.102 cm. Calculate the value of Young's modulus for the material of the bar.

Solution:

In Fig. 3.32, let AB be the initial position of the bar resting symmetrically on two knife-edges K_1 and K_2, 60 cm apart, with 20 cm of it projecting beyond the knife-edges on either side.

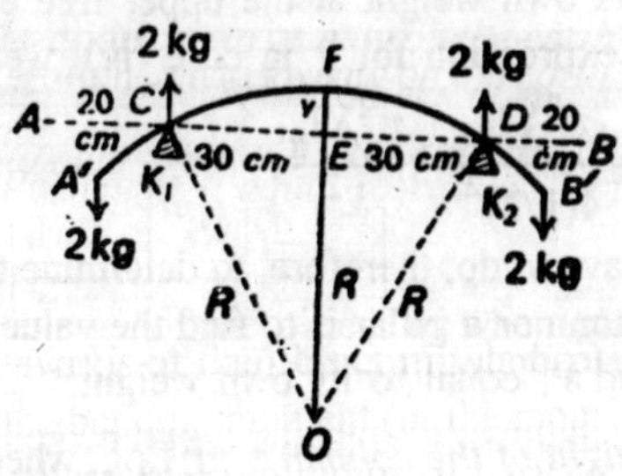

Fig. 3.32

When a weight of 2 kg is suspended from each end, the bar takes the form A'B', its portion CD between the knife edges bending into a circular arc of radius R and centre O, with its mid-point E elevated through a distance EF = y = 0.102 cm.

Clearly, *bending couple applied to the bar at either end = 2 × 1000 × 981 × 20 dynes-cm.*

The bar being in equilibrium, this must be just balanced by the *bending moment* (or the moment of the resistance to bending), equal to $Y\,I_g/R$, set up in the bar.

We, therefore, have $\frac{Y I_g}{R} = 2 \times 1000 \times 981 \times 20,$

whence, $Y = 2 \times 1000 \times 981 \times 20 \times \frac{R}{I_g}.$

To obtain the value or R we observe from the geometry of the figure that

$$CE \times ED = (20F - y)y.$$

Or, $30 \times 30 = (2R - y)\,y \approx 2Ry$

$R = 30 \times 30\ 2y$

Or, $30 \times 30/2 \times 0.102$ cm.

And I_g for the bar $= \frac{bd^3}{12} = \frac{b(b^3)}{12} = \frac{b^4}{12} = \frac{(2)^4}{12} = \frac{16}{12} = \frac{4}{3}(\text{cm})^4$

∴ *Young's modulus for the material of the bar, i.e.,*

$Y = 2 \times 1000 \times 981 \times 20 \times 30 \times 30 \times 3/2 \times 0.102 \times 4 = 1.3 \times 10^{11}$ dynes/cm^2.

Example 2:

A steel strip is clamped horizontally at one end. On applying a 500 gm load at the free end, the bending in equilibrium state is 5.0 cm. Calculate (i) the potential energy in the strip, (ii) the frequency of vibration if the load is disturbed from equilibrium. (Neglect mass of the strip itself).

Solution:

(i) We know that the depression of the free and loaded end of a cantilever (its own weight being neglected) is given by $y = WL^3/3YI_g$, whence the load or the force under which the cantilever remains in equilibrium, *i.e.*, $W = (3YI_g/L^3)y$.

This must naturally be equal in magnitude, though opposite in sign, to the elastic reaction set up in the strip.

∴ work done in depressing the loaded end through a further distance dy

$$= W dy = \left(\frac{3YI_g}{L^3}\right) y\,dy$$

Hence work done or potential energy stored up in depressing the strip through the whole distance y

$$= \int_0^y \frac{3YI_g}{L^3} y\,dy = \frac{3YI_g}{L^3}\frac{y^2}{2} = \frac{W}{y}\cdot\frac{y^2}{2} = \frac{1}{2}Wy$$

$$= \frac{1}{2} \times 500 \times 980 \times 5 = 12.25 \times 10^5 \text{ ergs.}$$

(ii) We have seen under 3.9, how the loaded end of a cantilever executes simple harmonic oscillations when disturbed from its equilibrium position and how the time-period of these oscillations is given by

$$T = 2\pi\sqrt{\frac{y}{g}}.$$

Since y = 5 cm and g = 980 cm/sec^2,

we have $T = 2\pi\sqrt{\dfrac{5}{980}} = 0.4487$ sec.

Hence *frequency of oscillation,* $n = \frac{1}{0.4487} = 2.228 \approx 2.23$/sec.

Example 3:

A uniform bar of circular cross-section, 120 cm long, rests on two knife-edges at its two ends. When it is centrally loaded with a weight of 3 kg, it is depressed through a distance of 1.5 cm. Calculate the critical load for it when used as a column with hinged ends.

Solution:

As we know, the depression of the mid-point of a bar or beam, supported at the ends and loaded centrally with a weight W is given by

$$y = \frac{WL^3}{48YI_g}, \text{ whence } YI_g = \frac{WL^3}{48y}.$$

Here, $W = 3 \times 1000 \times 980$ dynes, $L = 120$ cm and $y = 1.5$ cm.

So that, $YI_g = 3 \times 1000 \times 980 \times \frac{(120)^3}{48} \times 1.5$ dynes cm^2.

Now, Euler's formula for critical load for a column hinged at both ends is

$$P = \frac{\pi YI_g}{L^2}.$$

∴ *critical load for the column,*

i.e., $$P = \frac{\pi^2 \times 3 \times 1000 \times 980 \times (120)^3}{48 \times 1.5 \times (120)^2} \textit{ dynes}$$

$$= \frac{\pi^2 \times 3 \times 1000 \times 980 \times 120}{48 \times 1.5 \times 980 \times 1000} = \frac{\pi^2 \times 3 \times 120}{48 \times 1.5} = 50\text{kg}.$$

Example 4:

Calculate the working axial load for a column, 200 cm in length and 4 cm in diameter, when both ends of the column are (i) pin-joined, (ii) fixed. Factor of safety = 4 and Y for the material of the column = 20×10^{11} dynes/cm^2. (Take g = 1000 cm./sec^2).

Solution:

(i) The safe working load P_w for a column with *hinged or pin-jointed* ends is, as we know, given by the relation $P_w = \pi^2 YI_g/nL^2$, where n is the factor of safety.

Since $I_g = \frac{\pi r^4}{4} = 4\pi(\text{cm})^4$, we have

safe working load far the column,

$$P_w = \frac{\pi^2 \times 20 \times 10^{11} \times 4\pi}{4 \times (200)^2} = p^3 \times 5 \times 10^7 \text{ dynes.}$$

$$= \frac{\pi^2 \times 5 \times 10^7}{1000 \times 1000} = 1550\text{kg}.$$

(ii) The safe working load for a column with fixed ends is given by

$P_w = \frac{4\pi^2 YI_g}{nL^2}$, which is clearly 4 times that in case (i).

∴ *safe working load in this case, i.e.,* $P_w = 4 \times 1550 = 6200$ kg

Example 5:

Obtain an expression for the radius of curvature of a flat curve in terms of the slope of the curve and use the result to find the value o deflection in the case of a bar fixed horizontally at one end and loade at the other.

Solution:

Let P and Q be two points, distance δx apart on a flat curve OA (Fig. 3.33.) such that the radius of curvature of the portion PQ is R, wit C as its centre of curvature, and let PD and QE be tangents to the curv at P and Q and <PCQ = <DPE = θ.

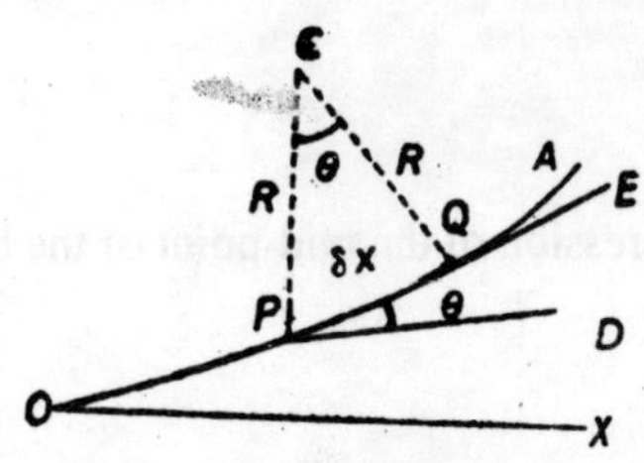

Fig. 3.33

Then, clearly $PQ = \delta x = R\theta$.

Since θ is the change in slope of the tangents from P to Q, we have

$$\theta = \left(\frac{dy}{dx}\right)Q - \left(\frac{dy}{dx}\right)P.$$

And since the rate of change of slope is d^2y/dx^2, the change in slope from P to Q is also equal to $\left(\frac{d^2y}{dx^2}\right)\delta x$,

We therefore, have

$$\frac{1}{R} = \frac{d^2y}{dx^2}\delta x. \text{ And}$$

$$\therefore \qquad \delta x = R\left(\frac{d^2y}{dx^2}\right)\delta x,$$

whence, $$R\left(\frac{d^2y}{dx^2}\right) = 1.$$

So that $$\frac{1}{R} = \frac{d^2y}{dx^2} = \textit{rate of change of slope.}$$

Thus, if the curvature (1/R) be small (as it usually is in the case of bent bars or beams) its value at any point on the curve is given by the rate of change of slope at that point.

Example 6:

(a) A rectangular bar, 2 cm in breadth and 1 cm in depth and 100 cm in length, is supported at the ends and a load of 2 kg is applied at its middle. Calculate the depression if the Young's modulus of the material of the bar to 20 × 10^{11} dynes/cm^2.

(b) In an experiment, the diameter of the rod was 1.26cm and the distance between the two knife edges, 70 cm. On putting a load of 900gm at the middle point, the depression was 0.025 cm. Calculate the Young's modulus for the substance.

Solution:

(a) In this case, the depression of the mid-point of the bar supported at its two ends is given by

$$y = \frac{WL^3}{48YI_g}.$$

Now, W = 2 × 1000 × 980 dynes.

L = 100 = 10^2 cm. Y = 20 × 10^{11} dynes/cm^2

and $I_g = bd^3/12 = 2 \times (1)^3/12 = 1/6$.

So that, depression of the mid-point of the bar,

i.e.,
$$y = \frac{2\times1000\times980\times(10^2)^3}{48\times20\times10^{11}\times(1/6)} = \frac{49}{400} = 0.1225\text{cm}.$$

(b) Here, again, $y = \frac{WL^3}{48YI_g}$.

And since $I_g = \frac{\pi r^4}{4}$, we have

$$y = \frac{WL^3\times4}{48Y\times\pi r^4} = \frac{WL^3}{12Y\pi r^4},$$

whence, $Y = \frac{WL^3}{12y\pi r^4}$.

Substituting the given values, therefore, we have *Young's modulus for the material of the rod,*

$$Y = \frac{900\times980\times(70)^3}{12\times0.025\times\pi(1.26/2)^4} = 2.039\times10^{12}\ \text{dynes/cm}^2.$$

Example 7:

Compare the loads required to produce equal depression for two beams of the same material, length and weight when one has a circular cross-section and the other has a square cross-section.

Solution:

Let W and W' be the loads required to produce the same depression y in the two cases respectively. Then, if I_g and I_g' be the respective geometrical moments of inertia of the two beams, we have

$$y = \frac{WL^3}{48YI_g} = \frac{W'L^3}{48YI'_g}$$

Or, $\frac{W}{W'} = \frac{I_g}{I'_g}$

As we know, *for the beam of circular cross-section of radius r,* $I_g = \frac{\pi r^4}{4}$ and for the beam of square cross-section, because b = d, $I'_g = \frac{b(b^3)}{12} = \frac{b^4}{12}$. Since the weights, and hence the volumes, of the two beams are equal (their material and hence their density being the same), we have

$$\pi r^2 L = b^2 L.$$

Or,

$$b^2 = \pi r^2.$$

$$\frac{W}{W'} = \frac{\frac{\pi r^4}{4}}{\frac{b^4}{12}} = \frac{3\pi r^4}{(b^2)^2} = \frac{3\pi r^4}{(\pi r^2)^2} = \frac{3\pi r^4}{\pi^2 r^4} = \frac{3}{\pi}.$$

Thus, *the loads in the two cases must be as 3 : π.*

EXERCISES

1. A solid cylindrical rod of radius 6 mm bends by 8 mm under a certain load. Deduce the bending when all other things remaining the same, the beam is replaced by a hollow cylindrical one with external radius 10 mm and internal, 8 mm.

   (Ans. 1.756 mm)

   [**Hint:** The length L and the load W being the same in either case, $y \propto 1/I_g$. If, therefore, y and y be the bondings in the two cases respectively, we have $y'/y = I_g/I'_g$. And, since $I_g = \pi r^4/4$ and $I'_g = \pi(r_1^4 - r_2^4)/4$, we have

   $$y' = \left(\frac{r^4}{r_1^4 - r_2^4}\right) y.]$$

2. A uniform beam is clamped at one end and loaded at the other. Obtain the relation between the load and depression at the loaded end when (a) the weight of the beam can be neglected, W when the weight of the beam cannot be neglected.

3. (a) The end of a strip cantilever depresses 10 mm under a certain load. Calculate the depression under the same load for another cantilever of the same material, two times in length, two times in breadth and three times in thickness. **(Ans.** 1.48 mm)

   (b) What should be the change in the load if the depression of the strip cantilever is to be maintained at 10 mm but its dimension are halved? **(Ans.** *The load should be halved)*

4. Give the theory of transverse vibrations of a bar fixed at one end and loaded at the other.

5. A light beam of circular cross section is clamped horizontally at one end and a heavy mass is attached at the other end. Find the depression at the loaded end.

   If the mass is pressed down a little and then released, show that it will perform S.H.M. Explain how from a knowledge of the period of oscillation, the mass and the dimensions of the bar, the value of Young's modulus for the material of the bar may be determined.

6. What is meant by a beam? Explain the terms: *neutral surface, neutral axis flexural rigidity* and *bending moment* of a beam.

7. A steel wire of 1.00 mm radius is bent in the form of a circular arc of radius 50 cm. Calculate (i) the bending moment, (ii) the maximum stress. (Given Young's modulus for steel = 2×10^{12} dynes/cm^2).

   (**Ans.** (i) 3.142×10^6 dyne-cm, (ii) 4×10^9 dynes/cm^2)

   [**Hint:** (ii) stress/strain = Y. ∴ stress, P, say = Y × strain. If R be the radius of the bent wire (or rather its neutral axis), strain at a distance z from the neural axis = z/R (3.5). Since strain is maximum on the surface of the wire, z = r, the radius of the wire. So that, maximum strain = r/R and ∴ *maximum stress* = Yr/R.]

8. Obtain, in terms of the bending moment M, an expression for (i) the longitudinal stress in a beam at a distance z from its neutral axis, (ii) the maximum stress in the case of (a) a cylindrical rod of radius r and (b) a cylindrical pipe of external and internal diameters d_1 and d_2 respectively.

   (**Ans.** (i) $\frac{M}{I_g}z$; (ii) (a) $\frac{4M}{\pi^3}$ (b) $\frac{32Md_1}{\pi(d_1^4 - d_2^4)}$)

   [**Hint:** (i) At distance z from the neutral axis, *stress* P = Yz/R. Now, *bending moment* M = YI_g/R, whence, Y/R = M/I_g.

   $$\therefore \quad P = \frac{M}{I_g}z.$$

   (ii) Since maximum stress occurs at the surface of a beam, z = r in case (a) and z = $d_1/2$ in case (b) Atso $I_g = \pi r^4/4$ in case (a) and I_g = it $\pi[d_1/2)^4 - (d_2/2)^4]/4$ in case (b).]

9. (a) What is a cantilever ? Obtain an expression for the depression at the free end of a thin light beam clamped horizontally at one end and loaded at the other.

(b) If in (a) above, the beam be of a rectangular cross-section, with breadth b and depth d, and if the depression of its free end be y under a load W, show that for the same load, the depression of the free end will be $y' = y\,(d^2/b^2)$ if the beam be used with d as breadth and b as depth.

[Hint: $I_g = bd^3/12$ in the first case and equal to $db^3/12$ in the second case.

Therefore, $y = \dfrac{WL^3}{3Y} \dfrac{bd^3}{12}$

and $y' = \dfrac{WL^3}{\dfrac{db^3}{12}}$

So that, $\dfrac{y'}{y} = \dfrac{d^2}{b^2}.$

Or, $y = y\left(\dfrac{d^2}{b^2}\right).$

10. A cantilever of length l is loaded at a point x. Calculate the depression of the tip of the cantilever. If the same load is shifted to the tip, what is the depression of the point x?

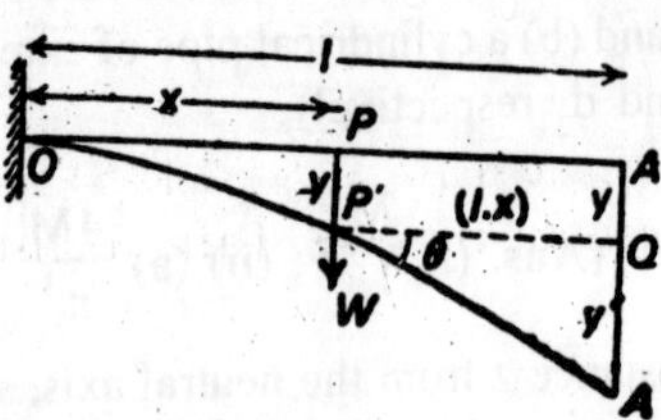

Fig. 3.34

(Ans. $\dfrac{Wx^2}{2YI_g}\left(l - \dfrac{x}{3}\right)$; *same.)*

[Hint : We have ***depression at the distance x from the fixed end,***

$$y = \frac{W}{YI_g}\left(\frac{lx^2}{2} - \frac{x^3}{6}\right)$$

$\therefore$ *slope of the curve at* $x_1 = \dfrac{dy}{dx} = \tan\theta = \dfrac{W}{YI_g}\left(lx - \dfrac{x^2}{2}\right)$.

Now, with the load W at P distant x_1 from the fixed end the depression at P is PP' = y = $Wx^3/3YI_g$ there being no depression beyond P', with the rest of the cantilever (P'A') tangential to the curve at P', making an angle θ with the horizontal, where tan θ

$$= \frac{dy}{dx} = \frac{W}{YI_g}\left(x^2 - \frac{x^2}{2}\right) = \frac{Wx^2}{3YI_g}.$$

$\therefore$ *displacement AA' of the tip of the cantilever* = AQ + QA' = y + y' = y + P'Q tan θ = y + (*l* – x) tan θ

$$= \frac{Wx^3}{3YI_g} + (l - x)\frac{Wx^2}{2YI_g}$$

$$= \frac{W}{YI_g}\left(\frac{lx^2}{2} - \frac{x^3}{6}\right) = \frac{Wx^2}{2YI_g}\left(l - \frac{x}{3}\right).$$

And, when load W is shifted to the tip of the cantilever, we have

depression at distance x from the fixed end $= \dfrac{W}{YI_g}\left(\dfrac{lx^2}{2} - \dfrac{x^3}{6}\right)$

$= \dfrac{Wx^2}{2YI_g}\left(l - \dfrac{x}{3}\right)$, *i.e.*, the *same as that of the tip in the first case.*

11. A cantilever of length 50 cm is depressed by 15.00 mm at the loaded end. Calculate the depression at a distance of 30 cm from the fixed end and deduce the formula used.

(**Ans.** 6.48 mm)

[**Hint:** At the loaded end y = 1.5 = $WL^3/3YI_g$, whence the value of WL^3/YI_g, can be obtained. Then, at distance x from the fixed end,

$y = \dfrac{W}{YI_g}\left(\dfrac{Lx^2}{2} - \dfrac{x^3}{6}\right)$ and may be easily evaluated.